Oceanography from Space

Oceanography from Space

Revisited

Edited by

Vittorio Barale
Joint Research Centre, European Commission, Ispra, Italy

J.F.R. Gower
Institute of Ocean Sciences, Fisheries and Oceans Canada, Sidney, BC, Canada

and

L. Alberotanza
Istituto di Scienze Marine (ISMAR), CNR, Venice, Italy

Springer

JRC
EUROPEAN COMMISSION

Editors
Vittorio Barale
Joint Research Centre
European Commission
TP272
Via E. Fermi 2749
21027 Ispra (VA)
Italia
vittorio.barale@jrc.ec.europa.eu

Luigi Alberotanza
Istituto di Scienze Marine (ISMAR)
CNR
Castello 1364
30122 Venezia
Italia
luigi.alberotanza@ve.ismar.cnr.it

Jim Gower
Institute of Ocean Sciences
Fisheries and Oceans Canada
9860 West Saanich Road
Sidney, B.C. V8L 4B2
Canada
Jim.Gower@dfo-mpo.gc.ca

ISBN 978-90-481-8680-8 e-ISBN 978-90-481-8681-5
DOI 10.1007/978-90-481-8681-5
Springer Dordrecht Heidelberg London New York

Library of Congress Control Number: 2010922544

Cover illustration: A graphical rendering of (part of) an arial view of San Giorgio Island, in front of San Marco's basin.

Printed on acid-free paper

Springer is part of Springer Science+Business Media (www.springer.com)

To all those sailors
Who dreamed before us
Of another way to sail the oceans

Preface

To all those sailors / Who dreamed before us / Of another way to sail the oceans.

The dedication of this Volume is meant to recall, and honour, the bold pioneers of ocean exploration, ancient as well as modern. As a marine scientist, dealing with the oceans through the complex tools, filters and mechanisms of contemporary research, I have always wondered what it was like, in centuries past, to look at that vast horizon with the naked eye, not knowing what was ahead, and yet to sail on. I have tried to imagine what ancient sailors felt, when "the unknown swirls around and engulfs the mind", as a forgotten author simply described the brave, perhaps reckless, act of facing such a hostile, menacing and yet fascinating adventure. Innovation has always been the key element, I think, for their success: another way, a better way, a more effective, safer and worthier way was the proper answer to the challenge.

The map of our world has been changed time and again, from the geographical as well as the social, economic and scientific points of view, by the new discoveries of those sailors. One of the positive qualities of human beings is without doubt the inborn desire to expand their horizons, to see what lies beyond, to learn and understand. Exploring the oceans not only brought fame and fortune to the sailors, it also opened the way to progress in all spheres of life: new lands, new markets and new people meant new ideas as well, new perceptions of the world around. In this sense, the "space" oceanographers who gathered in Venice, for the first "Oceans from Space" Symposium, in 1980, were like those pioneers who opened new trade routes, as what they discovered since then in reality surpassed many times what was expected in the beginning.

Both in ancient and in modern times, novel expeditions cost a lot. Merchants and rulers had a keen interest to invest into such risky enterprises as fragile ships sailing towards the unknown, because of the profits to be made with the goods they would bring back, and of the invaluable benefits brought by new discoveries. Oceanographers had similar, if less material goals, when they started to look at the expensive tools of space exploration, for opening new avenues of research. Satellite missions require huge investments, and are not immune from the risk of total failure, but they do bring back invaluable goods such as plentiful data, generated by large-scale, long-term systematic observations, as well as surprising, intriguing and promising new knowledge. The oceanographers who came back to Venice, for the second "Oceans from Space" Symposium, in 1990, were faced by difficulties of continuing to fund their enterprise, in spite of the extraordinary possibilities hinted by their early results.

It was only during the third edition of "Oceans from Space", in 2000, that the need for new, unprecedented commitments, in terms of both funding and research programmes, became evident not only to the scientific community, but also in the political circles where goals and priorities of public investments are set. Climate change, and its unexpected consequences, both in the environmental as well as socio-economic realms, was upon us, at that time, and action seemed urgent. It is still, as we get together once again, in Venice, the city that represents so well our complex, dangerous and yet captivating bond with the sea. The achievements, the understanding we gained over the past three decades of ocean observations, are reviewed and commented in the present collection of key-note lectures delivered at the fourth "Oceans from Space" Symposium, in 2010. But, as usual, the future outlooks offered by many of the authors focus on the open questions, on the new challenges that have been brought about in the course of the most recent explorations.

The chapters of this Volume provide an overview of the path followed so far, the instruments available today, and the plans for tomorrow, when observing the oceans with passive or active microwave, infrared and visible remote sensing. In some instances, they start to unveil the advantages deriving from the use of complementary techniques and the added value of their combined views. The breadth and complexity of the environmental themes and of the diverse techniques covered in the papers called for the contribution of several prominent figures in our scientific community, whose names and affiliations appear in the following list of Contributors. To all of them go my sincerest thanks, for the enthusiasm with which they responded to my call for help, and for the patience with which they endured my continuous reminders of incumbent deadlines. Further, special thanks are due to my colleagues at the Joint Research Centre of the European Commission, whose help as "fast reviewers" of the submitted papers has been highly appreciated, when time was running short, and is acknowledged here, now that their obscure, but invaluable, job is done. Their names are also recalled in the list of Contributors that follows. Finally, I would like thank my co-editors of this Volume, and dear friends, Jim Gower and Luigi Alberotanza. Without Jim, there would be no "Oceans from Space" conference series, period. Without Luigi, none of the "Oceans from Space" editions would have enjoyed the success that they did. It has been a privilege, and a great pleasure, to work with both of them.

Ispra, Italy Vittorio Barale

Contents

Contributors

Robert Arnone Oceanography Division, Naval Research Laboratory, Stennis Space Center, MS 39529, USA, arnone@nrlssc.navy.mil

Vittorio Barale* Institute for Environment and Sustainability, Joint Research Centre, European Commission, Ispra 21027, Italy, vittorio.barale@jrc.ec.europa.eu

Helen M. Beggs Australian Bureau of Meteorology, Centre for Australian Weather and Climate Research, Melbourne VIC 3001, Australia, H.Beggs@bom.gov.au

Jean-François Berthon* Institute for Environment and Sustainability, Joint Research Centre, European Commission, Ispra 21027, Italy, jean-francois.berthon@jrc.ec.europa.eu

Tess B. Brandon NOAA National Oceanographic Data Center, Silver Spring, MD 20910, USA, tess.brandon@noaa.gov

Marty Brewer Remote Sensing Systems, Santa Rosa, CA 95401, USA, brewer@remss.com

Brandon Casey Oceanography Division, Naval Research Laboratory, Stennis Space Center, Carriere, MS 39426, USA, casey@nrlssc.navy.mil

Kenneth S. Casey NOAA National Oceanographic Data Center, Silver Spring, MD 20910, USA, Kenneth.Casey@noaa.gov

Bertrand Chapron Laboratoire Océanographie Spatiale, Ifremer, Centre de Brest, Plouzané 29280, France, Bertrand.Chapron@ifremer.fr

Paolo Cipollini Ocean Observing and Climate Research Group, National Oceanography Centre, Southampton SO14 3ZH, UK, cipo@noc.soton.ac.uk

Fabrice Collard CLS, Radar Applications Division, Plouzané 29280, France, dr.fab@cls.fr

Peter Cornillon Graduate School of Oceanography, University of Rhode Island, Narragansett RI 02882, USA, pcornillon@gso.uri.edu

Gianfranco De Grandi* Institute for Environment and Sustainability, Joint Research Centre, European Commission, Ispra 21027, Italy, frank.de-grandi@jrc.ec.europa.eu

Craig J. Donlon ESTEC (EOP-SME), European Space Agency, Noordwijk 2201 AZ, The Netherlands, Craig.Donlon@esa.int

Mark Dowell* Institute for Environment and Sustainability, Joint Research Centre, European Commission, Ispra 21027, Italy, mark.dowell@jrc.ec.europa.eu

Jean-Noel Druon* Institute for the Protection and Security of the Citizen, Joint Research Centre, European Commission, Ispra 21027, Italy, jean-noel.druon@jrc.ec.europa.eu

Vyacheslav A. Dubina Satellite Oceanography Department, Pacific Oceanological Institute, Far Eastern Branch of the Russian Academy of Sciences, Vladivostok 690041, Russia, dubina@poi.dvo.ru

Robert Evans Rosenstiel School of Marine and Atmospheric Science, University of Miami, Miami, FL 33149, USA, revans@rsmas.miami.edu

Guido Ferraro Institute for the Protection and Security of the Citizen, Joint Research Centre, European Commission, Ispra 21027, Italy, guido.ferraro@jrc.ec.europa.eu

Jordi Font Department of Physical Oceanography, Institute of Marine Sciences, Barcelona 37-49, 08003, Spain, jfont@icm.csic.es

Lee-Lueng Fu Jet Propulsion Laboratory, California Institute of Technology, Pasadena, CA 91109-8099, USA, lee-lueng.fu@jpl.nasa.gov

Chelle L. Gentemann Remote Sensing Systems, Santa Rosa, CA 95401, USA, gentemann@remss.com

Howard R. Gordon Department of Physics, University of Miami, Coral Gables, FL 33124, USA, gordon@physics.miami.edu

Jim F.R. Gower Institute of Ocean Sciences, Fisheries and Oceans Canada, Sidney, BC V8L 4B2, Canada, Jim.Gower@dfo-mpo.gc.ca

Harm Greidanus Institute for the Protection and Security of the Citizen, Joint Research Centre, European Commission, Ispra 21027, Italy, harm.greidanus@jrc.ec.europa.eu

Kyle Hilburn Remote Sensing Systems, Santa Rosa, CA 95401, USA, hilburn@remss.com

Nicolas Hoepffner* Institute for Environment and Sustainability, Joint Research Centre, European Commission, Ispra 21027, Italy, nicolas.hoepffner@jrc.ec.europa.eu

Johnny Johannessen Nansen Environmental and Remote Sensing Center, Bergen 5006, Norway, johnny.johannessen@nersc.no

Kristina B. Katsaros Rosenstiel School of Marine and Atmospheric Science, Miami, FL, USA: and Northwest Research Associates, Bellevue, WA 98052, USA, katsaros@whidbey.net

Dong-Shang Ko Oceanography Division, Naval Research Laboratory, Stennis Space Center, Carriere, MS 39426, USA, ko@nrlssc.navy.mil

Vladimir Kudryavtsev Nansen Environmental and Remote Sensing Center, Bergen 5006, Norway, vladimir.kudryavtsev@nersc.no

Sherwin Ladner Oceanography Division, Naval Research Laboratory, Stennis Space Center, Carriere, MS 39426, USA, ladner@nrlssc.navy.mil

Gary Lagerloef Earth and Space Research, Seattle, WA 98121, USA, Lager@esr.org

W. Timothy Liu Jet Propulsion Laboratory, California Institute of Technology, Pasadena, CA 91109-8099, USA, w.t.liu@jpl.nasa.gov

Frederic Mélin* Institute for Environment and Sustainability, Joint Research Centre, European Commission, Ispra 21027, Italy, frederic.melin@jrc.ec.europa.eu

Peter J. Minnett Meteorology and Physical Oceanography, Rosenstiel School of Marine and Atmospheric Science, University of Miami, Miami, FL 33149-1098, USA, pminnett@rsmas.miami.edu

Leonid M. Mitnik Satellite Oceanography Department, Pacific Oceanological Institute, Far Eastern Branch of the Russian Academy of Sciences, Vladivostok 690041, Russia, mitnik@poi.dvo.ru

Leo Nykjaer* Institute for Environment and Sustainability, Joint Research Centre, European Commission, Ispra 21027, Italy, leo.nykjaer@jrc.ec.europa.eu

Rodrigo Perez-Garcia* Institute for Environment and Sustainability, Joint Research Centre, European Commission, Ispra 21027, Italy, rodrigo.perez-garcia@jrc.ec.europa.eu

Trevor Platt Coastal Ocean Sciences, Bedford Institute of Oceanography, Dartmouth, Nova Scotia B2Y 4A2, Canada, tplatt@Dal.Ca

Monica Posada Institute for the Protection and Security of the Citizen, Joint Research Centre, European Commission, Ispra 21027, Italy, monica.posada@jrc.ec.europa.eu

Marie-Helene Rio CLS-DOS, Parc Technologique du Canal, Ramonville Saint Agne 31526, France, mrio@cls.fr

Ian S. Robinson Ocean Observing and Climate Research Group, National Oceanography Centre, Southampton SO14 3ZH, UK, isr@noc.soton.ac.uk

Roland Romeiser Rosenstiel School of Marine and Atmospheric Science, University of Miami, Miami, FL 33149-1031, USA, rromeiser@rsmas.miami.edu

Hartmut Runge Remote Sensing Technology Institute, DLR, Wessling 82234, Germany, Hartmut.Runge@dlr.de

Shubha Sathyendranath Plymouth Marine Laboratory, Plymouth PL1 3DH, UK, shubha@dal.ca

Deborah Smith Remote Sensing Systems, Santa Rosa, CA 95401, USA, smith@remss.com

Walter H.F. Smith Laboratory for Satellite Altimetry, National Oceanic and Atmospheric Administration, Silver Spring, MD 20910, USA, Walter.HF.Smith@noaa.gov

Adolf Stips* Institute for Environment and Sustainability, Joint Research Centre, European Commission, Ispra 21027, Italy, adolf.stips@jrc.ec.europa.eu

Steffen Suchandt Remote Sensing Technology Institute, DLR, Wessling 82234, Germany, steffen.suchandt@dlr.de

Anna C.S. Sutcliffe Ocean Observing and Climate Research Group, National Oceanography Centre, Southampton SO14 3ZH, UK, anna.sutcliffe@noc.soton.ac.uk

Wenqing Tang Jet Propulsion Laboratory, California Institute of Technology, Pasadena, CA 91109-8099, USA, wqt@pacific.jpl.nasa.gov

Dario Tarchi* Institute for the Protection and Security of the Citizen, Joint Research Centre, European Commission, Ispra 21027, Italy, dario.tarchi@jrc.ec.europa.eu

Jean Verdebout* Institute for Environment and Sustainability, Joint Research Centre, European Commission, Ispra 21027, Italy, jean.verdebout@jrc.ec.europa.eu

Michel Verstraete* Institute for Environment and Sustainability, Joint Research Centre, European Commission, Ispra 21027, Italy, michel.verstraete@jrc.ec.europa.eu

Michele Vespe Institute for the Protection and Security of the Citizen, Joint Research Centre, European Commission, Ispra 21027, Italy, michele.vespe@jrc.ec.europa.eu

Kenneth J. Voss Physics Department, University of Miami, Coral Gables, FL 33124, USA, voss@physics.miami.edu

Frank J. Wentz Remote Sensing Systems, Santa Rosa, CA 95401, USA, frank.wentz@remss.com

Xiaosu Xie Jet Propulsion Laboratory, California Institute of Technology, Pasadena, CA 91109-8099, USA, xiaosu@pacific.jpl.nasa.gov

Giuseppe Zibordi Institute for Environment and Sustainability, Joint Research Centre, European Commission, Ispra 21027, Italy, giuseppe.zibordi@jrc.ec.europa.eu

---

*JRC EC Review Team

Key-note Lectures

presented at the

"Oceans from Space"

Symposium

Venice, Italy

26–30 April 2010

Ezio Balliano
Le pagine nell'acqua, 2009

Chapter 1
Oceans from Space, a Once-a-Decade Review of Progress: Satellite Oceanography in a Changing World

Jim F. R. Gower

1.1 Introduction

At the first Oceans from Space conference in Venice in 1980, we celebrated NASA's launch of two ocean satellites, Seasat and Nimbus 7, which showed the amazing capabilities of the new images and measurements from space. We saw that altimetry could measure currents and waves, that ocean color could measure surface chlorophyll and plankton blooms, that SAR could measure waves and ocean fronts, that scatterometers could measure surface wind, and that microwaves could map ice and measure sea surface temperatures through cloud. We looked forward to the launch of NOSS, the planned US National Ocean Satellite System, which would "operationalize" satellite oceanography, and which as a result would be much more expensive.

At the second conference in 1990, we were a sadder but wiser group, NOSS had failed to appear, the CZCS had eventually died, and we were seeing gaps in our data time series. At the same time, we had become uncomfortably aware that global climate change was likely upon us, and that we needed continuing time series of exactly these types of data.

By 2000 we were again happier, celebrating the new strengths of the global ocean satellite community. MODIS had been launched, the ERS satellites marked the start of ESA's major role in global earth observation. Japan had a brief success with ADEOS, Russia had contributed radar satellites, and Topex/Poseidon had collected a significant time series of altimetric data.

Today in 2010, we can continue to rejoice in our new data sources, but we now have no remaining doubt of the dangers inherent in allowing un-checked growth in carbon dioxide concentration in the earth's atmosphere. We see the need for global, stable, long-term time series of ocean satellite data to show us the changes that are occurring. Happily, we now have more satellites and a much greater capability

J.F.R. Gower (✉)
Institute of Ocean Sciences, Fisheries and Oceans Canada, Sidney, BC V8L 4B2, Canada
e-mail: Jim.Gower@dfo-mpo.gc.ca

V. Barale et al. (eds.), *Oceanography from Space*,
DOI 10.1007/978-90-481-8681-5_1,

for transmitting and processing data. This capability takes us beyond the problem often talked about in the 1980s of "Drinking from a Fire-Hose." I was never sure that the analogy made the problem of handling large data flow seem serious. Drinking from a fire-hose might be slightly messy, but one would never need to go thirsty.

1.2 Ocean Satellites Showing a Changing World

Climate change is giving us all something to think about. It has huge social, economic and ethical implications. Scientifically though, it is fascinating. It keeps me from retiring, and I am sure I am not alone in this. At the 2000 conference I said I would like to stay working long enough to see the world agree that human-induced climate change is happening, and to start to do something serious about it. I can now see that this fails to give me any well-defined retirement date. A consensus has been reached, but there will always be hold-outs. Action will be expensive and will benefit some people more than others. Wind farms are being built, and there are rumours that coal-fired power stations are being closed, though many more are being built. Today in Canada, we stress that the coal we are exporting is "metallurgical" (i.e. needed for steel production) and therefore in some sense "greener" than coal being burned for generation of electricity. We are having a harder time excusing our tar sands, but exports of this relatively "high carbon" oil continue to increase. It seems that the effects of climate change need to be much more strongly felt, by many more people, before serious and concerted action occurs.

Ocean satellites are providing a number of time series that demonstrate the problems we are facing. I have my favorites and maybe you do too. I would like to present a few here, and to suggest that we might consider awarding a Venice prize for the chosen "best". Some time series are longer than others and some, though short, already have fascinating implications.

1.3 The Keeling Curve: Origin of All Climate Change Time Series

For this conference I should emphasize ocean satellite data, but I need to start where recent, human-induced climate change begins, at the Keeling curve. This is a fascinating and frightening time series,[1] showing the CO_2 concentrations in the atmosphere as measured on Mauna Loa in Hawaii. Figure 1.1 shows the monthly averages, starting in 1957, curving upwards.

At the start of the series, values were near 315 parts per million (ppm), and we have now reached 390. The agreed pre-industrial concentration is 280, so the

[1] Available at http://scrippsco2.ucsd.edu/data/atmospheric_co2.html

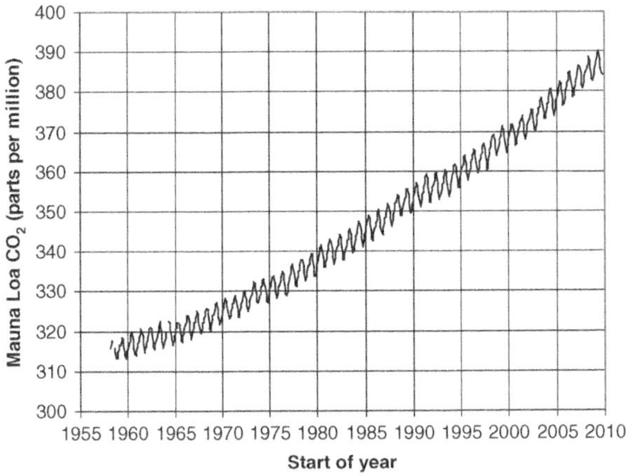

Fig. 1.1 The keeling curve: monthly averages of carbon dioxide in the northern hemisphere atmosphere, measured on Mauna Loa, Hawaii (20°N)

much-feared doubling refers to 560. The proposed world target of 350 ppm[2] was passed in about 1988 with no sign of slowing down.

When the annual cycle is removed, the curve shows a continuing and near-constant acceleration in CO_2 concentration, in spite of the stated aim of almost all national governments to reduce rates of emission. A good fit to the (12-month averaged) Keeling data between 1957 and 2009 is

$$CO_2(\text{ppm}) = 297 + 0.011(\text{year} - 1919)^2 \quad (1.1)$$

If this trend of the past 50 years continues into the future, the world will get to 410 ppm by 2020, the possible date of the next Venice conference, and we will see doubling of the pre-industrial value by 2073.

If the best fit, constant acceleration, curve of equation (1.1) is subtracted from the Keeling data, the residual differences show peaks in CO_2 concentrations which correlate with El-Nino (Fig. 1.2). They also show a single peak at 1990 which the Keeling research group at Scripps were unable to explain (Keeling et al., 1995).

At present, the Keeling curve continues its acceleration, even though the world entered a major economic recession in 2007, which would be expected to reduce emissions. A recent analysis predicted that the recession would change the present growth rate in emissions from their average rise of about 2.5% per year, to a drop of 3% for 2009. Such a drop is only by about 0.11 ppm in 1 year. It would need to continue for several years before it would be evident, given the variability shown in Fig. 1.2.

[2]See http://www.350.org

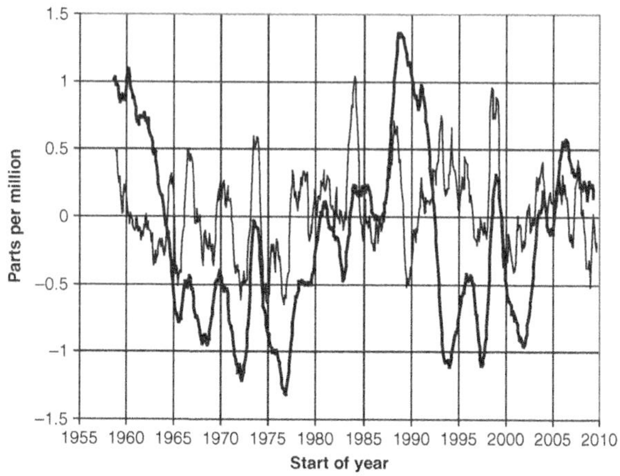

Fig. 1.2 Residual differences in carbon dioxide concentration between the keeling curve and the constant acceleration equation (1.1) (*heavy line*), compared with the Multi-variate El-Nino Index (Wolter and Timlin, 1998) (*light line*)

Figure 1.1 also shows a small but regular annual cycle. In May of every year the world breaks the "all-time" record for carbon dioxide concentration. Between then and October, plant growth on land in the northern hemisphere reduces the monthly averages.

The amplitude of this annual cycle has increased by about 20% over the 50 years (Fig. 1.3), suggesting that annual plant growth on land has measurably increased. Perhaps, here we are seeing a benefit from the increased levels of CO_2 in the atmosphere, though it seems doubtful that this will balance the associated losses.

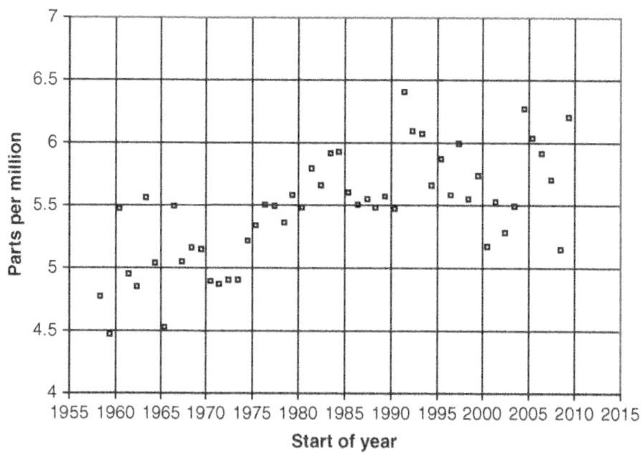

Fig. 1.3 Amounts of the annual draw-down evident in Fig. 1.1, from the northern spring (AMJ) to fall (SON) in each year. The data suggest that the amount of the drop is growing, probably indicating increased land productivity in a higher-CO_2 world

1.4 Ocean Colour and a Change in Global Productivity

As oceanographers, we need to ask what the ocean has contributed to this change. Has global ocean productivity (or biomass, as we effectively measure it) increased? If the total increase of ocean productivity is as large as that on land, then presumably the Keeling curve would not show an increase in its annual cycle amplitude, as the northern summer decrease would be offset by an increase in total productivity of the southern ocean.

The time series we have for this is poor, to say the least. It has been suggested that the average biomass deduced from the entire CZCS mission (1978–1986) was less than that more recently deduced by SeaWiFS for 1997–2004 (Antoine et al., 2005). Changes within the SeaWiFS data time series have also been reported (Gregg et al., 2005; Behrenfeld et al., 2006), and shown to be strongly affected by the El-Nino/La-Nina in 1997–1999. If these data are plotted as time series (Fig. 1.4), they lack the impressive impact of a multi-point plot, but they can be important nonetheless. The patterns of change seem related to an increase in area of oligotrophic waters in ocean basin gyres (Polovina, 2008).

In determining any trend in biomass, we come up against the continuing problem of providing long-term, stable, "climate quality" time series of global ocean optical data. Ocean color has already experienced the 10-year gap between CZCS and SeaWiFS. In an open letter to the ocean optics community Siegel et al. (2008) stated "It appears likely that the ocean biology and biogeochemistry communities will (again) face a multi-year gap in our climate data records." A problem which I would summarize in verse as:

We've needed one more SeaWiFS for many, many years
Instead we've bought 2 MODISs, a MERIS and a VIIRS

I should stress that both MODIS and MERIS have brought their own scientific and operational successes, but they were not designed as simple, stable instruments.

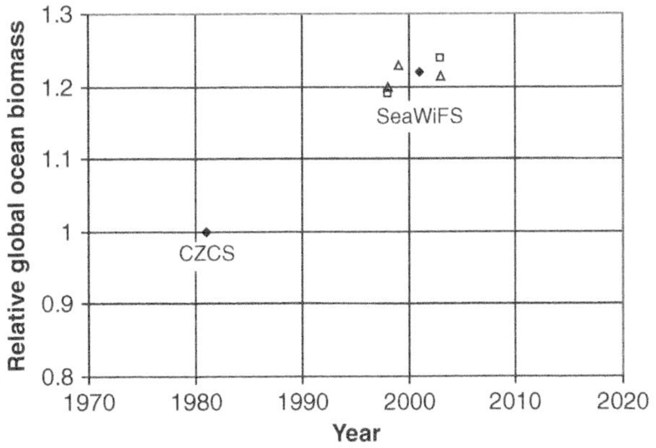

Fig. 1.4 Relative change in global ocean biomass from CZCS and SeaWiFS. *Filled diamonds*: Antoine et al. (2005). *Open squares*: Gregg et al. (2005). *Open triangles*: Behrenfeld et al. (2006)

It is possible that clever calibration may allow them to successfully continue the SeaWiFS time series. It is even possible that SeaWiFS may rise from the dead, and operate for a few more years (G. Feldman, personal communication).

1.5 Sea Surface Temperature (SST)

The Reynolds time series is based on ship and buoy data, using satellite data in data-sparse areas. This is only partly a "satellite" data time series and is becoming less so as surface data sources such as Argo, increase in density. Figure 1.5 shows the Hadley SST data series, in which the long-term warming trend is very clear, with short-term increases at the 1972 and 1997 El-Nino events, and a recent slight interruption of the steady warming, which is being made much of by warming skeptics ("cooling since 1998"), but which in fact looks typical of the variability in the record.

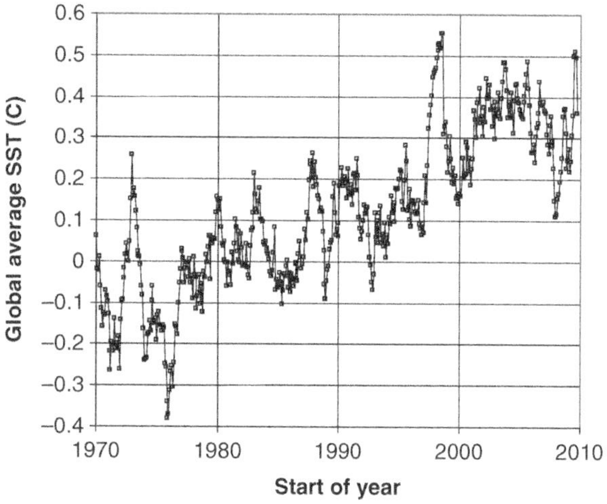

Fig. 1.5 Time series of global SST provided by the Hadley Centre, UK. Data show a warming trend of 0.014°C/year, with increases of about 0.3°C during the 1972 and 1997 El Ninos, but a smaller signal from the 1982 event

1.6 Polar Sea Ice Cover

The US National Snow and Ice Data Center (NSIDC) provides an archive of Arctic ice cover data based on satellite microwave radiometer observations of polar ice.[3] Figure 1.6 shows the anomaly time series deduced for each month by subtracting a fixed annual cycle. From the late 1970s to about the year 2000, the measured total ice areas oscillated with the seasons between about 5 and 15 million km^2. In Fig. 1.6, a slow drop of about 1 million km^2 is evident over this period, accelerating

[3] Available at ftp://sidads.colorado.edu/DATASETS/NOAA/G02135/

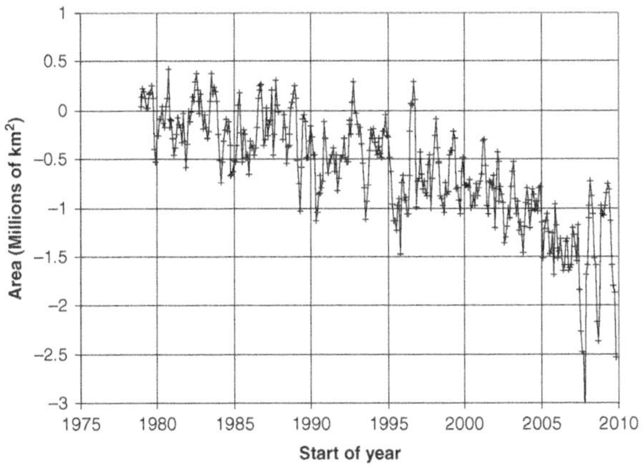

Fig. 1.6 Arctic sea ice area anomaly time series from the US National Snow and Ice Data Center, based on satellite microwave radiometer data, subtracting a fixed annual cycle for all years

after 2000, but recovering after 2007. The minimum in September 2007 represented a loss of almost half the usual late summer ice cover. In the three most recent years it appears that the annual cycle has changed to one with a larger amplitude. This may be related to the loss of multi-year ice in 2007. It is definitely a time series to watch.

1.7 The GRACE Satellite and Melting Ice Caps

The sea-ice loss in Fig. 1.6 is significant for northern navigation and Arctic warming, but will not affect sea level rise. Altimeter surveys of Greenland and Antarctica are attempting to show the mass loss from these two major ice caps.

Gravity measurements from GRACE (Gravity Recovery And Climate Experiment) suggest that these data can also provide useful time series. Recent data seem to show not only mass loss due to ice melting on Greenland and Antarctica, but an acceleration in this loss as shown in Fig. 1.7 (Velicogna, 2009). The short time series show average melt rates between 2002 and 2009 of 230 ± 33 Gt/year for Greenland and 143 ± 73 Gt/year for Antarctica, giving a total melt rate of 370 ± 80 Gt/year, equivalent to a sea level rise rate of 1.1 ± 0.2 mm/year.

The data also show apparent accelerations of 30 ± 11 Gt/year^2 for Greenland and 26 ± 14 Gt/year^2 for Antarctica. These imply melt rates in 2003 of 140 and 100, for a total of 240 Gt/year, or a sea level rise of 0.7 mm/year, increasing to melt rates in 2008 of 290 and 250, for a total of 540 Gt/year, or a sea level rise of 1.6 mm/year.

The time series in Fig. 1.7 are short, and we might hope to see significantly longer series at Venice 2020. Sadly, GRACE is due to die before then. It consists of a pair of satellites about 200 km apart, whose exact separation is precisely measured to show the effects of small changes in earth's gravity. For sensitivity, it is in a low orbit and needs frequent boosts to counteract atmospheric drag. It was designed for

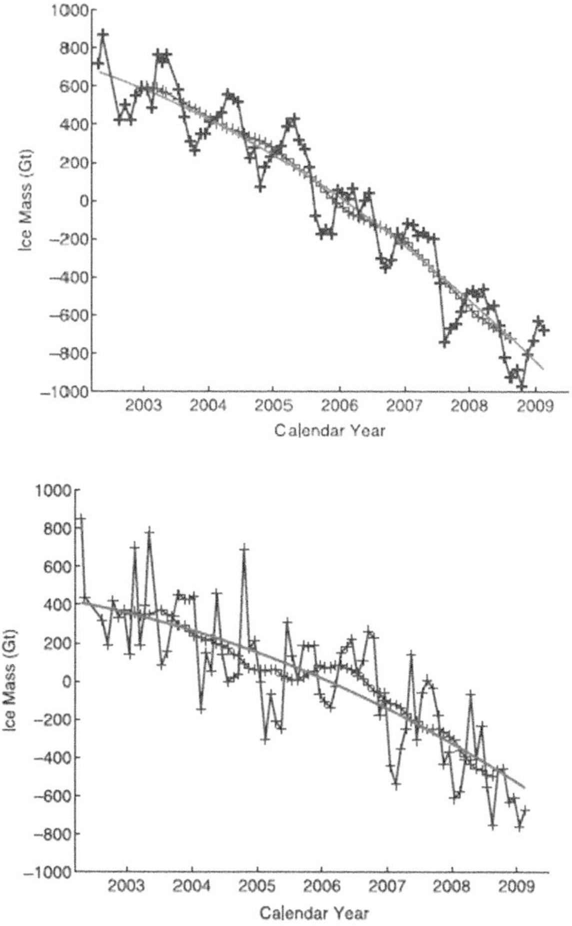

Fig. 1.7 Ice mass loss from Greenland (*top*) and Antartica (*bottom*) seen by the Gravity Recovery and Climate Experiment, GRACE (Velicogna, 2009). Each *plot* shows points, the result of smoothing by a 13 month window, and a quadratic fit

a 5-year life, is now in a 2-year extension, and will eventually run out of the fuel needed to keep it aloft.

1.8 Satellite Altimetry and Global Sea Level Rise

One of the most fascinating time series, especially for people living near a coast, is of global sea level measured by the Topex/Poseidon and Jason satellite altimeters (Leuliette et al., 2004). The data[4] (Fig. 1.8) show a very clear and relatively constant rate of increase of 3.1 $\pm$ 0.4 mm/year with an rms noise level on the individual points of only a few millimeters, much lower than is possible from shore-based tide gauges.

[4] Available from http://sealevel.colorado.edu/results.php

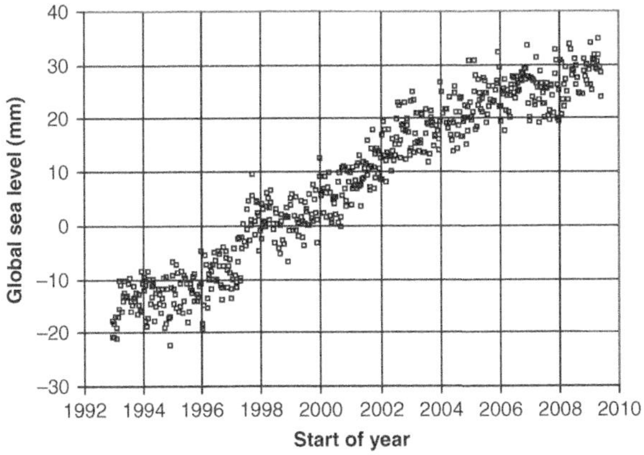

Fig. 1.8 The time series of global average sea surface height, based on 10-day repeat coverage of the Topex/Poseidon and Jason altimetry satellites, showing an average global rise rate of 3.1 mm/year

The series shows a peak of about a centimeter at the time of the 1997/98 El-Nino and a slower rise rate since the end of 2006. Sea surface heights are linked to the global sea surface temperature shown in Fig. 1.5 through the expansion of near-surface sea water. The 0.3°C average temperature increase associated with the 1997 El-Nino is equivalent to 7 mm of sea level rise if applied to the top 100 m of the ocean surface. The dip in temperatures between 2005 and 2009 explains some of the recent slowing in rise rate in Fig. 1.8. We note that GRACE data suggest that we should see an acceleration in global sea level rise. As yet there is no indication of this in the altimetry (Fig. 1.8), if anything the rise seems to have slowed since 2006. These observations need to be reconciled.

It is worth remembering that the global significance of plots like Figs. 1.5, 1.6 and 1.8 means that they will be inspected by many non-scientists who are more used to looking at time series of company share prices, indicating investment value. To such people, it is the recent trend which has special importance. Hence, they focus on "recent cooling" in Fig. 1.5 and "slow down" in Fig. 1.8. I'm not sure how they would interpret Fig. 1.6. Certainly, something in the Arctic has changed. Climate scientists have higher tolerance for "short-term fluctuations." Maybe this sometimes leads them to make bad investment decisions? Figure 1.8 is a relatively short time series, which will have lengthened significantly by the time of Venice 2020, either confirming the slow down, continuing the steady upward trend, or showing acceleration.

1.9 Venice Acqua Alta

And so we return to Venice. Any city close to sea level needs to take global sea level rise very seriously. At the first two conferences in Venice (1980 and 1990), I was not aware of any flooding from the sea. In 2000, we saw the problem. Tides in the Mediterranean are small and are often ignored by boaters. In Marseilles, for example, the tide range is only about 20 cm. In Venice at the head of the Adriatic, tides are larger, but the range is still less than a metre. The phenomenon of "acqua alta" or high water in Venice is related more to winds blowing up the Adriatic, piling water in a "surge" which can be at least a metre higher. Lower barometric pressure associated with a storm will also raise the level by up to 30 cm, 10 cm for each 10 mb drop in pressure.

Venice also has to take seriously any drop in the level of the land on which the city is built. Since 1930, the area surrounding Venice has sunk about 20 cm due to the extraction of arterial water. Thanks partly to scientific work by the ISDGM (Instituto per lo Studio della Dinamica delle Grandi Masse, now ISMAR, Istituto di Scienze Marine), the host institute of these conferences, this extraction has now been halted, but the loss of height remains (Carbognin and Gatto, personal communication[5]).

Floods have affected Venice throughout recorded history. Today, a tide of 90 cm above the standard Venice reference will start to cover St Mark's Square, but will cause few other problems. A tide of 110 cm will put 12% of the surface area of Venice under water. When this level is expected, forecast warnings are sounded (Fig. 1.9). At 130 cm, 70% of the city is flooded, and at 150 cm this rises to 96%. The floods do not usually last long, dropping as the high tide passes. The water is usually clear and drains away with relatively little damage, unlike a river flood,

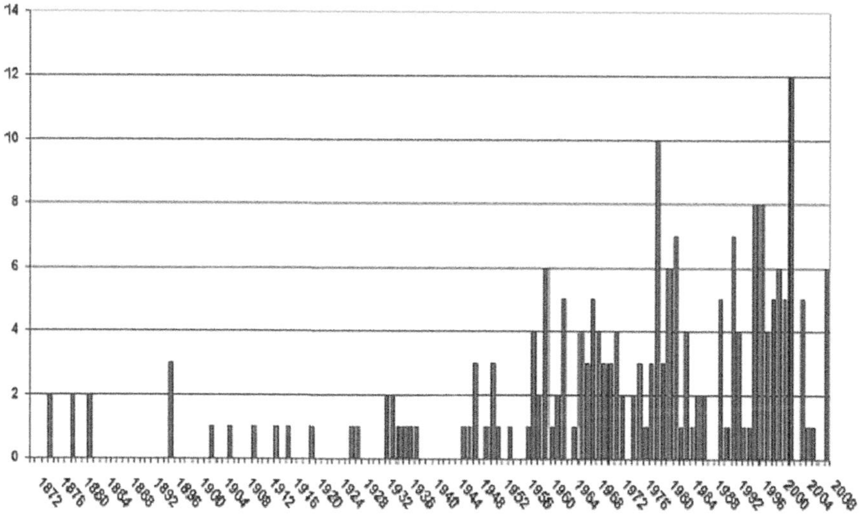

Fig. 1.9 The number of times in each year that water level at the Venice tide station has exceeded 110 cm, the level at which an alarm is sounded[6]

[5] See http://iahs.info/redbooks/a151/iahs_151_0321.pdf

[6] Data from www.comune.venezia.it/maree

which may bring mud and debris. I remember watching a small fish swim by the meeting hall in 2000. An added problem of the high water is that boats cannot pass under some bridges. Tourists now come to witness "acqua alta", but the city suffers.

The highest "acqua alta" in recent times was on 4 November 1966, reaching a level of 194 cm, enough to flood the entire city and do extensive damage. Since then people are more prepared, but levels have never come within 30 cm of this record. Recent highs were 156 cm on 1 December 2008, 147 cm on 16 November 2002 and 144 cm on 6 November 2000. Most extreme highs occur in November and December, but on 16 April 1936, water reached 136 cm. I see there will be a full moon at 12:21 UT on Wednesday 28 April 2010, during our conference. We must hope the barometer stays high and the winds light.

The project MOSE[7] is now underway, installing rising gates to block the gaps in the chain of offshore islands which include the Venice Lido. This is a huge and expensive project which has already faced delays and budget over-runs. It is now proceeding slowly and is due to be completed in about 2014.

1.10 Conclusions

What can we conclude? First, let us be selfish and conclude that we should all live long and scientifically productive lives, which will let us see how this dangerous adventure of the human race turns out. I hope we will all be here again at Oceans from Space, Venice 2020. For this, both Venice and ourselves need to survive. The extra 10 years will give us an improved perspective, and maybe the Keeling curve will have slowed. If not, the sea level rise may indeed have accelerated, and MOSE may well have already been proven inadequate.

At the 2000 conference I said I hoped to see humanity's first contact with intelligent life elsewhere in the universe. This may seem a strange dream, yet it is reasonably rational. To many specialists the mystery is why another civilization has not already contacted us. Maybe global warming tends to wipe out intelligent life? Maybe we'll see.

On a more down-to-earth note, we certainly conclude that the work covered by this conference is important, and that its importance is growing. We hope that governments will appreciate the need for improved monitoring of the global marine environment, and more importantly that they will act on the need to reduce emissions of carbon dioxide. I look for a significant reduction in the rise rate of the Keeling curve below the constant acceleration shown in Fig. 1.1 and Equation (1.1). I'm less sure how we hope to reduce the actual level to 350 ppm.[8]

Finally, let me also ask once again for any ideas on "best ocean satellite-based time series." There must be others. I'm working on statistics of bright blooms

---

[7] See http://en.wikipedia.org/wiki/MOSE_project

[8] As suggested by www.350.org

detected by the MCI index of MERIS. Can we show that blooms are changing or increasing in frequency (see, for example, Barale et al., 2008)? What other trends are apparent? Winds, waves, coral bleaching? Please point out candidates.

Acknowledgements Thanks to the Institute of Ocean Sciences (IOS) of the Canadian Department of Fisheries and Oceans and the Canadian Space Agency for continuing research support, to the Canadian Federal Government whose rules allow me to work past the age of 65, to Eric Lindstrom of NASA who recommended that I include GRACE results, and to my colleague Stephanie King who recommended that I leave out the "silly poem" (see Ocean Colour) and who helped prepare figures. Thanks also to colleagues at IOS and to V. Barale, JRC EC, Ispra (I), for useful comments and discussions.

References

Antoine D, Morel A, Gordon HR, Banzon VJ, Evans RH (2005) Bridging ocean color observations of the 1980s and 2000s in search of long-term trends. J Geophys Res 110:C06009, doi:10.1029/2004JC002620

Barale V, Jaquet JM, Ndiaye M (2008) Algal blooming patterns and trends in the Mediterranean Sea as derived from the SeaWiFS data set (1998–2003). Rem Sens Environ, special issue on: Earth Observations for Marine and Coastal Biodiversity and Ecosystems 112:3300–3313

Behrenfeld MJ, O'Malley RT, Siegel DA, McClain CR, Sarmiento JL, Feldman GC, Milligan AJ, Falkowski PG, Letelier RM, Boss ES (2006) Climate-driven trends in contemporary ocean productivity. Nature 444:752–755

Gregg WW, Casey NW, McClain CR (2005) Recent trends in global ocean chlorophyll. Geophys Res Lett 32:L03606, doi:10.1029/2004GL021808

Keeling CD, Whorf TP, Wahlen M, van der Plicht J (1995) Interannual extremes in the rate of rise of atmospheric carbon dioxide since 1980. Nature 375:666–670

Leuliette EW, Nerem RS, Mitchum GT (2004) Results of TOPEX/Poseidon and Jason calibration to construct a continuous record of mean sea level. Mar Geod 27:79–94

Polovina JJ, Howell EA, Abecassis M (2008) Ocean's least productive waters are expanding. Geophys Res Lett 35:L03618, doi:10.1029/2007GL031745

Siegel DA, Yoder JA, McClain CR (2008) Ocean Color Letter, at http://oceancolor.gsfc.nasa.gov/DOCS/OCB_OceanColorLetter.pdf

Velicogna I (2009) Increasing rates of mass loss from the Greenland and Antarctic ice sheets revealed by GRACE. Geophys Res Lett 36:L19503, doi:10.1029/2009GL040222

Wolter K, Timlin MS (1998) Measuring the strength of ENSO events – how does 1997/98 rank? Weather 53:315–324

Chapter 2
Passive Microwave Remote Sensing of the Ocean: An Overview

Chelle L. Gentemann, Frank J. Wentz, Marty Brewer, Kyle Hilburn, and Deborah Smith

2.1 Introduction

Global geophysical measurements from passive microwave radiometers provide key variables for scientists and forecasters. The daily measurements of Sea Surface Temperature (SST), wind speed, water vapor, cloud liquid water, rain rate, and, in the future, Sea Surface Salinity (SSS) over the oceans has provided data sets used to significantly improve our understanding of the Earth system. The data are used extensively in numerical weather prediction, hurricane forecasting, climate monitoring, ecosystem forecasting and fisheries; as well as for climate, weather, oceanographic, metorological and ecosystem research. The measurement accuracy is tied to the evolution of both the calibration methods and retrieval algorithms.

2.2 Background

Designed to measure rainfall, the first Passive MicroWave (PMW) radiometer was launched in December 1972 on the Nimbus-5 satellite. After a short gap, PMW radiometers have been continuously observing the oceans since the launch of Nimbus-7 in 1978. This instrument was followed by the Special Sensing Microwave Imager (SSM/I) series. More recently, several other PMW radiometers have been launched on National Aeronautics and Space Administration (NASA), Japan Aerospace eXploration Agency (JAXA), and European Space Agency (ESA) satellites (Table 2.1).

The Electrically Scanning Microwave Radiometer (ESMR) on Nimbus-5 had only one channel at 19.35 GHz and was capable of measuring both rainfall and sea ice detection.

From October 1978 to July 1987, the Nimbus-7 Scanning Multi-channel Microwave Radiometer (SMMR) measured at 6.6, 10.7, 18.0, 21.0, and 37 GHz in

C.L. Gentemann (✉)
Remote Sensing Systems, Santa Rosa, CA 95401, USA
e-mail: gentemann@remss.com

V. Barale et al. (eds.), *Oceanography from Space*,
DOI 10.1007/978-90-481-8681-5_2, © Springer Science+Business Media B.V. 2010

Table 2.1 PMW radiometer mission characteristics

Satellite	Sensor	Launch	Failure	Frequency (GHz)	Coverage
Nimbus-5	ESMR	12/1972	5/1977	19.4	Global
Nimbus-7	SMMR	10/1978	8/1987	6.6, 10.7, 18.0, 21.0, 37.0	Global
SEASAT	SMMR	6/1978	10/1978	6.6, 10.7, 18.0, 21.0, 37.0	Global
DMSP F08	SSM/I	7/1987	12/1991	19.4, 22.2, 37.0, 85.5	Global
DMSP F10	SSM/I	12/1990	11/1997	19.4, 22.2, 37.0, 85.5	Global
DMSP F11	SSM/I	12/1991	5/2000	19.4, 22.2, 37.0, 85.5	Global
DMSP F13	SSM/I	5/1995	11/2009	19.4, 22.2, 37.0, 85.5	Global
DMSP F14	SSM/I	5/1997	8/2008	19.4, 22.2, 37.0, 85.5	Global
DMSP F15	SSM/I	12/1999	Present	19.4, 22.2, 37.0, 85.5	Global
TRMM	TMI	12/1997	Present	10.7, 19.4, 21.3, 37.0, 85.5	40S-40N
ADEOS-II	AMSR	12/2002	10/2003	6.9, 10.7, 18.7, 23.8, 36.5, 89.0	Global
AQUA	AMSR-E	5/2002	Present	6.9, 10.7, 18.7, 23.8, 36.5, 89.0	Global
Coriolis	WindSat	6/2003	Present	6.8, 10.7, 18.7, 23.8, 37.0	Global
DMSP F16	SSMI/S	10/2003	Present	–	Global
DMSP F17	SSMI/S	11/2006	Present	–	Global
SMOS	MIRAS	11/2009	–	1.4	Global
DMSP F18	SSMI/S	10/2009	Present	–	Global
GPM	GMI	(7/2013)	–	10.7, 18.7, 23.8, 36.5, 89.0	65S-65N
SAC-D	Aquarius	(5/2010)	–	1.4	Global
GCOM-W	AMSR2	(2/2012)	–	6.9, 7.3, 10.7, 18.7, 23.8, 36.5, 89.0	Global
C2	MIS	(5/2016)	–	6.8, 10.7, 18.7, 23.8, 37.0, 89.0	Global

both the horizontal and vertical polarizations (Gloersen et al., 1984). SMMR geophysical retrievals were compromised by non-negligible switch leakages (Han and Kim, 1988), rendering the SMMR measurements useful for detection of sea ice but not accurate enough for geophysical retrievals.

The Defense Meteorological Satellite Program (DMSP) satellite series launched the first SSM/I on F-08 in June 1987. This was followed by SSM/Is on F-09 to F-15. The DMSP satellites orbit the earth in 102 min, at approximately 833 km with an inclination of 98.8° (Hollinger et al., 1990). The F-series alternate between early and late morning Local Equator Crossing Times (LECTs). The SSM/I instrument measures at 19.4, 22.2, 37.0, and 85.5 GHz. Both vertical and horizontal polarizations are measured for all channels except the 22.2 GHz which only measures the vertical. SSM/I was the first satellite PMW radiometer to have external calibration accomplished by viewing a mirror that reflects cold space and a hot reference absorber once each scan, every 1.9 s. The cold space is a known 2.7 K while the hot absorber temperature is monitored with thermistors. The frequent calibration minimizes receiver gain fluctuation contributions to the signal but does not correct radiometer nonlinearity (if it exists). This well-calibrated instrument's measurements are used to determine wind speed, water vapor, cloud liquid water, rain rates, and sea ice concentration over global oceans.

In December 1997, NASA launched the Tropical Rainfall Measuring Mission (TRMM) carrying the TRMM Microwave Imager (TMI), a PMW radiometer measuring at 10.7, 19.4, 21.3, 37.0, and 85.5 GHz. Similar to SSM/I, all channels measure both vertical and horizontal polarizations, except the 21.3 GHz which only measures in the vertical (Kummerow et al., 1998). Designed to measure the tropics and sample the diurnal cycle, the satellite was launched with an orbital inclination of 35° at an altitude of 350 km (later changed to 400 km to extend satellite life). This equatorial orbit yields coverage from 39N to 39S. The satellite is sun-asynchronous, processing through the diurnal cycle every 23 days. Again, similar to SSM/I, the feed horns and main reflector rotate, with a period of 1.9 s, about an axis parallel to the local spacecraft nadir. The stationary hot reference absorber and cold calibration reflector are positioned so that they pass between the feed horns and main reflector once per scan. The temperature of the warm load is monitored by three thermistors while the cold reflector views the cosmic microwave (MW) background at 2.7 K. At fairly regular intervals the platform yaws from forward (aft) viewing direction to aft (forward). Each scan consists of 104 discrete samples spaced by 8 km. In addition to the geophysical variables measured by SSM/I, TMI is able to measure SST. TMI suffered calibration problems due to an emissive reflector, for which corrections were developed and implemented.

NASA's AQUA satellite carries the JAXA's Advanced Microwave Scanning Radiometer – Earth Observing System (AMSR-E). The AQUA satellite was launched in May 2002 into a polar, sun-synchronous orbit at an altitude of 705 km, with a LECT of 1:30 AM/PM. AMSR-E has 12 channels corresponding to 6 frequencies: 6.9, 10.7, 18.7, 23.8, 36.5, and 89.0 GHz, all except 23.8 measure both vertical and horizontal polarizations (Parkinson, 2003). The calibration is completed similar to SSM/I and TMI using a cold reflector and hot absorber with eight thermistors. The AMSR-E hot absorber has large thermal gradients not well measured by the thermistors. A correction for this error in the calibration reference point has been developed and implemented. In addition to the geophysical variables measured by SSM/I, AMSR-E is able to measure SSTs. Almost global coverage is attainable in 2 days (Fig. 2.1).

The Naval Research Laboratory (NRL) launched the Coriolis satellite in January 2003. The sun-synchronous orbit is at an altitude of 840 km with a LECT at 6:00 AM/ PM (Gaiser et al., 2004). Coriolis carries the WindSat instrument, a fully polarimetric PMW radiometer intended to retrieve wind direction in addition to wind speed. The fully polarimetric channels are at 10.7, 18.7, and 37.0 GHz, but the instrument also has channels at 6.8 and 23.8 that only measure the vertical and horizontal polarizations. Calibration is similar to SSM/I with a cold reflector and hot absorber measured by six thermistors.

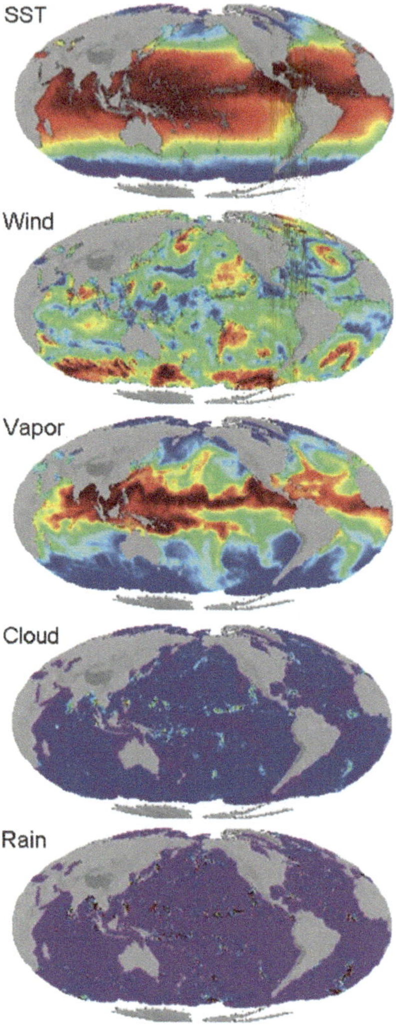

Fig. 2.1 AMSR-E geophysical retrievals 1–2 October 2009. Small amounts of missing data due to rain events are visible in the SST and wind retrievals

DMSP satellites F16 and forward carry the Special Sensor Microwave Imager/Sounder (SSMIS). F16 was launched in October 2003 into a sun-synchronous orbit at an altitude of 830 km and a LECT of 8 AM/PM. SSMIS has 24 channels, several of which are similar to the SSM/I set (19.35, 22.2, and 37.0 GHz). The additional channels are intended for atmospheric sounding. The calibration is completed similar to SSM/I using a cold reflector and hot absorber. SSMIS has two main problems, an emissive antenna and non-uniform hot absorber. Corrections for these issues have been developed and implemented.

Future PMW radiometers include JAXA's Global Change Observation Mission – Water (GCOM-W) AMSR2, the National Polar Orbiting Earth observing System of Systems (NPOESS) C2 satellite will carry the Microwave Imager Sounder (MIS), and NASA's Global Precipitation Mission (GPM) will carry the GPM Microwave Imager (GMI). For all these instruments, the planned calibration is similar to SSM/I using a cold reflector and hot absorber.

GCOM-W is to be launched in February 2012 into NASA's A-Train satellite formation in a sun-synchronous orbit with an altitude of 700 km and a LECT of 1:30 AM/PM. AMSR2 is similar to AMSR-E but has an improved hot absorber and an additional channel at 7.3 GHz to minimize Radio Frequency Interference (RFI). With a launch date set for February 2012, it is hoped that the AQUA AMSR-E remains healthy until then to allow for satellite inter-calibration.

Two other future instruments, the European Space Agency's Soil Moisture and Ocean Salinity (SMOS) Microwave Imaging Radiometer using Aperture Synthesis (MIRAS) and the Satélite de Aplicaciones Científicas-D (SAC-D) Aquarius are intended to measure ocean salinity and only have a single channel at 1.4 GHz. SMOS launched in November 2009 into a sun-synchronous orbit at 800 km with an LECT of 6:00 AM/PM. Aquarius is scheduled to be launched in May 2010 into a sun-synchronous orbit at 650 km with a LECT of 6:00 AM/PM. Both of these instruments are designed to provide measurements of ocean salinity.

2.3 Calibration

To create a climate quality, inter-calibrated dataset of PMW geophysical retrievals, it is necessary to start the process using radiometer counts and work towards calibrated geophysical retrievals. Table 2.2 describes the steps to produce a calibrated brightness temperature (TB). First, it is necessary to reverse engineer the antenna temperatures (TAs) or TBs back to radiometer counts. Often there are small provider added corrections or adjustments put into the TA or TBs which are sometimes undocumented. For example, SSMI/S had five TB version changes in the first 2 years of data. Therefore, the first step is to reverse these steps and remove any corrections. Starting from radiometer counts, the first two steps in the calibration procedure are crucial to accurately determining other errors.

To ensure that any subsequent collocations or comparisons that are performed are correct, it is necessary to do a geolocation analysis. The correction to the

Table 2.2 Calibration steps for PMW radiometers

	Geolocation analysis	Attitude adjustment	Along-scan correction	Absolute calibration	Hot load correction	Antenna emissivity
SSM/I	NRL/RSS	No	Yes	APC	No[a]	0
TMI	Goddard	Dynamic	Yes	APC	No	3.5%
AMSRE	RSS	Fixed	Yes	APC	Yes	0
AMSRA	RSS	Dynamic	Yes	APC	Yes	0
WindSat	NRL/RSS	Fixed	Yes	APC	Yes	0
SSMIS	RSS	No	Yes	APC	Yes	0.5–3.5%

[a]Errors due to hot load are removed when doing the zonal TB inter-calibration

geolocation is different than a correction for erroneous satellite pointing information (roll/pitch/yaw), which is a correction for the mounting of the instrument on the satellite. Pointing is usually off by about 0.1° from the satellite specified roll/pitch/yaw. The geolocation correction uses ascending minus descending TA to ensure that islands do not "move". The geolocation analysis has been performed by a number of groups, NRL and Remote Sensing Systems (RSS) both contributed to SSM/I, TMI was completed by Goddard, and other instruments as specified in Table 2.2.

Corrections from this point onward are determined by comparisons between the satellite TA measurements and TAs simulated using a radiative transfer model (RTM). Using collocated environmental information, RTM simulated TBs are determined. These TBs are then transformed into TAs using the instrument, channel specific antenna patterns.

After the pointing is corrected, the spacecraft reported roll/pitch/yaw are then examined for errors using comparisons of the observed minus RTM TAs. Spacecraft pointing is determined by a number of different methods, the preferred being a star tracker. Another method is a horizon balancing sensor. For SSM/I no pointing information was given, so it was assumed to be correct. TMI has a dynamic pointing correction that changes within an orbit because the horizon sensor used prior to the orbit boost is not as accurate as a star tracker. After boost, the horizon sensor was disabled and pointing was determined from two on-board gyroscopes, also not as accurate as a star tracker. AMSR-E had no pointing problems, as the AQUA had a star tracker. The AMSR on ADEOS-II needed a dynamic correction, while WindSAT needed a simple fixed correction to the roll/pitch/yaw.

Once instrument mounting errors and satellite attitude errors have been corrected for, an along-scan correction is completed. It is very important to complete the first two corrections first because TA is dependent on incidence angle. Not correcting for pointing errors would result in an erroneous cross-scan biasing. As the mirror rotates, at the edge of the earth scene the view will begin to contain obstructions such as the satellite itself or part of the cold mirror. Additionally, during the scan, the antenna side-lobe pattern may result in contributions from different parts of the spacecraft. Therefore, the difference between the TA and RTM simulated TAs are again used to examine the data for along-scan biases. This correction is needed for every instrument.

The antenna pattern correction (APC) is then completed. Pre-launch, an APC is determined, consisting of the spill over and cross-polarization values. After launch, the spill over and cross-polarization values are adjusted so that the measured TAs match the simulated TAs. This correction is needed for all instruments. Next, a correction for the hot load thermal gradients and antenna emissivity are developed. These are only needed for specific instruments. The determination of TB from counts for PMW radiometers is completed using two known temperatures to infer the scene temperature. For each scan, the feedhorns view a mirror that reflects cold space, a known 2.7 K, a hot absorber, measured by several thermistors, and Earth scenes. Assuming a linear response, the Earth scene temperatures are then determined by fitting a slope to the two known measurements as shown in Fig. 2.2. This

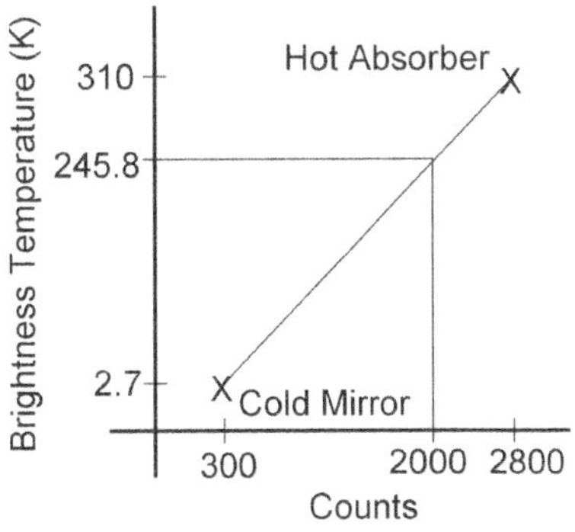

Fig. 2.2 Calculation of earth scene brightness temperatures using the radiometer counts and calibration points (cold mirror and hot absorber) known temperatures

2-point calibration system continuously compensates for variations in the radiometer gain and noise temperatures. This seemingly simple calibration methodology is fraught with subtle difficulties. The cold mirror is relatively trouble-free, as long as lunar contamination is flagged. Occasionally, the cold mirror will not reflect deep space, but the moon instead. These data must be removed.

The hot absorber has been more problematic as the thermistors often do not adequately measure thermal gradients across the hot absorber. For example, a hot load correction is required for AMSR-E because of a design flaw in the AMSR-E hot load. The hot reference load acts as a blackbody emitter and its temperature is measured by precision thermistors. Unfortunately, during the course of an orbit, large thermal gradients develop within the hot load due to solar heating making it difficult to determine the average effective temperature from the thermistor readings. The thermistors themselves measure these gradients and may vary by up to 15 K between themselves at any time for AMSR-E. Several other instruments have had similar, but smaller, issues. RTM simulations are used to determine an effective hot load temperature which is a regression of the measured hot load thermistor temperatures. The follow-on instrument, AMSR2 on GCOM-W, has an improved hot absorber design that should mitigate these issues.

Finally, the main reflector is assumed to be a perfect reflector with an emissivity of 0.0, but this is not always the case. For example, a bias recognized in the TMI measurements was attributed to the degradation of the primary antenna. Atomic oxygen present at TMI's low altitude (350 km) led to rapid oxidization of the thin, vapor-deposited aluminum coating on the graphite primary antenna, resulting in a much higher antenna emissivity than expected. The measured radiation is comprised of the reflected earth scene and antenna emissions. Emissivity of the antenna was deduced during the calibration procedure to be 3.5%. The antenna emissivity correction utilizes additional information from instrument thermistors to estimate the antenna temperature, thereby reducing the effect of the temporal variance. This emissivity is constant for all the TMI channels. SSMI/S has an emissive antenna where the emissivity appears to increase as a function of frequency, changing from 0.5 to 3.5%.

2.4 Retrieval Algorithm

Geophysical retrievals from PMW radiometers are commonly determined using a radiative transfer model to derive a regression algorithm (Wentz and Meissner, 2000). A schematic of the derivation of the regression coefficients is shown in Fig. 2.3. A large ensemble of ocean-atmosphere scenes is first assembled. The specification of the atmospheres comes from quality-controlled radiosonde flights launched from small islands (Wentz, 1997). One half of these radiosonde flights are used for deriving the regression coefficients, and the other half is withheld for testing the algorithm. A cloud layer of various columnar water densities ranging from 0 to 0.3 mm is superimposed on the radiosonde profiles. Underneath these simulated atmospheres, we place a rough ocean surface. SST is randomly varied from 0 to 30°C, the wind speed is randomly varied from 0 to 20 m/s, and the wind direction is randomly varied from 0 to 360°.

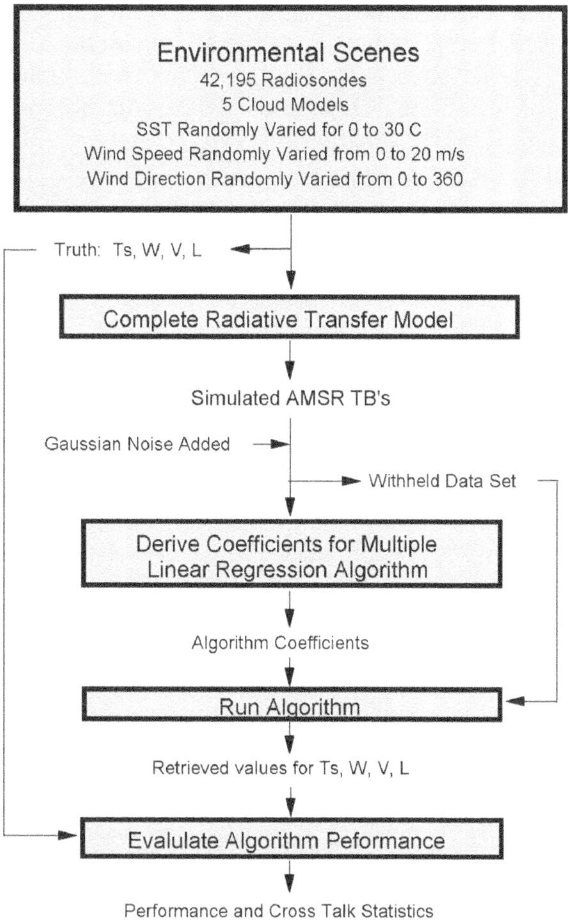

Fig. 2.3 Derivation of regression coefficients

Atmospheric brightness temperatures and transmittance are computed from these scenes and noise, commensurate with measurement error which depends on spatial resolution, is added. The noise-added simulated brightness temperatures along with the known environmental scene are used to generate multiple linear regression coefficients. Algorithm testing is undertaken by repeating the process using the withheld scenes.

2.5 Geophysical Retrievals

2.5.1 Wind Speed

Ocean surface winds are crucial to transferring heat, gases, energy and momentum between the atmosphere and ocean. Winds also determine the large scale ocean circulation and transport, power global weather patterns, and play a key role in marine ecosystems. Hurricanes, typhoons, and mid-latitude winter storms all contain high wind speeds that threaten international shipping and the lives and property of people along the coasts. Ocean surface winds change rapidly in both time and space and satellite sampling and accuracy make these observations the most useful wind data available for research and forecasting over the global oceans.

Surface wind speeds (at 10 m height, without directions) are routinely estimated from passive microwave radiometers (SSM/I, AMSR-E, TMI, SSMIS) on a spatial scale of roughly 25 km. Wind speeds in the range of 0–30 m/s are simultaneously retrieved along with SST, water vapor, cloud liquid water and rain rates using an algorithm that exploits the polarization signature of wind induced sea surface emissivity (Wentz, 1997). Radiometer winds are quite accurate under typical ocean conditions when no rain is present, however when even a little rain exists, the wind speeds are unusable. Validations of radiometer winds in rain-free conditions have been performed. Comparisons with ocean buoy and weather model winds show root-mean-square differences near (model winds) or less than 1 m/s (buoy winds) in rain-free conditions (Mears et al., 2001; Meissner et al., 2001). Since 1996, there have been three or more radiometers in polar orbits simultaneously, resulting in good spatial and temporal sampling, yielding over 95% Earth ocean surface coverage in a given day.

WindSat is a passive fully-polarimetric microwave radiometer designed to measure ocean surface vector winds. It has been found to have wind accuracies close to that of scatterometers for winds between 6 and 20 m/s, with significant wind direction uncertainty below 6 m/s (Bettenhausen et al., 2006). WindSat vector winds have been poor in rainy conditions until recently when a new WindSat algorithm has been developed that improves WindSat winds even in rain (Meissner and Wentz, 2009). The quality of these new winds is similar to QuikScat in all but very heavy rain and very low winds. Excellent agreement (to within 0.5 m/s) is found between passive radiometer wind speeds, polarimetric radiometer wind vectors and scatterometer vector winds despite the different measuring methods of each instrument (Wentz and Meissner, 2007). Only a few small regions of difference exist that seem to be related to the 37 GHz observations of the ocean surface and atmosphere.

Combined surface wind data sets have recently become more available and are very useful in atmospheric and oceanographic research due to the lack of data gaps. One example, the Cross Correlated Multi-Platform (CCMP) winds (Atlas et al., 2009), use carefully inter-calibrated PMW wind speeds from radiometers and wind vectors from scatterometers. Simple interpolation schemes are unable to adequately represent fast-moving storms in mid-latitude regions when making a merged wind product with no gaps. An advanced 4-dimensional variational analysis method is used in the CCMP to merge the satellite winds with the European Center for Medium-range Weather Forecasting (ECMWF) Re-Analysis (ERA)-40 model wind vectors, providing a gridded wind product consisting of an analyzed wind field every 6 h for 20 years.

The satellite winds used in the CCMP include over 20 years of SSMI winds. A recent study showed that these carefully inter-calibrated SSM/I winds have no spurious trends. Wentz et al. (2007) found agreement between ocean buoy trends and the SSM/I trends for many buoy types and different ocean regions. The overall difference in wind trend (SSM/I minus buoy) is –0.02 m/s/decade. This gives one confidence in using the passive microwave winds in climate studies.

2.5.2 Water Vapor

Over 99% of the atmospheric moisture is in the form of water vapor, and this vapor is the principal source of the atmospheric energy that drives the development of weather systems on short time scales and influences the climate on longer time scales. Tropospheric water vapor measurements are an important component to the hydrological cycle and global warming (Held and Soden, 2006; Trenberth et al., 2005). The microwave measurement of water vapor can also be used as a proxy to detect global warming of the lower troposphere with a signal-to-noise ratio that is five times better than the AMSU method of measuring the temperature change (Wentz and Schabel, 2000).

Satellite microwave measurements near the 22.2 GHz vapor absorption line provide the most accurate means to determine the total amount of vapor in the atmosphere. Quality controlled radiosondes from stations on small islands or ships are used for validation of the columnar water vapor retrievals. Simulations show that retrievals are accurate to 0.1 mm total columnar water vapor. Comparisons of AMSR-E water vapor retrievals with ship based radiosondes show an error of 2.2–0.5 mm (Szczodrak et al., 2006) which includes errors due to differences between a radiosonde point measurement and the larger AMSR-E footprint.

2.5.3 Cloud Liquid Water

Cloud water links the hydrological and radiative components of the climate system. Cloud water can be retrieved from passive microwave measurements because of its strong spectral signature and polarization signature (Wentz, 1997). Passive microwave observations provide a direct estimate of the total absorption along the sensor viewing path. At 18 and 37 GHz, clouds are semi-transparent allowing for measurement of the total columnar absorption. The absorption is related to the total amount of liquid water in the viewing path, after accounting for oxygen and water vapor absorption.

Validation of columnar cloud liquid water is a difficult undertaking. The spatial variability of clouds makes comparisons between upward looking ground based radiometers and the large footprint size of the downward looking satellite retrievals problematic. The upward looking ground-based radiometers also have very limited geographic distribution, making meaningful validation over global conditions impossible. Generally, validation is completed using a statistical histogram method (Wentz, 1997).

2.5.4 Rain Rate

Rainfall is the key hydrological parameter, so much so that changes in the spatial distribution of rainfall have led to the collapse of civilizations (Haug et al., 2003; O'Conner and Kiker, 2004). Rain is one of the most difficult parameters to accurately retrieve using remote sensing because of its extreme variability in space and time over a variety of scales. The most accurate and physically-based rain retrieval techniques take advantage of the interactions between microwave radiation and water, and both passive and active microwave remote sensing techniques can be used to derive rain rates over both ocean and land.

PMW observations respond to the presence of rain in the instrument field-of-view with two primary signals: an emission signal and a scattering signal (Petty, 1994). The ocean surface is roughly 50% emissive, so it serves as a cold background around 150 K against which to observe rain. Since the ocean is an expansive flat surface, the emission is strongly polarized. For typical incidence angles and clear skies, vertical polarization brightness temperatures are larger than horizontal polarization brightness temperatures by as much as 100 K. The emission depends on the sea surface temperature, salinity, and surface roughness.

Emission from small round rain and cloud drops is unpolarized, and the liquid emission strongly decreases the polarization seen by the sensor. Heavy rain can bring the difference between vertical and horizontal polarization brightness temperatures down to zero. The emission signal has a strong spectral signature that increases with frequency – that is, higher microwave frequencies are more affected by rain. The strength of the emission signal depends on the total amount of liquid water below the freezing level, and this is related to the surface rain rate. The primary factors governing this relationship are: the height of the freezing level, the relative portioning of cloud and rain water, and the horizontal inhomogeneity – the beamfilling effect (Hilburn and Wentz, 2008; Wentz and Spencer, 1998). The scattering signal measures a decrease in brightness temperatures due to the presence of ice above the freezing level (Spencer et al., 1989). Usually the scattering signal is used over a warm background, and is especially useful over land. The relationship of the scattering signal to surface rain rate is less direct than it is for the emission signal.

The relationship between the emission signal and the rain rate is strongly nonlinear. Since rain is horizontally inhomogeneous over satellite footprints (which may range in diameter from 6 to 56 km), the measurement represents an average over the satellite footprint. Averaging a highly variable observable quantity, when the observable quantity is nonlinearly related to the desired quantity, results in a bias in the desired quantity. This is the beamfilling effect, and it causes rain rates to be underestimated by PMW radiances.

Different sensors have systematically different spatial resolutions and the probability distribution function of liquid water in the footprint changes systematically with the size of the footprint. For example, an infinitely small satellite footprint would model the variability of liquid in the footprint with a delta function, whereas a satellite footprint the size of the Earth models that variability with the global rain probability distribution function – typically taken to be a mixed log-normal distribution. Fortunately, real satellite footprints do not vary that much. The spatial resolution of SSM/I rain retrievals is nominally 32 km, and the spatial resolution of AMSR rain retrievals is nominally 12 km. This means that SSM/I rain retrievals require a larger beamfilling correction than AMSR rain retrievals, because SSM/I retrievals have more spatial averaging.

Hilburn and Wentz (2008) developed a new beamfilling correction by simulating lower resolution SSM/I data with higher resolution AMSR data. Rain retrievals were computed from the simulated SSM/I data at several resolutions and compared to the AMSR rain retrievals at the highest possible resolution to deduce how the variability of liquid water changes systematically with footprint size. When the new correction was applied to satellite data, rain rates agreed to within 3% (after removing sampling biases due to the different local times-of-day for each satellite). New inter-calibrated rain rate retrievals have been successfully used to close the water cycle (Wentz et al., 2007), show excellent agreement with rain gauges on ocean buoys (Bowman et al., 2009), and correlate well with the TRMM Precipitation Radar (Cecil and Wingo, 2009).

2.5.5 Sea Ice

PMW retrievals of sea ice form one of the most important climate data records in existence. The time series of sea ice, from 1979 – present, has provided measurements of ice concentration and classification of sea ice types (multiyear or first-year ice) on a daily basis. The PMW sea ice retrievals are vital because of their ability to see through clouds. Large ice shelf breakup events, such as the Larsen Ice shelf breakup, have been witnessed and monitored using PMW retrievals. Sea ice is important to the global climate as it acts to regulate heat, moisture, and salinity in the polar ocean. The recent increase in summer Arctic sea ice acts as a positive feedback for global warming by changing the albedo.

There are two common retrieval algorithms for sea ice, the NASA team algorithm and the bootstrap algorithm. Both algorithms use the polarization and gradient ratios to determine ice concentration and type. At 19 GHz the difference between the vertical and horizontal polarizations is small for sea ice (both first-year and multiyear) and large for ocean. The two polarizations are different for first-year versus multi-year ice at 37 GHz (Cavalieri et al., 1984). The primary error in the NASA team algorithm is due to the effects of surface glazing and layering on these channel ratios (Comiso et al., 1997). Newer team algorithms use the 89 GHz gradient ratio to minimize these errors (Markus and Cavalieri, 2000). The bootstrap algorithm uses the polarization and gradient ratios, combining different channels, such as the 19 and 37 vertical polarization ratio (Comiso et al., 1997). Both algorithms use different methodologies to filter weather effects.

Validation of the sea ice retrievals has been completed through inter-comparison between different algorithms and comparison to visible and infrared satellite measurements. The NASA team algorithm and bootstrap algorithm generally agree with each other but differ by 10–35% in areas within the ice pack (Comiso and Steffen, 2001).

2.5.6 Sea Surface Temperature

Sea surface temperature is a key climate and weather measurement routinely made each day by satellite infrared (IR) and PMW radiometers, in-situ moored and drifting buoys, and ships of opportunity. These measurements are used to create daily spatially-complete global maps of SST that are then used for weather prediction, ocean forecasts, and in coastal applications such as fisheries forecasts, pollution monitoring, and tourism. They are also widely used by oceanography, meteorology, and climate scientists for research. Prior to 1998, SSTs were only available globally from IR satellite retrievals, but with the launch of TMI, PMW retrievals became possible. While IR SSTs have a higher resolution than PMW SSTs (1–4 km as compared to 25 km), their retrieval is prevented by clouds giving PMW SSTs improved coverage since they are retrieved through clouds.

Between 4 and 11 GHz the vertically polarized TB of the sea-surface has an appreciable sensitivity to SST. In addition to SST, TB depends on the sea-surface roughness and on the atmospheric temperature and moisture profile. Fortunately, the spectral and polarimetric signatures of the surface-roughness and the atmosphere are quite distinct from the SST signature, and the influence of these effects can be removed given simultaneous measurements at multiple frequencies and polarizations. Both TMI and AMSR-E measure multiple frequencies that are more than sufficient to remove the surface-roughness and atmospheric effects. Sea-surface roughness, which is tightly correlated with the local wind, is usually parameterized in terms of the near-surface wind speed and direction. The additional 7 GHz channel present on AMSR-E and not TMI, provides improved estimates of sea-surface roughness and improved accuracy for SSTs less than 12°C (Gentemann et al., in press). All channels are used to simultaneously retrieve SST, wind speed, columnar

Table 2.3 Nighttime satellite – buoy SST errors, bias and standard deviation (STD)

	TOGA TAO/TRITON			PIRATA		
Satellite	Collocations	Bias	STD	Collocations	Bias	STD
TMI	84,072	−0.09	0.67	11,669	−0.09	0.60
AMSR-E	21,461	−0.03	0.41	2,837	−0.00	0.35

water vapor, cloud liquid water, and rain rate (Wentz and Meissner, 2000). SST retrieval is prevented only in regions with sun-glitter, rain, and near land. Since only a small number of retrievals are unsuccessful, almost complete global coverage is achieved daily. Any errors in retrieved wind speed, water vapor, or cloud liquid water can result in errors in retrieved SST.

Buoy measurements from the Tropical Atmosphere Ocean/Triangle Trans-Ocean Buoy Network (TAO/TRITON) and the Pilot Research Moored Array in the Tropical Atlantic (PIRATA) are used to validate the PMW SSTs. Table 2.3 shows the mean difference, mean satellite minus buoy SST difference and standard deviation (STD) for each of the buoy arrays. Comparisons with TMI data from 1 January 1998 to 9 June 2005 show that the TAO and PIRATA arrays have very small mean biases, –0.09 and –0.09°C, and STD of 0.67 and 0.60°C respectively. Comparisons with AMSR-E data (1 May 2002–9 June 2005) show the TAO and PIRATA arrays have very small biases (−0.03 and −0.01°C) and STD (0.41 and 0.35°C, respectively).

2.5.7 Sea Surface Salinity

The first measurements of SSS from space will be from the SMOS and Aquarius. SSS is important to ocean circulation, the global hydrological cycle, and climate. Monitoring SSS will provide information on geophysical processes that affect SSS and the global hydrological cycle, such as the sea ice freeze/thaw cycle, evaporation and precipitation over the ocean, and land runoff. The Aquarius mission will attempt to measure SSS with a 150 km spatial resolution and a measurement error of < 0.2 PSS-78 (Practical Salinity Scale of 1978) (Lagerloef et al., 2008).

At 1.4 GHz, retrievals are sufficiently sensitive to SSS to allow for accurate retrieval of SSS. The retrievals depend on the dielectric constant of sea water, the wind-induced sea-surface emissivity and scattering characteristics, atmospheric absorption, particularly that due to rain, and Faraday rotation. Additional contributions from near-land emissions, galactic background radiation reflection, and reflected solar radiation present increased difficulties.

2.6 Erroneous Retrievals

2.6.1 Rain Contamination

The retrievals for SST, wind speed, and vapor must be flagged as bad data in the presence of rain. This is usually done by looking at the simultaneous retrieval of rain rate. Occasionally, sub-pixel rain cells contaminate these retrievals but are not flagged as rain. These can be seen as anomalously warm or cold SSTs or anomalously high wind values, usually only affecting 1–2 pixels in a region where other data nearby has been flagged as rain contaminated. In working with PMW data, area-rain flagging is necessary to remove these anomalously affected cells near rain. Only then can climatological results be trusted.

2.6.2 Near Land Emission

Near land, the lobes to the side of the main beam can result in side-lobe contamination. This contamination results in geographic dependent retrieval errors unless the data are flagged as erroneous. This contamination impacts all the geophysical retrievals from PMW radiometers to differing extents depending on the land emission signal at the frequencies included in the various retrieval algorithms. For example, because the 10.7 GHz channels is affected more by land emissions, the land contamination at 10.7 GHz is larger than at 6.9 GHz, resulting in a warm bias and small increase in standard deviation for both TMI and AMSR-E measurements near land, but the effect is larger in the TMI retrievals.

To estimate the side-lobe contamination in the TMI PMW SST retrievals we have compared contemporaneous Visible Infrared Radiometer Scanner (VIRS) IR SST retrievals in coastal regions, using data from January 1998 to December 1998. VIRS is an infrared radiometer carried on the TRMM satellite alongside TMI. VIRS SSTs were determined to have a standard deviation of 0.7°C when compared to Reynolds Optimal Interpolated SSTs (Ricciardulli and Wentz, 2004).

To investigate how the effect of land contamination on the TMI SSTs diminishes away from land, the distance from land for each data point was calculated. The effect of land contamination can be seen in the mean difference, TMI minus VIRS SST (Fig. 2.4). The mean difference away from land is roughly 0.12°C, which is approximated by the difference expected between a skin (VIRS) and subskin (TMI)

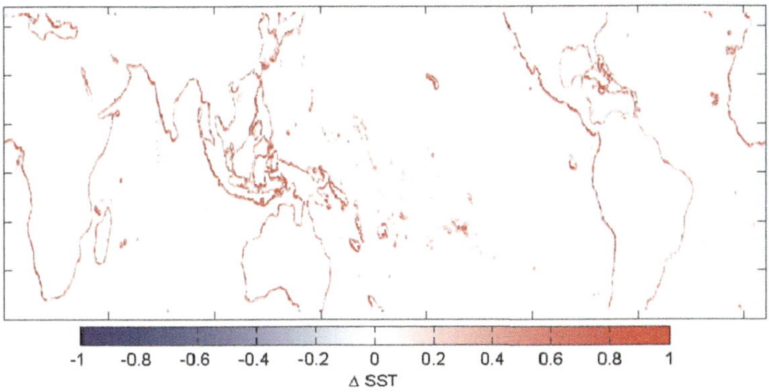

Fig. 2.4 Estimate of bias due to side-lobe contamination near land for 10.7 GHz SST retrievals

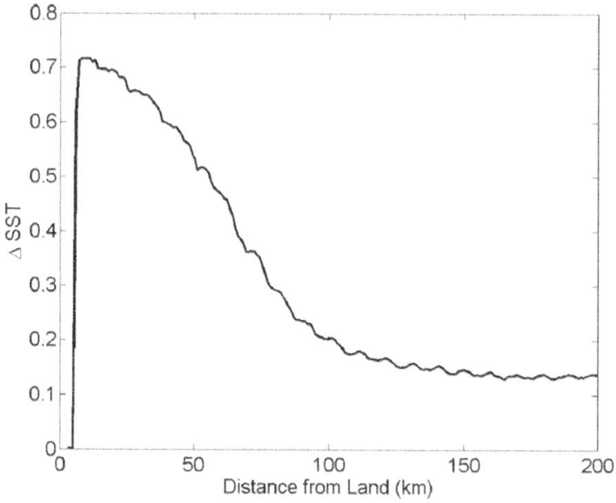

Fig. 2.5 Land contamination bias derived from TMI VIRS comparisons. This global average shows that by removing data within 100–150 km of land, side-lobe contamination will be removed

measurement of SST. As the distance to land decreases, the mean difference increases, with a maximum magnitude of 0.72 K, indicating that the bias due to land contamination is on the order of 0.6 K. From Fig. 2.5, it is clear that biasing exists mostly for retrievals less than 100–150 km from land. These results are specific to the 10.7 GHz SST retrieval from TMI. Although AMSR-E also has land contamination, the impact is less at 6.9 GHz, the primary channel used for AMSR-E SSTs.

2.6.3 Radio Frequency Interference

RFI is arguably the fastest growing source of errors in microwave SSTs and wind speeds. The RFI impact on water vapor, cloud liquid water, and rain rate is less, but growing. RFI errors are largely caused by media broadcasting activities (including television and radio) from commercial satellites in geostationary orbits. Geostationary RFI results when signals broadcast from these communication satellites reflect off the Earth's ocean surface into a PMW instrument's field of view. Ground-based instrumentation in the microwave range is also producing RFI, some sources of which have been identified and characterized. Both these types of anthropogenic RFI are increasing in magnitude and extent. While it is relatively straightforward to identify and flag data affected by large RFI contamination, less-obvious RFI contamination can be difficult to identify. The spatial and temporal nature of the RFI removal must be carefully monitored to avoid spurious trends in climate data records. The addition of new communication satellites, more power, more ground coverage, and the use of more frequencies near PMW instrument measurement bandwidths signify that sources of RFI will continue to change and increase in the future.

The RFI errors resulting from geostationary broadcast sources are primarily dependent on communication broadcast frequency, power and direction, PMW instrument bandwidth, signal glint angle, and ocean surface roughness. The observation bandwidths of PMW instruments are typically wider than the protected bands allocated for PMW remote sensing. Thus, PMW instruments can receive RFI from legal activity using nearby frequency bands allocated for communication and other commercial uses. AMSR-E and WindSat are the two PMW instruments most affected by RFI, while SSM/I and TMI both appear to be relatively unaffected. This is likely because the lower frequency channels of AMSR-E and WindSat, particularly the 10.7 and 18.7 GHz measurement channels, are sensitive to frequencies used extensively for media broadcasting. WindSat has more significant RFI than AMSR-E due to wider observation bandwidth. Observing more bandwidth tends to yield less noise, but also leads to more interference from frequencies further from the channel's center observation frequency. For example, at 18.7 GHz, WindSat receives interference from DirecTV nationwide broadcast beams. AMSR-E, with narrower bandwidth at 18.7 GHz, does not appear to be significantly affected by nationwide broadcast frequencies, but does receive RFI from DirecTV spot beams, which broadcast at frequencies closer to the center observation frequency of the 18.7 GHz channel.

Power and direction are also important factors affecting RFI. Satellite media broadcasts appear to direct most signal power very carefully to specific markets. Powerful signals can result in large RFI errors within certain regions. To serve smaller but growing geographically dispersed markets, media satellites also broadcast wide, low power beams to cover much larger, less populated regions. These lower power beams induce more subtle RFI effects that can be difficult to detect and remove. Assuming the Earth observation point is within the footprint of a geostationary broadcast, the magnitude of RFI is highly dependent on the glint angle, or how close the observation reflection vector comes to pointing at the RFI source.

RFI induced errors in AMSR-E ocean products were investigated over the entire 7 year mission data set. The effects of the different sources of RFI are listed in Table 2.4, including which PMW passes are affected and the time period of interference. Because most geostationary broadcast power is directed toward the northern hemisphere, many broadcast beams only reflect into the descending pass AMSR-E field-of-view.

From the start of the AMSR-E mission in 2002, HotBird, which is positioned over 13.0° East longitude, and Astra, located at 19.2° East, have steadily increased RFI in European waters over time. DirecTV-10 at 102.8° West and DirecTV-11 at 99.2° West have produced RFI in American waters sinc, and Atlantic Bird 4A at 7.2° West has been contributing to Mediterranean Sea RFI since 2009. Also from beginning of mission in 2002, SkyBrazil has directed power toward the southern hemisphere, therefore reflecting into ascending passes of ASMR-E and producing RFI off the coasts of southern Brazil and Argentina.

Ground-based RFI sources are also growing stronger and more numerous over time. Unlike the Geostationary RFI, the ground-based RFI affects both ascending and descending swaths, though to different extents. This is likely due to differing levels of RFI activity at the AMSR-E local observation times of 1:30 AM or 1:30 PM. Although errors caused by these ground-based sources cover fairly small regions, the size and intensity of these RFI effects have been increasing over the

Table 2.4 Sources of RFI

Source	Region affected	Frequency (GHz)	Effect on data (↓ decreases)	Affected overpass	Period of interference
HotBird	Europe	10.7	↓ SSTs ↑ Winds	Descending	Pre 2002 – present
Astra	Europe	10.7	↓ SSTs ↑ Winds	Descending	Pre 2002 – present
Atl.Bird 4A	Mediterranean	10.7	↓ SSTs ↑ Winds	Descending	April 2009 – present
DirecTV-10	USA	18.7	↓ SSTs ↑ Wind ↓vapor ↑cloud ↑rain	Descending	September 2007 – present
DirecTV-11	USA	18.7	↓ SSTs ↓ Wind ↓vapor ↑cloud ↑rain	Descending	July 2008 – present
SkyBrazil	SE American Coast	10.7	↓ SSTs ↑ Winds	Ascending	Pre 2002 – present
Ground-based	Ascension Island	6.9	↓ SSTs No wind effect	Both	Pre 2002 – present
Ground-based	Gulf of Aden	10.7	↓ SSTs ↑ Winds	Both	March 2009 – present
Ground-based	Coastal Netherlands Coastal Norway	6.9	↑ SSTs No wind effect	Both	2004 – present
Ground-based	Mumbai	6.9	↑ SSTs No wind effect	Both	2003 – present

years. Ground-based RFI sources can operate intermittently, sometimes even sporadically. The most prominent regions include coastal Netherlands and Norway, coastal Mumbai, the Gulf of Aden through the waters south of Oman, and waters around Ascension Island.

Regions of RFI are located by differencing AMSR-E SSTs derived using all SST channels (6.9–36.5 GHz) from those derived without 6.9 GHz (10.7–36.5 GHz), as well as by differencing winds derived using all wind channels (10.7–36.5 GHz) from those derived without 10.7 GHz (18.7–36.5 GHz). An example is shown in Fig. 2.6.

Since most geostationary sources affect the AMSR-E descending passes, this plot shows the wind (North America) and SST (Europe) descending orbit difference maps. The wind RFI around North America caused by DirecTV outlines US coastal waters and the Great Lakes (both pictured), with some subtle effects detected as far as Hawaii and possibly the Canary Islands off the coast of Africa (neither shown). The SST RFI around Europe shows consistently increasing extent and intensity over the years.

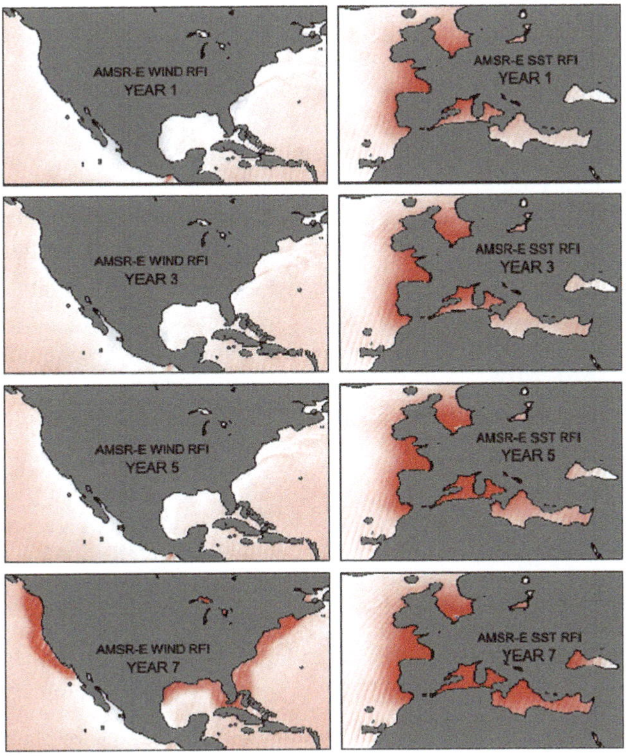

Fig. 2.6 RFI induced wind (*left*) and SST (*right*) errors shown in descending pass difference plots for years 1, 3, 5 and 7 of the ASMR-E mission (starting July, 2002, 2004, 2006, 2008) over North America and Europe where the RFI has increased most in coverage and intensity over the years. The striping is caused by the shifting orbital pattern of the most intense geostationary glint angles

A ground RFI source off the Netherland coast has concurrently increased power to become more prominent as seen by the small distinctive dot forming over the years. The ground source produces SST errors of opposite sign compared to the geostationary RFI in the region. In this small region, two prominent sources of RFI error tend to cancel each other, potentially complicating detection and removal. The striping visible in Fig. 2.6 is not due to any cross-swath problem with the SSTs or wind speeds, but is due to the glint angle geometry which results in a heavily stripped glint angle pattern caused by AMSR-E's ground track repeat pattern every 233 orbits.

Glint angles and broadcast footprints are together highly predictive of potential RFI bias. Therefore, to remove RFI errors from the AMSR-E SST and wind products we calculate the signal glint angles using the longitude of the geostationary orbits. These glint angles, together with analysis of broadcast footprints, are used to remove retrievals with high probability of RFI error.

2.7 Conclusions

PMW retrievals of wind speed, water vapor, cloud liquid water, rain rate, sea ice, and SST have provided key information for research, climate, and operational applications. For research and operational applications, the daily global coverage provided by PMW retrievals are a significant advance over the pre-satellite era which relied on ship and buoy observations. For climate monitoring, the careful inter-calibration of the PMW radiometers and consistent (single algorithm) processing of the entire data set has provided an accurate 22 year time series of PMW retrievals.

Acknowledgements The AMSR-E SSTs are from Remote Sensing Systems, processed using the version 5 algorithm, and available at www.remss.com . This work was funded by the NASA grants NNG04HZ29C, NNG07HW15C, NNH08CC60C, and NNH09CF43C.

References

Atlas RM, Hoffman RN, Ardizzone J, Leidner SM, Jusem JC (2009) Development of a new cross-calibrated, multi-platform (CCMP) ocean surface wind product. Paper presented at AMS 13th Conference on Integrated Observing and Assimilation Systems for Atmosphere, Oceans and Land Surface (IOAS-AOLS), Phoenix, AZ, USA

Bettenhausen MH, Smith CK, Bevilacqua RM, Wang NY, Gaiser PW, Cox SK (2006) A nonlinear optimization algorithm for windsat wind vector retrievals. IEEE Trans Geosci Rem Sen 44(3):597–608

Bowman KP, Homeyer CR, Stone DG (2009) A comparison of oceanic precipitation estimates in the tropics and subtropics. J Appl Meteorol Climatol 48:1335–1344

Cavalieri DJ, Gloersen P, Campbell JW (1984) Determination of sea ice parameters with the nimbus 7 SMMR. J Geophys Res 89:5355–5369

Cecil DJ, Wingo M (2009) Comparison of TRMM rain-rate retrievals in tropical cyclones. J Meteorol Soc Jpn 87A:369–380, doi:10.2151/jmsj.87A.369

Comiso JC, Cavalieri DJ, Parkinson CL, Gloersen P (1997) Passive microwave algorithms for sea ice concentration: a comparison of two techniques. Rem Sens Environ 60(3): 357–384

Comiso JC, Steffen K (2001) Studies of Antarctic sea ice concentrations from satellite data and their applications. J Geophys Res 106(C12):31361–31385

Gaiser PW, St. Germain KM, Twarog E, Poe G, Purdy W, Richardson D, Grossman W, Jones WL, Spencer D, Golba G, Cleveland J, Choy L, Bevilacqua RM, Chang P (2004) The windsat spaceborne polarimetric microwave radiometer: sensor description and early orbit performance. IEEE Trans Geosci Rem Sen 42(11):2347–2361

Gentemann CL, Meissner T, Wentz FJ Accuracy of satellite sea surface temperatures at 7 and 11 GHz. IEEE Trans Geosci Rem Sen (in press)

Gloersen P, Cavalieri DJ, Chang ATC, Wilheit TT, Campbell WJ, Johannessen OM, Katsaros KB, Kunzi KF, Ross DB, Staelin D, Windsor EPL, Barath FT, Gudmandsen P, Langham E, Ramseier RO (1984) Summary of results from the first NIMBUS 7 SMMR observations. J Geophys Res 89:5335–5344

Han D, Kim ST (1988) Effects of switch leakages upon Nimbus-7 SMMR calibration. NASA Memorandum 100705, Goddard Space Flight Center, Greenbelt, MD, USA, 17pp

Haug GH, Gunther D, Peterson LC, Sigman DM, Hughen KA, Aeschlimann B (2003) Climate and the collapse of maya civilization. Science 299:1731–1735

Held IM, Soden BJ (2006) Robust responses of the hydrological cycle to global warming. J Climate 19:5686–5699

Hilburn KA, Wentz FJ (2008) Intercalibrated passive microwave rain products from the unified microwave ocean retrieval algorithm (UMORA). J Appl Meteorol Climatol 47:778–794

Hollinger J, Peirce JL, Poe G (1990) SSM/I instrument evaluation. IEEE Trans Geosci Rem Sens 28(5):781–790

Kummerow CD, Barnes W, Kozu T, Shieu J, Simpson JJ (1998) The tropical rainfall measuring mission (TRMM) sensor package. J Atmos Ocean Technol 15(3):809–817

Lagerloef GSE, Colomb FR, Le Vine D, Wentz FJ, Yueh SH, Ruf CS, Lilly J, Gunn Y, Chao A, deCharon A, Feldman G, Swift C (2008) The aquarius/SAC-D mission: designed to meet the salinity remote-sensing challenge. Oceanography 21:68–81

Markus T, Cavalieri DJ (2000) An enhancement of the NASA team sea ice algorithm. IEEE Trans Geosci Rem Sens 38:1387–1398

Mears CA, Smith DK, Wentz FJ (2001) Comparison of special sensor microwave imager and buoy-measured wind speeds from 1987–1997. J Geophys Res 106(C6):11719–11729

Meissner T, Smith DK, Wentz FJ (2001) A 10-year intercomparison between collocated special sensor microwave imager oceanic surface wind speed retrievals and global analyses. J Geophys Res 106(C6):11731–11742

Meissner T, Wentz FJ (2009) Wind vector retrievals under rain with passive satellite microwave radiometers. IEEE Trans Geosci Rem Sens 47(9):3065–3083

O'Conner T, Kiker GA (2004) Collapse of the mapungubwe society: vulnerability of pastoralism to increasing aridity. Clim Change 66:49–66

Parkinson CL (2003) Aqua: an earth-observing satellite mission to examine water and other climate variables. IEEE Trans Geosci Rem Sens 41(2):173–183

Petty GW (1994) Physical retrievals of over-ocean rain rate from multichannel microwave imagery. Part I: Theoretical characteristics of normalized polarization and scattering indices. Meteorol Atmos Phys 54:79–99

Ricciardulli L, Wentz FJ (2004) Uncertainties in sea surface retrievals from space: comparison of microwave and infrared observations from TRMM. J Geophys Res 109:C12013, doi:10.1029/2003JC002247

Spencer RW (1989) Precipitation retrieval over land and ocean with the SSM/I: Identification and characteristics of the scattering signal. J Atmos Ocean Technol 6:254–273

Szczodrak M, Minnett PJ, Gentemann CL (2006) Comprison of AMSR-E retrievals of total water vapor over ocean. Paper presented at 2006 Ocean Sciences Meeting, Honolulu, HI, USA

Trenberth KE, Fasullo J, Smith L (2005) Trends and variability in column-integrated atmospheric water vapor. Clim Dyn 24:741–758

Wentz FJ (1997) A well calibrated ocean algorithm for special sensor microwave/imager. J Geophys Res 102(C4):8703–8718

Wentz FJ, Meissner T (2000) AMSR Ocean Algorithm, Version-2, Tech report 121599A-1, Remote Sensing Systems, Santa Rosa, CA, USA, 66pp. Available online

Wentz FJ, Meissner T (2007) AMSR_E ocean algorithms. 051707, Remote Sensing Systems, Santa Rosa, CA, USA, 6pp

Wentz FJ, Ricciardulli L, Hilburn KA, Mears CA (2007) How much more rain will global warming bring? Science 317:233–235

Wentz FJ, Schabel MC (2000) Precise climate monitoring using complementary satellite data sets. Nature 403(6768):414–416

Wentz FJ, Spencer RW (1998) SSM/I rain retrievals within a unified all-weather ocean algorithm. J Atmos Sci 55(9):1613–1627

Chapter 3
SMOS and Aquarius/SAC-D Missions: The Era of Spaceborne Salinity Measurements is About to Begin

Gary Lagerloef and Jordi Font

3.1 Introduction

During the Oceans from Space Venice 2000 meeting a decade ago, a friendly wager was made among a few participants. The potential for salinity measurement from space was a topic of lively discussion at a Special Session on Salinity Remote Sensing. There were some promising developments presented indicating that this capability would be achieved during decade ahead. The European Space Agency (ESA) pathfinder Soil Moisture Ocean Salinity (SMOS) mission had been selected in 1999. Meanwhile, NASA scientists were actively planning a pathfinder salinity mission proposal for a mid-2001 deadline, eventually to be named Aquarius. It was then estimated that these missions would be launched as early as mid-decade. Some participants were quite skeptical. After all, the scientific accuracy required is ~0.2 pss (practical salinity scale 1978), which equates to about ½ pinch of salt in a bottle of wine. So the wager was made, with a bottle of wine at stake, whether or not satellite-based salinity measurements would be presented at Oceans for Space 2010.

The wager's outcome is still unresolved, and will be decided at the 2010 meeting, although the odds may slightly favor the proponents. SMOS was launched successfully into orbit November 2nd, 2009. After a lengthy checkout period, preliminary data will be available to the science validation team about 2 months prior to the Oceans from Space 2010 meeting. Every effort will be made to have some preliminary results to show. Meanwhile, the Aquarius/SAC-D mission, a partnership between NASA and Argentina is now planned for launch in the Fall 2010, several months after the Oceans from Space 2010 meeting.

The past decade clearly has seen major progress, both technically and scientifically, toward meeting the challenge of measuring ocean salinity from space. This paper presents some of the history of this decade of progress, by describing the development of the SMOS and Aquarius/SAC-D missions and their scientific capabilities. The timely completion and launch of these satellite missions indicates that

G. Lagerloef (✉)
Earth and Space Research, Seattle, WA 98121, USA
e-mail: Lager@esr.org

V. Barale et al. (eds.), *Oceanography from Space*,
DOI 10.1007/978-90-481-8681-5_3,

the year 2010 marks the turning point that will begin the era of space borne salinity measurements.

3.2 Scientific Background – Links Between the Ocean Circulation, Water Cycle and Climate

The 1990s saw a confluence of two important developments for sea surface salinity (SSS) remote sensing. Scientifically, there was increasing awareness of the connection between surface salinity, ocean circulation and climate variability (e.g. Broecker, 1991). Meanwhile, advances in microwave radiometer technology were making it feasible to measure SSS at the levels of accuracy, spatial and temporal resolution needed to address important scientific questions (Lagerloef et al., 1995). By the time of Oceans from Space 2000, the key scientific themes had been identified for applying satellite SSS data study the links to ocean circulation and climate. These included tropical air sea interactions and El Niño, high-latitude convection and salinity anomalies, mid-latitude subduction processes, and the relation of salinity to changes in the global water cycle. The mission concepts were evolving to address these topics with measurement capabilities of ~0.2 pss accuracy on ~100–200 km and 10–30 day resolutions.

Scientific interest continued to grow during the past decade, reflected, for example, in the *Journal of Geophysical Research* special section on ocean salinity (Lagerloef, 2002). This included about two dozen peer-reviewed papers addressing the role of SSS on upper ocean dynamics, air-sea interaction and climate based on observational and modeling studies. Ocean salinity's critical importance to understanding and predicting climate variability was further documented in the report of the US CLIVAR Salinity Working Group (US CLIVAR, 2007) and the Intergovernmental Panel on Climate Change (IPCC, 2007).

These assessments identified salinity variability as a key index of the marine hydrologic cycle. SSS is a tracer for varying evaporation and precipitation, runoff and ice processes. These have important consequences for oceanic currents and mixing dynamics that influence the ocean's capacity to absorb, transport and store heat, freshwater and carbon dioxide. The assessments also reviewed clear observational evidence of decades-long changes, for example, of decreasing salinity in the subpolar North Atlantic and Southern Ocean, while the near surface salinity in the subtropics was increasing.

More recent studies reveal new features in these trends and links to water cycle, circulation and anthropogenic climate change. Stott et al. (2008) attributed to human influence the recent increases in the observed salinity in the Atlantic (20–50°N). Gordon and Giulivi (2008) found opposing trends, increasing since the late-1980s, in the sub-tropical gyres of North Atlantic and North Pacific, with the latter experiencing a relative freshening (Fig. 3.1). The authors attributed this to increasing atmospheric transport of fresh water from Atlantic to Pacific via the trade winds across Central America. The North Atlantic and Nordic Seas upper ocean freshening trend of the 1960s–1990s has reversed over the last decade (Holliday et al., 2008). This may be attributed to changes in the ocean circulation (Hakkinnen and

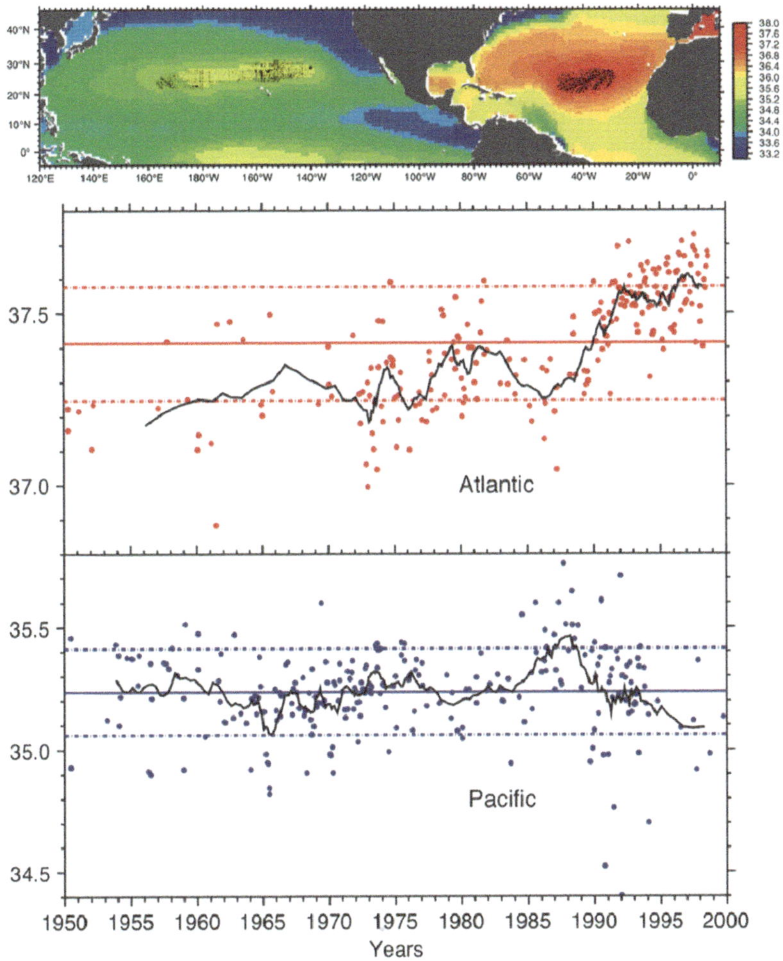

Fig. 3.1 Recently documented inversely correlated surface salinity trends in Atlantic and Pacific subtropical gyres, consistent with increasing atmospheric water transport from the Atlantic to Pacific (from Gordon and Giulivi, 2008)

Rhines, 2009), whereby warm and salty subtropical waters increased their penetration toward the Nordic seas. This suggests that salinity trends are related to changes in ocean circulation as well as the hydrologic cycle, and these different linkages need to be resolved.

Many of the fundamental processes involving salinity in the modulation of upper-ocean circulation and mixing remain poorly understood in both tropical and high-latitude regions. Nor are they adequately represented in climate models, and yet model studies do indicate that expanded monitoring of salinity (both satellite and in-situ) will measurably improve climate forecasts on inter-annual to decadal timescales (US Clivar, 2007).

3.3 Basic Principles and Issue for Salinity Remote Sensing

Salinity remote sensing is possible because the microwave emission of the sea surface at a given radio frequency depends partly on the dielectric coefficient of sea water, which in turn is partly related to salinity and temperature (Klein and Swift, 1977; Meissner and Wentz, 2003). The total power of the emission at horizontal (H) and vertical (V) polarization is measured remotely with a microwave radiometer. The output is given in terms of a parameter called brightness temperature (T_{BH} and T_{BV}) at each polarization, which are respectively the products of the surface emissivities (e_H and e_V) and the absolute temperature of the sea surface (T):

$$\begin{aligned} T_{BH} &= e_H T \\ T_{BV} &= e_V T \end{aligned} \tag{3.1}$$

The polarized emissivity for a flat sea (no wind/wave roughness), given by Equation (3.2), is governed by Fresnel reflection, the dielectric coefficient (ε), and the viewing angle from nadir (θ) (e.g. Swift and McIntosh, 1983).

$$\begin{aligned} e_H &= 1 - \left[\frac{\cos\theta - (\varepsilon - \sin^2\theta)^{1/2}}{\cos\theta + (\varepsilon - \sin^2\theta)^{1/2}}\right]^2 \\ e_V &= 1 - \left[\frac{\varepsilon\cos\theta - (\varepsilon - \sin^2\theta)^{1/2}}{\varepsilon\cos\theta + (\varepsilon - \sin^2\theta)^{1/2}}\right]^2 \end{aligned} \tag{3.2}$$

The above equations apply to all emitting surfaces, including seawater. At $\theta = 0$, both polarizations are the same.

Salinity (S) and temperature (T) enter the formulation through the complex dielectric coefficient ε, which depends on the microwave radio frequency (f), the electrical conductivity of sea water $C(S,T)$ and other factors, some of which are also dependent on (S,T). Klein and Swift (1977) is perhaps the most commonly applied model and is theoretically based on a simplified Debye equation and fitted with laboratory measurements of the dielectric coefficient. However, there are uncertainties in this model and additional studies have been carried out in recent years, including Blanch and Aguasca (2004), Meissner and Wentz (2003), Strogryn (1997), and Ellison et al. (1998), which all have inconsistencies relative to one-another. Differences among these models were evaluated by Wilson et al. (2004) in comparison to carefully controlled H and V polarization microwave brightness temperature measurements at $f = 1.413\,\text{GHz}$ (the L-band frequency to be used for measuring salinity). The Klein–Swift and the Meissner–Wentz models generally agreed the best with the brightness temperature observations, with uncertainties of between 0.02 and 0.07 K, whereas the others showed significant trends over the expected ranges of temperature and salinity. While these uncertainties are of similar magnitude to other terms in the error budgets (see below), this model function error is an important concern, and because the dielectric coefficient is a fundamental physical property of seawater, it should be know as accurately as possible. A

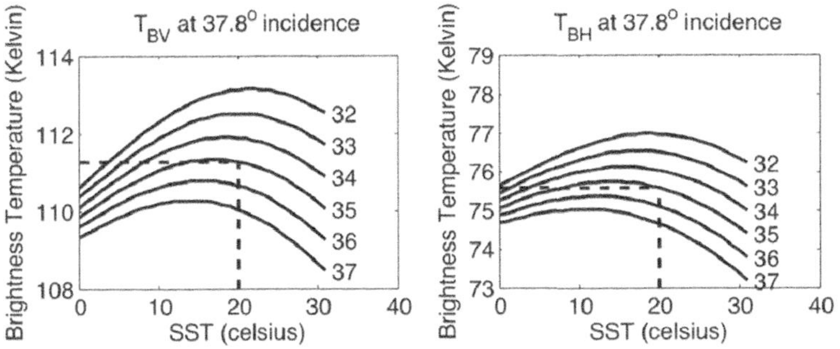

Fig. 3.2 Brightness Temperatures T_{BV} (*left*) and T_{BH} (*right*) as a function of SSS (contours) and SST (abscissa) for typical ocean surface conditions, and incidence angle of 37.8° (as an example from one of the Aquarius/SAC-D satellite viewing angles). Salinity can be determined from either polarization when SST is known (*dashed lines*). Calculations based on the dielectric model Klein and Swift (1977)

new set of laboratory measurements is presently being carried out (Lang, 2008). The dielectric model, combined with in-situ SSS and SST validation measurements will provide a consistent calibration reference for present and future salinity satellite missions.

Satellite remote sensing of salinity is done in the protected L-band frequency centered at 1.413 GHz to avoid radio interference. At this band, the brightness temperature change relative to a change in salinity, although small, is nevertheless enough to make SSS remote sensing possible, given sufficiently sensitive radiometric measurements. Figure 3.2 shows the relationship between T_{BH} and T_{BV} with SST and SSS for a particular viewing angle. The contour lines are for salinities ranging from 32 to 37 psu. It is easy to see that a unique value of salinity can be retrieved when SST and either T_{BH} or T_{BV} are known. This is the essence of how salinity remote sensing is achieved, although it is more complicated in practice. The SMOS and Aquarius instruments approach the salinity retrieval in very different ways, based on sensor design, as will be explained further below.

The dynamic range of brightness temperature is about 5 K over the range of typical open ocean surface salinity and temperature conditions. At a given temperature, brightness temperature decreases as salinity increases, whereas the tendency with respect to temperature changes sign. The salinity contours are spread farther apart for V polarization than for H, and therefore V is slightly more sensitive to salinity changes. The sensitivity is strongly affected by temperature, being largest at the highest temperatures and yielding better measurement precision in warm versus cold ocean conditions. Corrected brightness temperature will need to be measured to 0.02–0.08 K precision to achieve 0.1 pss salinity resolution. The difference in sensitivity between polarizations also increases with incidence angle (not shown). Temporal and spatial averaging can reduce random error. The degraded measurement precision in higher latitudes will be partly offset by averaging with the greater sampling frequency from a polar orbiting satellite.

Other sources of error in the SSS retrieval include numerous effects that change the brightness temperatures from the idealized flat surface emmission in Equations (3.1) and (3.2) to what is actually measured by a satellite radiometer in orbit. These include surface reflections of astronomical L-band radiation sources such as the cosmic background, galactic core, sun and moon. Attenuation in the atmosphere and ionosphere, including Faraday rotation (Yueh, 2000) must also be corrected. Clouds are transparent at L-band and pose not problem, but attenuation during very heavy rain can be significant and those data will need to be flagged. The T_B variation with respect to temperature falls generally between $\pm 0.15\,\mathrm{K}^{\circ}\mathrm{C}^{-1}$ and near zero over a broad S and T range. Knowledge of the surface temperature to within a few tenths °C will be adequate to correct T_B for temperature effects and can be obtained using data from other satellite systems. In general these terms are well understood and will be corrected with appropriate models and ancillary data. See Lagerloef et al. (2008) for these terms tabulated for the Aquarius error analysis. The optical depth for this microwave frequency in seawater is about 1–2 cm, and the remotely sensed measurement depends on the T and S in that surface layer thickness, which poses a potential problem when comparing satellite data to in-situ measurements within a few meters of the surface.

The error source posing the most significant problem, however, is the change in emissivity from surface roughness due to wind, including sea state, wave breaking and foam. The change in T_B due to wind is much smaller at L-band than at higher frequencies (Hollinger, 1971), but nevertheless it is still the largest error source for salinity remote sensing. The Wind and Salinity Experiment (WISE) field study early in the decade (Camps et al., 2004; Gabarró et al., 2004) measured the L-band response wind, wave height and foam at a range of incidence angles and developed empirical formulas relative to those variables. These results show that the T_{BH} response is much larger than T_{BV} for incidence angles from 25 to 65° and is typically 0.2–0.4 K/m/s of wind. This implies large corrections for even moderate winds of a few m/s. Recent airborne measurements show that the T_{BV} response is larger than indicated by the WISE data, and that there is a detectable modulation due to the wind direction in both polarizations (S. Yueh, 2009, personal communication). Clearly there remains considerable uncertainty in correcting the wind and roughness effect, and this will be addressed once the satellites are on orbit through the calibration and validation activities. The Aquarius instrument will use radar backscatter to help make this correction, where as SMOS will derive a wind correction through a complex inversion algorithm that will rely on a model such as WISE that covers the full range of SMOS incidence angles.

3.4 Soil Moisture Ocean Salinity (SMOS) Mission

3.4.1 Early Configuration and Evolution of Design

ESA organized in 1995 a consultative meeting on "Soil Moisture and Ocean Salinity, Measurement Requirements and Radiometer Techniques" to analyze the feasibility of low microwave passive remote sensing for the measurement of these two geophysical variables (Kerr et al., 1995). Besides confirming L-band radiometry as a viable option, it was concluded that the most promising technique to address the requirements for a simultaneous acquisition of both parameters was interferometric aperture synthesis radiometry, a concept developed in the 1950s to obtain high resolution radio images of celestial bodies and that had been demonstrated to be useful for Earth observation (Ruf et al., 1988). The interferometry design, inspired from the very large baseline antenna concept, consists of deploying small receivers in space, then reconstructing a brightness temperature (T_B) field through Fourier synthesis in a snapshot basis with a resolution corresponding to the spacing between the outmost receivers (Fig. 3.3).

A synthetic aperture radiometer measures all cross-correlation products between the signal pairs collected by the array elements (Corbella et al., 2004) and the total

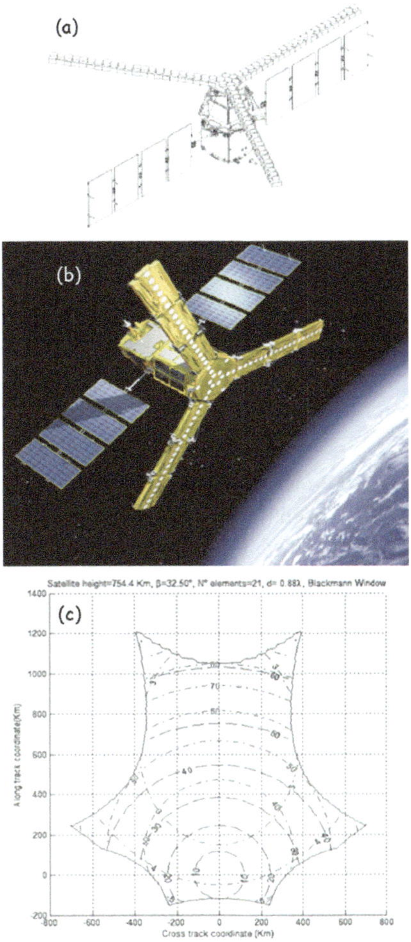

Fig. 3.3 (**a**) The initially proposed SMOS L-band plus C-band satellite in flight configuration (SMOS proposal to ESA). (**b**) Artist's view of the final SMOS configuration, with the 3 antenna arms of the MIRAS instrument and the PROTEUS platform with its solar panels deployed (ESA). (**c**) The SMOS instantaneous AF-FOV (*irregular curved hexagon*) with variable pixel characteristics: incidence angle (*dashed lines*) ranges from 0 to 65°, spatial resolution (*dash-dotted lines*) from 32 to 100 km, and expected radiometric sensitivity (*dash-dotted*) from 2.5 K at boresight to 5 K (generated by the SMOS end-to-end Performance Simulator)

power of the scene is also measured using at least one real aperture radiometer connected to one of the antennas. The relatively high spatial resolution (less than 50 km) and the short revisit time (1–3 days) imposed by soil moisture science objectives, are more feasible at this low frequency with such a new approach than with a classical pushbroom technique. Le Vine et al. (2000) were able to generate an SSS map using the Electronically Steered Thinned Array Radiometer (ESTAR), the first one-dimensional synthetic aperture radiometer flown on an aircraft.

Early in the 1990s ESA started preparing the specifications for a polarimetric two-dimensional synthetic aperture radiometer, improving the ESTAR design. The result was MIRAS (Microwave Imaging Radiometer with Aperture Synthesis), with a large number of antennas along a Y-shaped 3 arms structure (Martín-Neira and Goutoule, 1997). The optimum sampling strategy of the spatial frequency plane is on a hexagonal grid, instead of the rectangular one commonly used in signal or image processing (Camps, 1996). This strategy allows an increase of the maximum antenna separation without suffering from aliasing effects in the image reconstruction process, or alternatively for the same antenna spacing enlarging the alias-free field-of-view (AF-FOV) as compared to rectangular sampling. For a given number of elements, the array structure that provides the largest spatial frequency coverage (best angular resolution) is a Y structure.

The two-dimensional MIRAS interferometer allows to measure T_B at large incidences, for two polarisations. Moreover, the instrument records instantaneously a whole scene; as the satellite moves, a given point within the 2D FOV is observed from different view angles. One then obtains a series of independent measurements, which allows retrieving surface parameters with much improved accuracy. The concept is fully scalable and allows achieving very fine spatial resolution without moving parts. A first MIRAS feasibility study was carried out by France in 1992–1996 and the development of receivers (LICEF, Lightweight Cost-Effective Front-end) was started in 1995 by Spain. In 1998 EADS-CASA Espacio took the lead of the technology development through the MIRAS Demonstrator Pilot Project. The first measurements of the MIRAS prototype led to the finding of the Corbella equation (Corbella et al., 2004) that introduced a fundamental modification in the visibility equation used in radioastronomy.

In summer 1998 ESA launched the first call for Earth Explorer Opportunity Missions within its new Living Planet program. Taking advantage of the recent technological developments, a large group of land and ocean researchers, together with microwave technologists, submitted a SMOS (Soil Moisture and Ocean Salinity) proposal (Kerr et al., 2001; Font et al., 2004). The previous year a mission using MIRAS had been proposed to the French Space Agency (CNES) under the name of RAMSES (Radiométrie Appliquée à la Mesure de la Salinité et de l'Eau dans le Sol), but although being initially selected its implementation was finally discarded. ESA considered SMOS was a risky proposal, due to its new technological concept, never flown before on a satellite, but the maturity and the innovative character of this concept, as well as the timeliness and relevance of the proposed objectives for Earth observation qualified the proposal for being selected in May 1999 as the second Earth Explorer Opportunity Mission out of 27 submitted proposals. SMOS was established as a co-operative ESA-lead mission with contributions from the French CNES and the Spanish Center for Technological and Industrial Development (CDTI).

The initial configuration proposed an instrument working at 1.4 GHz with 25 equally spaced antenna elements in each one of its 4.5 m long arms, plus 4 additional receivers in the central hub. In order to fit in the launcher fairing, each arm was proposed to be folded in five sections in stowed configuration. The instrument would be installed on a PROTEUS generic platform provided by CNES with an antenna plane tilted 20–30° with respect to nadir to guarantee an incidence angle range within [0°, 50°]. The orbit was proposed to be sun-synchronous with Equator crossing at 6 AM (ascending) and 6 PM (descending) to minimize the perturbation on L-band signal (air, vegetation and soil temperature almost identical) and making the Faraday effect minimum. Raw measuring performances were expected to be: 30 to more than 90 km for ground resolution, 0.8–2 K for radiometric sensitivity, 1–3 days for temporal sampling, depending upon latitude, nature of the target and location within the instrument FOV.

An important aspect in the SMOS proposal was the need to use novel calibration techniques, combining both on board reference noise sources of known power level and external constant T_B targets, to ensure a high stability of the measurements. Concerning the operation mode, each 300 ms an image was to be taken, successively in horizontal and vertical polarization. The huge amount of data generated forced to propose some pre-processing on board, with averages of 5 images to obtain one to be formatted and sent to the platform. The resulting equivalent integration time is 1.5s/polarization, so 2 images are available every 3s. No mention was made by then of the possibility of full polarization capability. At the moment of preparing the proposal it was considered that the retrieval of ocean salinity required an independent measurement of sea surface temperature, to be provided through a secondary frequency. The preliminary analysis indicated that a C-band channel could potentially be useful. However, this option was soon discarded due to the mass and power limitations imposed by the use of the PROTEUS platform, suited for a minisatellite, but not allowing simultaneous operations of the two instruments.

The SMOS Phase A development started in 2000, Phase B in 2002, and Phase C/D in 2003 with a launch expected for 2007, that was later delayed until taking place in November 2, 2009. A configuration optimization analysis (Waldteufel et al., 2003) concluded, mainly driven by the more restringing soil moisture requirements in terms of resolution and coverage, that the number of elements per arm should be 21 (six on each one of three folding sections, plus three in the hub), the satellite steering angle 30°, the orbit height around 755 km, the tilt of the antenna plane close to 33°, and the spacing between antenna elements 0.875 wavelengths. During the detailed mission design it appeared that, in spite of the efforts made in successive improvements on the receivers and other components design, the PROTEUS capability was really at the limit with almost no margin, so it was decided to remove three of the receivers in the hub. The final number of antenna elements is 69 and 72 receivers, 66 LICEFs and six noise injection radiometers, are connected to them (McMullan et al., 2008).

The final configuration includes an orbit of 100 min duration with mean altitude of 758 km and inclination of 98.44°; low-Earth, polar, Sun-synchronous, quasi-circular, dusk-dawn, 149-day repeat cycle, 3-day sub-cycle. Global coverage, 80° N/S latitude, with a nominal swath of 1,050 km (3-day coverage) and narrow swath of 640 km (7-day, better radiometric accuracy and large number of incidence angles). The instrument operates at a frequency of 1,413 MHz with 1.2s integration time. Two possible observation modes are implemented: dual-polarization, where horizontal and vertical T_B are recorded in consecutive snapshots, and full-polarization, where the third and fourth Stokes parameters are also acquired in a more complex observation sequence (Martín-Neira et al., 2002). The satellite mass is 658 kg (platform: 275 kg, payload: 355 kg, fuel: 28 kg). The Data Processing Centre is at ESAC, Spain, long-term archive in Kiruna, Sweden, and User Services via ESA's Centre for Earth Observation ESRIN.

3.4.2 Key Science Requirements

SMOS is known as the ESA's Water Mission (Drinkwater et al., 2009) and its main objective is to demonstrate the feasibility of using spaceborne radiometric interferometry for Earth observation to provide global and continuous coverage of soil moisture and ocean salinity with resolution and accuracy adequate to fulfill the mission science requirements. A significant increase of the present knowledge of the spatial distribution and temporal evolution of these two geophysical variables, key to the Earth's global water cycle, is expected to improve the efficiency of our present systems for weather forecast, climate evolution analysis, prevention of natural catastrophic events impact, as well as water resources management.

For ocean salinity (Font et al., 2004), SMOS aims at meeting the salinity remote sensing objectives as defined by the Salinity and Sea Ice Working Group (Lagerloef, 2001): improving seasonal to interannual climate prediction, improving ocean rainfall estimates and global hydrologic budgets, and monitoring large scale salinity events and thermohaline convection. The mission expects being able to observe phenomena like barrier layer effects on tropical Pacific heat flux, halosteric adjustment of heat storage from sea level, North Atlantic thermohaline circulation, surface freshwater flux balance, among other relevant for large-scale and climatic studies. This requires an obtainable accuracy of 0.1–0.4 pss over 100–300 km in 10–30 days. Then the scientific requirement put to the mission was to obtain at least one mean value per 100 km square every month with an accuracy of 0.1 pss. This is a challenging requirement that may have to be relaxed depending on the finally obtained performances for the instrument and the salinity retrieval algorithm.

3.4.3 Basic SMOS Algorithm Approach

The SMOS approach to retrieve the salinity field from the reconstructed T_B images at each orbit uses the multiangular nature of the observations. Due to the shape of the FOV, as closer to the satellite sub track, a single spot in the ocean is seen in more successive snapshots and under more different angles. A maximum of above 60 horizontal and 60 vertical T_B measurements is obtained in the centre, decreasing to half of it at 300 km to both sides. The retrieval algorithm performs a minimization loop using a cost function where the recorded T_B is compared to a T_B modeled value for each one of the available angular measurements, until an optimal fit is reached.

The forward model, or geophysical model function (GMF), that provides the T_B values corresponding to specific seawater characteristics and viewing geometry is a key component of the retrieval algorithm. It has to simulate the emission from the top ocean layer, plus any other radiation at the same frequency coming from external sources (e.g. the cosmic background) and scattered on the ocean surface to the concerned direction, and finally the transformation the overall radiation leaving the surface suffers until reaching the SMOS antenna plane (from atmosphere attenuation and upward emission until Faraday polarization rotation in the ionosphere). The emissivity of a flat sea as function of temperature, salinity, viewing angle, frequency and polarization is quite well modeled using the geometric optics theory (Klein and Swift, 1977), but the different processes that impact on the L-band emission of a roughened surface were not fully described in the several theoretical formulations available at the moment of starting the development of SMOS algorithms. It was necessary to design several new components of the GMF for the SMOS Level 2 Ocean Salinity Processor (L2OP, Zine et al., 2008).

The effect of surface roughness on the T_B is the main geophysical source of error. Unlike Aquarius, SMOS does not have any means to acquire simultaneous independent information of this roughness to be used in the GMF. In addition to this, the available data reporting rough sea surface emissivity dependencies with wind speed does not allow to discriminate the best adapted formulation among the several theoretical models proposed (Font et al., 2006). The SMOS L2OP implements the approach of the polarized ocean T_B being the addition of two terms, one corresponding to the flat sea emission and the other one a correction to it due to the impact of the surface roughness. For this correction three different options were considered, to be tested, improved or even discarded once SMOS data are available. Two of them are theoretical formulations (statistical description of the sea surface plus electromagnetic scattering model) based on the two-scale model (Dinnat et al., 2002) and the small slope approximation (Johnson and Zhang, 1999), and the third one (Gabarró et al., 2004) is an experimental fit, using different roughness descriptors, from data acquired during the WISE trials carried out as part of the SMOS science definition studies (Camps et al., 2004). All these roughness models require the use of external information on wind speed, significant wave height, wave age, etc. to describe the sea state. They are provided to the L2OP by global operational forecasts (atmospheric and ocean wave models) from the European Centre for Medium range Weather Forecasts (ECMWF) that also deliver other parameters, as sea surface temperature, needed by different modules of the retrieval algorithm. These ECMWF variables are introduced as first guess values in the cost function, and during the minimization process they are also tuned like SSS, initially obtained from climatology, until reaching the optimum fit between modeled and measured T_B. This multi-parameter convergence is possible thanks to the over-determination produced by the existence of independent measurements (different incidence angles) of the T_B emitted by the ocean surface, which SSS does not change during the satellite overpass.

3.4.4 Expected SMOS Performance, Error Analysis

The determination of ocean salinity by SMOS has a major drawback, compared to the retrieval of soil moisture, in the much lower sensitivity of T_B to salinity changes, especially at low temperatures. This makes the instrument performance more critical for salinity retrieval. The science requirements established for the mission translated into quite strict radiometric requirements for MIRAS, but tests made after completing the instrument development indicated these were met with considerable margin. The Table 3.1 summarizes the MIRAS radiometric requirements and performances expressed in RMS of T_B as measured in May–June 2007 at the Maxwell Electromagnetic Chamber in ESTEC (European Space Technology Centre, ESA, The Netherlands). The values presented in the table provide the worst case, whenever several measurements were available (Font et al., 2010).

Besides the mentioned low range of T_B values that correspond to the whole range of salinity values in the world oceans, other key problems are impacting the quality of SMOS retrieval of ocean salinity. First, although the radiometric performance of the instrument is better than initially expected, the process of image reconstruction from the correlations of the measurements made by the individual antenna elements is introducing some errors (not well known yet) that would not exist if the measurement was directly made by a physical aperture antenna. Second, there is a need for simultaneous auxiliary information on the sea surface properties (temperature, roughness ...) to be estimated from external sources, as they are not directly measured by SMOS itself. The inaccuracy of this information (in terms of bias and noise in the auxiliary fields provided by the 3-hourly ECMWF forecasts) impacts on the retrieved salinity, in spite of these being only taken as reference values in the convergence retrieval procedure. And third, the imperfections in the different modules that constitute the GMF are also introducing degradations in the retrieval quality

Table 3.1 MIRAS system radiometric requirements and instrument-measured performances at boresight and at the edge of the FOV (32°)

	Required	Measured
Systematic error	1.5 K RMS (0°) 2.5 K RMS (32°)	0.9 K RMS in AF-FOV
Land ($T_{Bland} = 220$ K) Radiometric sensitivity	3.5 K RMS (0°) 5.8 K RMS (32°)	2.23 K RMS (0°) 3.95 K RMS (32°)
Ocean ($T_{Bocean} = 150$ K) Radiometric sensitivity	2.5 K RMS (0°) 4.1 K RMS (32°)	1.88 K RMS (0°) 3.32 K RMS (32°)
Stability (1.2s interval)	4.1 K RMS (<32°)	4.03 K RMS
Stability (6 d interval)	0.03 K	<0.02 K

with respect the one that would be obtained if a perfect model existed. This includes not only the mentioned problem of the surface roughness effect (how the roughness is described with the available environmental parameters and how the roughness modifies the ocean emission, with the additional impact caused by the presence of foam at high winds), but also the determination of other GMF components as the contribution to the signal of the polarized emission of galactic bodies reflected on the roughened sea surface.

The particularities of the imaging capability of an interferometric radiometer contribute to the performance of the salinity determination. We have mentioned the weakness due to the need of performing an image reconstruction step, but SMOS has remarkable strengths compared to measurements made by real aperture antennas. We have highlighted before the multi-angular observation that allows taking advantage of the sensitivity of T_B to the incidence angle to increase the robustness of the inversion. Another fundamental feature is the high angular resolution that allows imaging pixels of the order of 30 km and as a consequence identifying different elements within the FOV. This allows locating the pixels that include a direct, or more likely reflected, image of the Sun. With the SMOS orientation, the Sun is present in 97% of the snapshots; and considering the very high T_B of the Sun L-band emission, it is necessary to discard the few affected angular measurements instead of attempting a correction.

Concerning the observation modes, if MIRAS is operating in dual-polarization every 1.2 s either horizontal or vertical T_B is acquired in consecutive snapshots. Under full-polarization, some time is dedicated to acquire the cross-polarized components and then less data is available for each polarization and there is less noise reduction. However, the additional information can be used to avoid the singularities of the transformation from the antenna to the Earth reference frame, to allow improved RFI detection, to eventually identify azimuthal signals, or to estimate the Faraday rotation. Both modes are to be tested during SMOS Commissioning Phase (6 months after launch) to decide what is the nominal configuration to be used for operations. The salinity retrieval can be performed using the two polarized T_B separately or applying all the calculations to the first Stokes parameter, the sum of both polarizations. Doing the latter the number of independent measurements to integrate in the inversion is halved, then the noise reduction diminishes, but the problem of polarization mixing by Faraday rotation is avoided. Other advantages of this approach is that the uncertainties in the T_B associated to angular dependencies of the sea water dielectric constant model and in the roughness correction term are reduced, as well as the above mentioned singularities in the geometric transformation disappear.

Idealized tests of the SMOS L2OP performance have been done under different configurations and environmental conditions (Zine et al., 2008). Simulated scenes are used to compute the polarized T_B in the SMOS swath along an orbit. Then radiometric noise is added according to the expected MIRAS performance, and the processor is run with different errors and biases for the auxiliary parameters. These tests show that the retrieved SSS values from one satellite overpass will be affected by considerable noise, both from radiometric origin and from uncertainties in the algorithm and auxiliary data. This error is of the order of 0.5–0.7 pss in the centre of the swath and degrades to about 1.5 on its borders. These results improve with high SST scenes, but can be as bad as 1.2 pss (centre) and 2.4 pss (borders) for SST = 5°C. Introducing biases on the auxiliary parameters produces also a bias on the retrieved SSS that can reach up to 0.7 pss (centre) and 0.9 pss (border) when wind speed is biased by 2 m/s. More realistic conditions (e.g. using all the measured antenna patterns for the different MIRAS elements) have been considered in some SMOS system end-to-end performance tests, and these indicate that the above described accuracy of the salinity determination can be degraded by about 50%.

It appears evident that the quality of salinity retrieval obtained from a SMOS orbit (in grid points situated in an Icosahedral Snyder Equal Area projection, ISEA4H9, 15 km characteristic length scale, as interpolated during image reconstruction) will not meet the mission scientific requirements. This is expected to be significantly improved by performing spatio-temporal averages in the generation of global gridded maps (Boutin et al., 2003). A salinity error budget analysis (Sabia et al., 2010) made in an open ocean region using different combinations of configurations and auxiliary data uncertainties, concluded that an average of SMOS products over 30 days and 2° × 2° boxes would generate a SSS map with an error of 0.22 pss, very close to the mission requirements. In further processing steps this can be improved by introducing balancing terms in the cost function (Gabarró et al., 2009) and by bias reduction through external calibration techniques using other sources of salinity data.

3.5 Aquarius/SAC-D Mission

3.5.1 Early Configuration and Evolution of Design

The first significant step toward a NASA salinity mission began with the Salinity Sea Ice Working Group (SSIWG), established in early 1998. The SSIWG included participation from United States as well as European scientists and engineers. During that year the SMOS mission was also being formulated for proposal to ESA (see above). The SSIWG became an international, voluntary and open forum, and held workshops in 1998, 1999 and 2000 focusing on a range of scientific and technical issues covered by the charter (see www.esr.org/ssiwg/mainssiwg.html). The SSIWG provided the basic scientific framework and objectives for salinity remote sensing and outlined basic measurement requirements (Lagerloef et al., 2008). An analyses by Yueh et al. (2001) provided more rigorous assessment of the technical issues and feasibility. During this time, parallel efforts continued in both Europe (with SMOS) and in the United States. The NASA effort focused on satellite sensor concepts and mission designs to measure salinity as a primary objective. Another team in the US pursued a separate mission concept to measure soil moisture with science requirements that demanded much higher spatial and temporal resolution, but much less radiometric accuracy, than needed for salinity.

In 2000, NASA was preparing to release an Announcement of Opportunity (AO) for the Earth System Science Pathfinder (ESSP) program. That same year, the Goddard Space Flight Center (GSFC) and Jet Propulsion Laboratory (JPL) agreed to develop jointly an ESSP mission concept and proposal for an Ocean Salinity Measurement Mission (OSMM). They considered 3 mission concepts, including a (1) single large aperture feed horn, (2) a three-beam pushbroom design, and (3) a conically scanning antenna system. In December, 2000, the 3-beam pushbroom concept was selected after considerable technical evaluation. This offered the best trade-offs between system accuracy, sampling statistics and engineering complexity.

In January 2001, the OSMM concept was named *Aquarius* after the celestial constellation of the same name. In ancient middle-east mythology, Aquarius was the water-bearer whose appearance coincided with the rainy season, and in ancient Egypt, the flooding of the Nile. The image is of a man pouring water on the earth from a large urn. It is a suitable name for an earth science mission to explore how the water cycle, ocean circulation and climate interact in an era of likely anthropogenic climate change. The Aquarius Step 1 (science and mission concept) proposal was submitted to NASA in July 2001 under the ESSP AO. Of the 18 Step 1 proposed ESSP missions, Aquarius and five others were selected to proceed with a Step 2 (technical implementation and cost) proposal.

In November 2001, the Aquarius team agreed with the Argentina Comisión Nacional de Actividades Espaciales (CONAE) to propose a joint mission for Step 2. CONAE would furnish the satellite, mission operations and complementary sensors at no cost to NASA. The mission would become the fourth Satélite de Aplicaciones Científicas (SAC) developed by Argentina in partnership with the US. The SAC-D science objectives are local measurements over Argentina and contribute to global investigations of atmosphere, oceans and effects of human and natural processes on the environment, as per the Argentine National Space Program strategic plan.

The Aquarius NASA proposal was to provide the salinity sensor, science, launch vehicle and other implementation costs. The Step 2 proposal was completed in January 2002, and Aquarius was selected in July 2002 as one of two primary missions to proceed with further definition studies, along with one alternate. The project was then directed to do a 1-year risk reduction study, during which time the NASA-CONAE team re-configured the observatory design to allow CONAE and third party instruments to be included and resembles the final configuration we have today (Fig. 3.4). The mission formulation phase (Phase B) began in December 2003, leading to the system requirements reviews in August and September 2004, and preliminary design reviews in June–August 2005. The NASA mission confirmation review in September 2005 marked the start of the implementation phase (Phase C/D). The joint NASA-CONAE mission critical design review took place in Buenos Aires, July 2008, and the present launch date is scheduled for autumn 2010.

A schematic of the mission profile is shown in Fig. 3.5. The design incorporates several key functional elements. The pushbroom footprint pattern has a ~390 km wide swath consisting of 3 elliptical beam footprints of sizes 76 km × 94 km, 84 km × 120 km and 96 km × 156 km. The measurements will be relatively low spatial resolution, as was stressed in Lagerloef et al. (1995, 2008) and most suited

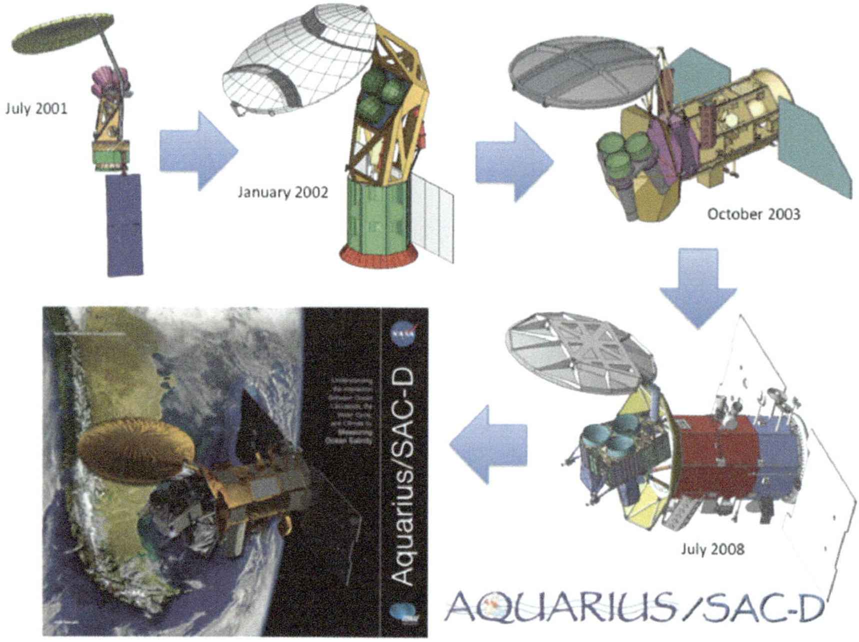

Fig. 3.4 The evolution of the Aquarius/SAC-D mission design: Step 1 Proposal (July 2001); Step 2 Proposal (January 2002); After reconfiguration design (October 2003); Mission Critical Design Review (July 2008); Artist concept of on-orbit observatory over Patagonia

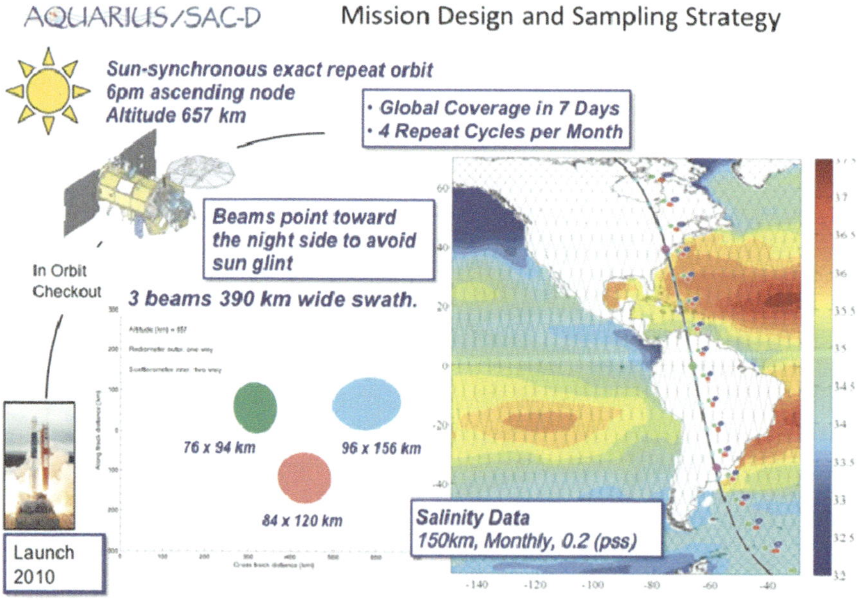

Fig. 3.5 A schematic of the Aquarius/SAC-D mission profile, showing the 3-beam push broom swath (*center*) and the approximate orbit geometry over the western hemisphere (*right*)

for resolving the basin scale SSS field. This will filter much of the eddy and frontal scales, yet provide much greater detail than is derived from historical data (World Ocean Atlas 2005), as illustrated by Lagerloef et al. (2008). The three Aquarius microwave radiometers will measure microwave brightness temperature in vertical and horizontal polarizations (T_{BH} and T_{BV} respectively), as well as polarimetric channels to correct for the Faraday rotation of the signal as it passes through the ionosphere (Yueh, 2000). These sensors are aligned with an offset 2.5 m aperture antenna reflector to generate the 3 fixed beams at incidence angles of 28.7, 37.8 and 45.6° relative to the ocean surface and form the 3 distinct footprints aligned across the swath (more technical details in Le Vine et al., 2007). The Aquarius microwave radiometers have very demanding requirements for low noise and calibration stability, and will be the most accurate ever developed for Earth remote sensing.

The satellite will be placed in a sun-synchronous polar orbit crossing the equator northward (ascending) at 6 PM. The sensor will be viewing away from the sun to avoid solar contamination of the science measurement. The 7-day repeat orbit track spacing at the equator is equal to the swath width. This ensures that the sampling pattern gives total area coverage (no significant swath gaps) and sufficient repeat observations to allow the errors to be reduced by monthly averages. The primary Aquarius microwave sensor combines an L-band microwave radiometer of unprecedented accuracy with an integrated L-band radar to provide a measurement correction for surface roughness, which as noted above, is the significant error source.

The CONAE Microwave Radiometer (MWR) will make complementary measurements of rain, wind and sea ice with 23.8 and 36.5 GHz channels in an overlapping swath pattern (Lagerloef et al., 2008). The MWR data will be used by the Aquarius data processing for rain and sea ice flags, and as supplementary rain and surface wind speed correction algorithms. The New InfraRed Scanner Technology (NIRST) camera is a narrow swath imager intended to detect forest fires and other thermal events on land. It can be tilted to observe preferred targets, and on occasion will be used to map SST within one Aquarius footprint.

3.5.2 Key Science Requirements

The principal scientific requirement is to make global SSS measurements over the open oceans with 150 km spatial resolution, and to achieve a measurement error less than 0.2 (pss) on a 30 day time scale, taking into account all sensor and geophysical random errors and biases. For comparison, the Global Ocean Data Assimilation Experiment (GODAE) requirement is one sample every 10 days/200 km^2 and SSS error of 0.1. Presently, the Global Ocean Observing System (GOOS) provides about 40% of this global sampling (exclusive of the sea ice covered regions and continental shelf) with in-situ observations (principally Argo). The requirement applies to the open ocean, sufficiently far from land or ice boundaries so that the warmer land and ice brightness temperatures, as compared to the ocean surface, do not contaminate the radiometric measurement. Generally this boundary zone is about 2–3 times the footprint diameter. The baseline mission is designed to operate for up to 3 years, with potential extended durations for 2 years or longer.

3.5.3 Basic Aquarius Algorithm Approach

The Aquarius retrieval algorithm is being developed using very thorough simulation to test and quantify errors in the retrievals prior to launch. Starting with the dielectric model of equations (3.1) and (3.2), the Aquarius science simulator generates a forward computation of the top of the atmosphere brightness temperatures based on an ocean model SSS and SST field. The simulator adds all the geophysical radiative sources described in Section 3.3, convolves the Aquarius antenna gain patterns and thus derives the brightness temperature input to the three individual radiometers (called antenna temperatures). Estimated errors for the sensor and geophysical corrections are added, and then the inverse calculation is performed to compare with the input SSS field. A 30-day simulation and retrieval analysis shows worst-case salinity errors ~0.5 in high latitudes and <0.2 in the latitude range 40N–40S for point measurements (5.76s integration time), which would be further reduced by monthly averaging. See also Lagerloef et al. (2008, 2010) and Kim et al. (2010) for more simulator details. A new 1-year simulation is now being computed and will be available at the time of the Oceans from Space 2010 meeting.

The Aquarius baseline retrieval algorithm utilizes both polarization channels in the basic retrieval model (Lagerloef et al., 2008)

$$S = a_0 + a_1 T_V + a_2 T_H + a_3 W + ... \tag{3.3}$$

The coefficients a_0, a_1, a_2 and a_3 are independent functions of SST and the individual beam incidence angle θ. These are presently derived by regression analysis with the simulated data and will be tuned with surface calibration during the mission. The sensitivity to wind roughness is less for T_V than for T_H, (Camps et al., 2004). This allows the possible tuning of the coefficients a_1 and a_2 to off set the roughness effect to some degree. W represents an independent wind parameterization which can be input based on the radar scatterometer data or from an ancillary data source. The present simulator uses ancillary wind fields, and the "at launch" processor will do likewise, until both the radiometer and radar sensors have been calibrated in-orbit and the correction algorithms tuned accordingly. The model can also be expanded to include non-linearities and additional ancillary terms such as wave height, rain rate or wind direction.

3.5.4 Expected Aquarius/SAC-D Performance, Error Analysis

A careful error analysis has been maintained for the Aquarius measurement system throughout the design and construction phase. This includes the measurement errors inherent in the sensor (NEDT noise and calibration stability), the roughness correction retrieval error from the radar, and a residual uncertainty from number of geophysical error sources based on the maturity of the models and the uncertainties in the associated ancillary data. Error terms are tabulated based on an individual observation consisting of a 5.76s data integration along track for each of the 3 Aquarius radiometer beams. The allocation is given in Table 3.2, where the sources are combined as a root sum square. Both the allocation and the current best estimate (CBE) are shown, and the margin is defined as the additional rss error that could

Table 3.2 Top portion shows the allocated sensor and geophysical errors (left column) in terms of v-polarized brightness temperature (T_{BV}) in Kelvin (K), and the Current Best Estimate (CBE) (right column) for each measurement with 5.76s integration time. The radiometer CBE data are based on the Aquarius instrument pre-launch calibration data. The margin is the additional RSS error that could be included before exceeding the allocation. Bottom portion shows the monthly RMS salinity error by latitude band based on the total error allocation of 0.38 K per sample. These include the measurement sensitivity as it varies with SST and the number of samples averaged per latitude band. The global RMS error allocation over all the latitude bands is 0.2 and CBE 0.15 (pss)

			3 beam RMS (K)	
Error sources			Allocation	CBE
Radiometer			0.15	0.13
Antenna			0.08	0.01
System pointing			0.05	0.02
Roughness			0.28	0.13
Solar			0.05	0.02
Galactic			0.05	0.04
Rain (Total liquid water)			0.02	0.01
Ionosphere			0.06	0.05
Atmosphere – other			0.05	0.02
SST			0.10	0.07
Antenna gain near land & ice			0.10	0.10
Model function			0.08	0.07
			Baseline mission 3 beams RMS (K)	
Brightness temperature error per observation			Allocation	CBE
Total RSS (K)			0.38	0.24
Margin RSS (K)			0.30	
			Baseline mission monthly salinity error (psu)	
Latitude range	Mean sensitivity (dTv/dS)	Mean # samples in 28 days	Allocation	CBE
0–10	0.756	10.9	0.15	0.10
11–20	0.731	11.3	0.16	0.10
21–30	0.671	12.1	0.16	0.10
31–40	0.567	13.5	0.18	0.12
41–50	0.455	15.9	0.21	0.13
51–60	0.357	20.3	0.24	0.15
61–70	0.271	30.2	0.26	0.16
Global RMS (pss)			0.20	0.13
Margin RSS (pss)			0.15	

be added to the CBE without exceeding the allocation. Table 3.2 also provides the separation of monthly average error by latitude range, according to the prevailing sea surface temperature. As noted earlier, the sensitivity decreases as temperature decreases, increasing the errors in higher latitudes.

The sample rate increases in higher latitudes from the polar orbiting orientation of the satellite, which is reflected in the mean number of samples per month shown in the table. The largest single error source in the table is the roughness effect due to wind and waves, as described above, and is the reason why the Aquarius instrument includes an integrated L-band (1.26 GHz) radar scatterometer to measure simultaneous oceanic backscatter in the footprint as noted above. The Aquarius roughness error allocation is presently based on limited airborne radiometer and radar combined measurements (Wilson et al., 2001) and is expected to improve with additional airborne data collected in March 2009, which are still being analyzed. More improvements will be derived once the satellite is on orbit. The remaining geophysical error sources are the estimated uncertainty residuals after the best known correction models have been applied. The degree of understanding has been the result of rigorous studies of the ionosphere, galactic reflections, sun and so forth (Le Vine and Abraham, 2002, 2004; Le Vine et al., 2005).

The estimates in Table 3.2 are based on the assumption that all the errors are uncorrelated. In nature, some of these errors are likely to exhibit long correlation scales, especially among the various geophysical effects, and with slow variations in the sensor calibration. The assumptions are tested with the Aquarius science simulator as noted above. Those results, also shown in Lagerloef et al. (2008, 2010), indicate retrieval errors within the science requirement of 0.2 pss monthly average with substantial margin.

3.6 Summary: A Look to the Future Follow-on Possibilities

The simultaneous flights of SMOS and Aquarius/SAC-D give the oceanographic community a rare opportunity to test and evaluate two very different technical approaches: phased-array versus real aperture radiometry. In addition, Aquarius will be the first L-band integrated passive-active (radiometer and radar scatterometer) sensor in space, providing added value for land and ice data analysis as well. Both missions, as pathfinders, have as part of their objectives to demonstrate the technical feasibility and scientific value of the data as a foundation for future salinity missions. Also, as pathfinders, they do not try to do too much – the focus is to provide rather coarse resolution averaged data, consistent with climatological scales (∼100–200 km, monthly), for the open ocean and removed from land and ice boundaries.

Once SSS measurements on these scales are demonstrated, calibrated and validated, the demand will grow to obtain ∼10 km resolution, near coastal measurements and higher accuracy. The higher spatial resolution and near-coast measurement issues can only be solved by flying ∼25 m aperture antennas. Will phased-array or real-aperture be the optimal approach? Can radiometric accuracy

<0.1 K be achieved with such large hardware dimensions? These are the types of over-arching technical challenges for the future, for which SMOS and Aquarius will provide vital technical data to consider the best options.

The other key consideration for future missions, of course, is data continuity. Aquarius and SMOS will provide unprecedented benchmark measurements of an essential climate variable at a time when the planet appears to be experiencing dramatic climate change. As we analyze the satellite SSS data in the next few years, the scientific benefit and need to maintain a climate data record for decades to come will become obvious.

Two opportunities for SSS measurement continuity beyond SMOS and Aquarius/SAC-D are likely. ESA has begun considering possible improvements on the basic MIRAS design to propose a series of operational satellites (SMOSops, SMOS Operational System) that could follow SMOS with similar characteristics. Options are also being studied to augment MIRAS to provide a simultaneous roughness measurement. Several preparatory studies have been carried out in 2007–2008 to analyze some of these possible improvements, although no decision will be taken until an evaluation of the SMOS mission achievements can be made.

In the United States, NASA is developing the Soil Moisture Mapping mission (SMAP) for possible launch in 2015. This is a conically scanning system with a large 6 m offset mesh antenna with a radiometer footprint size of $\sim$40 km. It offers the only viable option in NASA for follow-on salinity measurements, although it is primarily designed for soil moisture measurement. Like Aquarius, SMAP will have an integrated L-band radiometer and radar sensor which will provide simultaneous brightness and roughness measurements over the ocean. The sample rate will be much higher than for Aquarius, although the radiometric accuracy will not be as good. A simulation study is planned for early 2010 to analyze how well the SMAP design could perform to meet the Aquarius science measurement requirements.

We believe that Oceans from Space 2010 is witness to the start of a new era in ocean remote sensing; one that will include ocean salinity as a fully functional component of the array of space borne ocean measurements. The next decade will bring important new discoveries and oceanographic insights. What shall we wager for Oceans from Space 2020 regarding ocean salinity? Will we have a decade of satellite SSS data to analyze? Will ENSO forecast skill be measurably improved as a result? Will the global marine freshwater balance be known to within a few percent? Will upper ocean mixing process be much better understood as a result, and new parameterizations be imbedded in our best ocean general circulation and climate models? Will we have found some significant ocean features and phenomena that we did not anticipate? All of these speculations are certainly worth about ½ pinch of salt in a bottle of wine.

Acknowledgements This chapter is partly a contribution to the SMOS Barcelona Expert Centre on Radiometric Calibration and Ocean Salinity (SMOS-BEC) funded through grant ESP2007-05667-C04 from the Spanish Ministry of Science and Innovation.

References

Blanch S, Aguasca A (2004) Seawater dielectric permittivity model from measurements at L-band. Paper presented at International Geoscience and Remote Sensing Symposium IGARSS, 20–24 September 2004

Boutin J, Waldteufel P, Martin N, Caudal G, Dinnat EP (2003) Surface salinity retrieved from SMOS measurements over the global ocean: imprecisions due to sea surface roughness and temperature uncertainties. J Atmos Ocean Technol 21:1432–1447

Broecker WS (1991) The great ocean conveyor. Oceanography 4:79–89

Camps A (1996) Application of interferometric radiometry to Earth observation. PhD Thesis, Univ Politècnica de Catalunya, http://www.tdx.cesca.es/TDX-1020104-091741/

Camps A, Font J, Vall-llossera M, Gabarró C, Corbella I, Duffo N, Torres F, Blanch S, Aguasca A, Villarino R, Enrique L, Miranda J, Arenas J, Julià A, Etcheto J, Caselles V, Weill A, Boutin J, Contardo S, Niclós R, Rivas R, Reising SC, Wursteisen P, Berger M, Martín-Neira M (2004) The WISE 2000 and 2001 field experiments in support of the SMOS mission: sea surface L-band brightness temperature observations and their application to multi-angular salinity retrieval. IEEE Trans Geosci Rem Sens 42:804–823

Corbella I, Duffo N, Vall-llossera M, Camps A, Torres F (2004) The visibility function in interferometric aperture synthesis radiometry. IEEE Trans Geosci Rem Sens 42:1677–1682

Dinnat EP, Boutin J, Caudal G, Etcheto J, Waldteufel P (2002) Influence of sea surface emissivity model parameters at L-band for the estimation of salinity. Int J Remote Sens 23:5117–5122

Drinkwater M, Kerr YH, Font J, Berger M (2009) Exploring the water cycle of the blue planet: the soil moisture and ocean salinity (SMOS) mission. ESA Bull 137:6–15

Ellison W, Balana A, Delbos G, Lamkaouchi K, Eymard L, Guillou C, Prigent C (1998) New permittivity measurements of sea water. Radio Sci 33(3):639–648

Font J, Lagerloef GSE, Le Vine DM, Camps A, Zanifé OZ (2004) The determination of surface salinity with the European SMOS space mission. IEEE Trans Geosci Rem Sens 42: 2196–2205

Font J, Boutin J, Reul N, Waldteufel P, Gabarró C, Zine S, Tenerelli J, Petitcolin F, Vergely JL (2006) An iterative convergence algorithm to retrieve sea surface salinity from SMOS L-band radiometric measurements. Proceedings of the IEEE International Geoscience and Remote Sensing Symposium 2006 (IGARSS 2006), Denver, pp. 1697–1701

Font J, Camps A, Borges A, Martín-Neira M, Boutin J, Reul N, Kerr YH, Hahne A, Mecklenburg S (2010) SMOS: the challenging sea surface salinity measurement from space. Proc IEEE (in press)

Gabarró C, Font J, Camps A, Vall-llossera M, Julià A (2004) A new empirical model of sea surface microwave emissivity for salinity remote sensing. Geophys Res Lett 31(L0):1309, doi:10.1029/2003GL018964

Gabarró C, Portabella M, Talone M, Font J (2009) Towards an optimal SMOS ocean salinity inversion algorithm. IEEE Trans Geosci Rem Sens 6:509–513

Gordon AL, Giulivi CF (2008) Sea surface salinity trends over 50 years within the Subtropical North Atlantic. Oceanography 21:20–29

Hakkinen S, Rhines PB (2009) Shifting surface currents in the northern North Atlantic Ocean. J Geophys Res 114(C04005), doi:10.1029/2008JC004883

Hollinger JP (1971) Passive microwave measurements of sea surface roughness. IEEE Trans Geosci Electron GE-9(3):165–169

Holliday NP, Hughes SL, Bacon S, Beszczynska-Modieller A, Hansen B, Lavín A, Loeng H, Mork KA, Østerhus S, Sherwin T, Walczowski W (2008) Reversal of the 1960s to 1990s freshening trend in the northeast North Atlantic and Nordic Seas. Geophys Res Lett 35(L03614), doi:10.1029/2007GL032675

IPCC (2007) Climate change 2007: the physical science basis. In: Solomon S, Qin D, Manning M, Chen Z, Marquis M, Avery KB, Tignor M, Miller HL (eds.) Fourth assessment report, Cambridge University Press, Cambridge, UK, pp. 996

Johnson JT, Zhang M (1999) Theoretical study of the small slope approximation for ocean polarimetric thermal emission. Wave Random Complex 37:2305–2316

Kerr YH, Fukami K, Skou N, Srokosz MA, Lagerloef GSE, Goutoule JM, Le Vine DM, Martín-Neira M, Marczewski W, Laursen B, Gazdewich J, Barà J, Camps A (1995) Proceedings of the Consultative Meeting on Soil Moisture and Ocean Salinity Measurement Requirements and Radiometer Techniques (SMOS), ESA WPP-87, ESTEC, Noordwijk, The Netherlands

Kerr YH, Waldteufel P, Wigneron JP, Martinuzzi JM, Font J, Berger M (2001) Soil moisture from space: the soil moisture and ocean salinity (SMOS) mission. IEEE Trans Geosci Rem Sens 39:1729–1735

Kim SB, Wentz F, LeVine D, Lagerloef GSE (2010) Simulation of sea surface salinity retrieval with the Aquarius L-band radiometer. IEEE Trans Geosci Rem Sens (in press)

Klein LA, Swift CT (1977) An improved model for the dielectric constant of sea water at microwave frequencies. IEEE J Ocean Eng 2:104–111.

Lagerloef GSE, Swift C, Le Vine D (1995) Sea surface salinity: the next remote sensing challenge. Oceanography 8:44–50

Lagerloef GSE (2001) Satellite measurements of salinity. In: Steele J, Thorpe S, Turekian K (eds.) Encyclopedia of Ocean Sciences, Academic Press, London, pp. 2511–2516

Lagerloef GSE (2002) Introduction to the special section: the role of surface salinity on upper ocean dynamics, air sea interaction and climate. J Geophys Res 107(C12):8000, doi:10.1029/2002JC001669

Lagerloef GSE, Colomb F, Le Vine D, Wentz F, Yueh S, Ruf C, Lilly J, Gunn J, Chao Y, deCharon A, Feldman G, Swift C (2008) The Aquarius/SAC-D mission: designed to meet the salinity remote-sensing challenge. Oceanography 20:68–81

Lagerloef G, Boutin J, Chao Y, Delcroix T, Font J, Niiler P, Reul N, Riser S, Schmitt R, Stammer D, Wentz F (2010) Resolving the global surface salinity field and variations by blending satellite and in situ observations. In: Hall J, Harrison DE, Stammer D (eds.) Proceedings of OceanObs'09: Sustained Ocean Observations and Information for Society (Vol. 2), Venice, Italy, 21–25 September 2009, ESA Publication WPP-306

Lang RH, Tarkocin Y, Utku C, Le Vine DM (2008) Recent results on the accurate measurements of the dielectric constant of seawater at 1.413 GHz. Geoscience and Remote Sensing Symposium. IGARSS 2008. IEEE International, Vol. IV, Boston, MA, pp. 950–953

Le Vine DM, Zaitzeff JB, D'Sa EJ, Miller JL, Swift C, Goodberlet M (2000) Sea surface salinity: toward an operational remote-sensing system. In: Halpern D (ed.) Satellites, Oceanography and Society, Elsevier Oceanography Series 63, Amsterdam, pp. 321–335

Le Vine DM, Lagerloef GSE, Colomb R, Yueh S, Pellerano F (2007) Aquarius: an instrument to monitor sea surface salinity from space. IEEE Trans Geosci Rem Sens 45: 2040–2050

Le Vine DM, Abraham S (2002) The effect of the ionosphere on remote sensing of sea surface salinity from space: absorption and emission at L-band. IEEE Trans Geosci Rem Sens 40:771–782

Le Vine DM, Abraham S (2004) Galactic noise and passive microwave remote sensing from space at L-band. IEEE Trans Geosci Rem Sens 42:119–129

Le Vine DM, Abraham S, Wentz F, Lagerloef GSE (2005) Impact of the sun on remote sensing of sea surface salinity from space. Proc Internat Geosci Rem Sens Symp 1:288–291, doi:10.1109/IGARSS.2005.1526164

Martín-Neira M, Goutoule JM (1997) A two-dimensional aperture-synthesis radiometer for soil moisture and ocean salinity observations. ESA Bull 92:95–104

Martín-Neira M, Ribó S, Martín-Polegre AJ (2002) Polarimetric mode of MIRAS. IEEE Trans Geosci Rem Sens 40:1755–1768

McMullan KD, Brown MA, Martín-Neira M, Ritts W, Ekholm S, Marti J, Lemanczyk J (2008) SMOS: the payload. IEEE Trans Geosci Rem Sens 46:594–605

Meissner T, Wentz FJ (2003) The complex dielectric constant of pure and sea water from microwave satellite observations. IEEE Trans Geosci Remote Sens 42:1836–1849

Ruf CS, Swift CT, Tanner AB, Le Vine DM (1988) Interferometric synthetic aperture microwave radiometry for the remote sensing of the Earth. IEEE Trans Geosc Rem Sens 26: 597–611

Sabia R, Camps A, Talone M, Vall-llossera M, Font J (2010) Determination of the sea surface salinity error budget in the soil moisture and ocean salinity mission. IEEE Trans Geosci Rem Sens doi: 10.1109/TGRS.2009.2034648

Stott PA, Sutton RT, Smith DM (2008) Detection and attribution of Atlantic salinity changes. Geophys Res Lett 35:L21702, doi:10.1029/2008GL035874

Stogryn A (1997) Equations for the permittivity of sea water. GenCorp, Aerojet Electron Syst Rep, Azusa, CA

Swift CT, McIntosh RE (1983) Considerations for microwave remote sensing of ocean-surface salinity. IEEE Trans Geosci Rem Sens 21:480–491

US CLIVAR Salinity Working Group (2007) Report of the US CLIVAR Salinity Working Group. US CLIVAR Report No. 2007-1, US CLIVAR Office, Washington, DC, available online at: http://www.sclivar.org/Pubs/Salinity_final_report.pdf

Waldteufel P, Boutin J, Kerr YH (2003) Selecting an optimal configuration for the soil moisture and ocean salinity mission. Radio Sci 38:8051

Wilson W, Yueh S, Dinardo SJ, Chazanoff S, Kitiyakara A, Li FK, Rahmat-Samii Y (2001) Passive Active L- and S-band (PALS) microwave sensor for ocean salinity and soil moisture measurements. IEEE Trans Geosci Remote Sens 39:1039–1048

Wilson WJ, Yueh S, Dinardo SJ, Li FK (2004) High-stability L-band radiometer measurements of saltwater. IEEE Trans Geosci Remote Sens 42:1829–1835

Yueh SH (2000) Estimates of faraday rotation with passive microwave polarimetry for microwave remote sensing of earth surfaces. IEEE Trans Geosci Rem Sens 38(5):2434–2438

Yueh SH, West R, Wilson WJ, Li FK, Njoku EG, Rahmat-Samii Y (2001) Error sources and feasibility for microwave remote sensing of ocean surface salinity. IEEE Trans Geosci Rem Sens 39:1049–1060

Zine S, Boutin J, Font J, Reul N, Waldteufel P, Gabarró C, Tenerelli J, Petitcolin F, Vergely JL, Talone M, Delwart S (2008) Overview of the SMOS sea surface salinity prototype processor. IEEE Trans Geosci Rem Sens 46:621–645

Chapter 4
Discoveries About Tropical Cyclones Provided by Microwave Remote Sensing

Kristina B. Katsaros

4.1 Introduction

Since the first "Oceans from Space" meeting in Venice, in the spring of 1980, microwave remote sensing of storms over the oceans has developed in numerous ways. The first seeds to use microwave passive and active instruments were sown with the SEASAT satellite launched in 1978 (see e.g. Katsaros and Brown, 1991). It carried the Scanning Multichannel Microwave Radiometer (SSMR), the SEASAT scatterometer, an altimeter, and a Synthetic Aperture Radar (SAR). I review below how these types of instruments in their later incarnations have contributed to our understanding of tropical cyclones, also known as hurricanes and typhoons and by other names in different ocean basins. In addition to the four types of SEASAT instruments, since 1998 we have had the rain radar on the Tropical Rainfall Measuring Mission (TRMM; see Simpson et al., 1988; Kummerow et al., 2000), which has added a new whole dimension to observing precipitation in these storms. In Section 4.2, I give a short synopsis and examples of how each of these instrument types have benefited Tropical Cyclones (TC) research and provided practical applications. Often, employing data from more than one sensor contributes to interpreting the information inherent in each of them. I have not attempted to review the use of microwave sounders. In Section 4.3, I shall present some thoughts on how these advances can be amplified in the near future. Section 4.4 gives some conclusions.

4.2 Five Types of Microwave Sensors and Tropical Cyclones

This review cover, in turn, microwave radiometers, rain radars, scatterometry, altimetry and the use of Synthetic Aperture Radars. Table 4.1 lists these instruments in a generic way, giving typical microwave frequencies, swath widths and footprint sizes.

K.B. Katsaros (✉)
Rosenstiel School of Marine and Atmospheric Science, Miami, FL, USA: and Northwest Research Associates, Bellevue, WA 98052, USA
e-mail: katsaros@whidbey.net

V. Barale et al. (eds.), *Oceanography from Space*,
DOI 10.1007/978-90-481-8681-5_4,

Table 4.1 Some characteristics of the five types of microwave instruments discussed in relation to TC research and applications. See text for definitions

Instrument	Frequency	Swath Width	Resolution	Satellite/sensor
Radiometer	6.6, 10, 18, 21, 37, 89 GHz	500–1500 km	300 to 15 km	SEASAT SMMR SSM/I TMI
Rain Radar	C-band, 13.8 GHz	215 km	4.3 km	TRMM radar
Scatterometer	Ku-band	500–1800 km	12.5, 25, 50 km	SASS NSCAT QuikSCAT/ SEAWINDS
	C-band	500 km (1000 km)	25–50 km	AMI ASCAT
Altimeter	Ku-band, 14.6 GHz	5–7 km	5–7 km	ERS1/2 TOPEX-Poseidon Jason 1/2
SAR/ SCAN-SAR Wide	C-band	Variable SCANSAR, 500 km	100 m	SEASAT ERS1/2 RADARSAT1/2 ENVISAT

4.2.1 Microwave Radiometry

The SMMR on SEASAT was the prototype that allowed us to learn how the five channels at approximately, 6, 10, 18, 21, and 37 GHz functioned and how to interpret and calibrate the output. The radiometer operated with a conical scan in a sun-synchronous polar orbit, so it could observe a swath of 1,400 km twice/day with ground resolutions ranging from 300 to 25 km depending on frequency. Of interest in the context of TCs are the water parameters that can be obtained. Early on we learned that the total column water vapor agreed very well with values obtained by integrating a radiosonde humidity profile. This was mostly based on the 21 GHz hydrogen line signal, but the liquid water content and rain in the column plus effects of sea surface temperature and wind caused roughness of the sea had to be accounted for. A very similar sensor, the Scanning Multichannel Microwave/Imager, SSM/I has operated on a series of satellites in the Defense Meteorological Satellite Program (DMSP), since 1988 (Alishouse et al., 1990a, b) and other sensors have followed: the Advanced Microwave Scanning Radiometer, AMSR and one called AMSR-E on two recent satellites, Aqua and Terra plus the TRMM Microwave Imager, TMI, and others launched by China and India that are not yet widely available. These sensors provide the water vapor content of the air, which after a long time of learning how to incorporate an integrated quantity is now assimilated into atmospheric numerical models. The liquid water content derived from these radiometers provides total column liquid water, which can be interpreted in terms of precipitation. On the

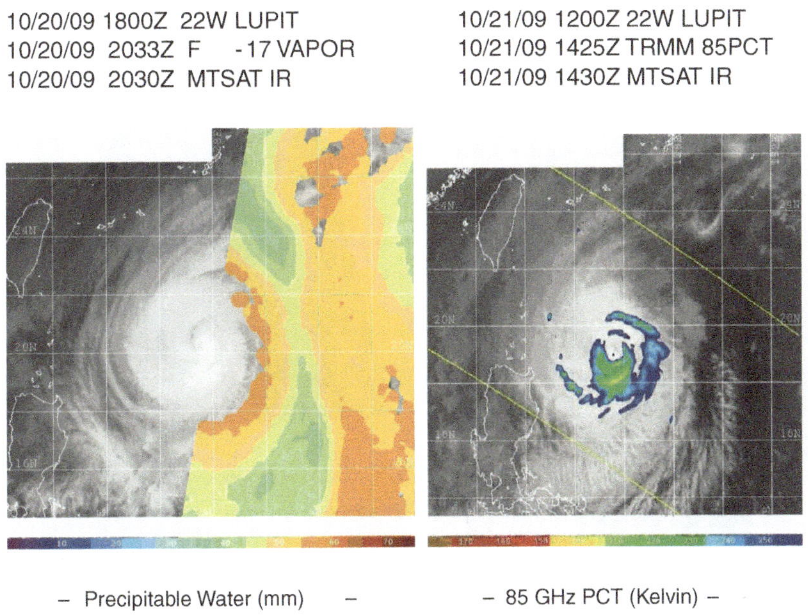

Fig. 4.1 *Left panel*: Column integrated water vapor in the *outer regions* of typhoon *Lupit* on October 20, 2009. Maximum values in the range of 70 kg/m^2. *Right panel*: Precipitation in typhoon *Lupit* on October 21, 2009, provided by the 85 GHZ algorithm using TRMM radiometer. Infrared image from the geostationary meteorological satellite (MTSAT) provides the background. Obtained from the Naval Research Laboratory (NRL), Monterey Marine Meteorology Division (Code 7500) Tropical Cyclone Page[1]

later microwave radiometers, notably SSM/I, a higher frequency, 85 GHz, provides information about ice in the upper regions of a cloud, which is very helpful for identifying raining cloud systems; Fig. 4.1 shows the very high humidity content in the outer regions of typhoon *Lupit*, October 20, 2009 and the precipitation pattern, based on high cold and glaciated clouds, October 21, 2009.

Early in the life of the first SSM/I in space, Zhao (1994) carried out a study of the practical uses of passive microwave data for identifying the stage of TC development (early, intensifying and decaying). She even found evidence of the re-development/intensification with the replacement of the eyewall as first reported by Willoughby et al. (1982). Figure 4.2 illustrates the type of images available using SSM/I data for typhoon Elsie in the West Pacific. Figure 4.2a is based on the water vapor algorithm. In Fig. 4.2b we see the "normalized polarization algorithm" at 37 GHz (P37), which isolates the precipitation effects from variations in atmospheric water vapor content and the effects of surface wind speed (roughness). Figure 4.2c shows "the scattering index" employing the 85 GHz signals in two polarizations and the brightness temperature of the vertically polarized signal at 85 GHz (S85). The scattering index is indicative of the scattering of microwave

[1] http://www.nrlmry.navy.mil/sat_products.html

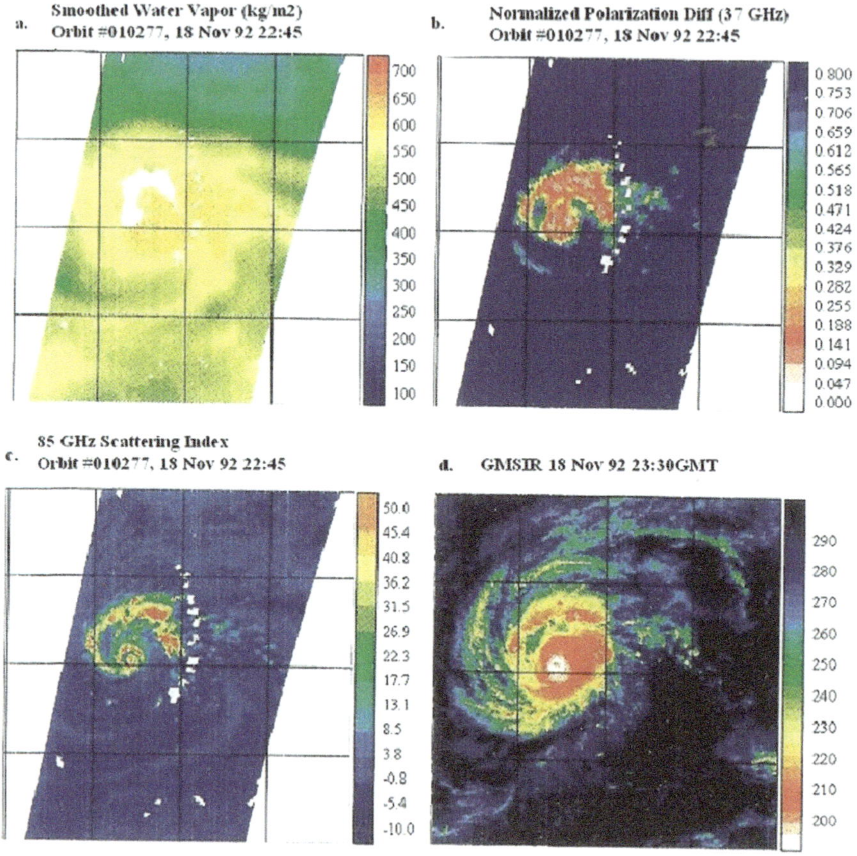

Fig. 4.2 SSM/I images of water vapor, P37, S85 and infrared image from GMS for typhoon *Hunt* at the intensifying typhoon stage (from Zhao, 1994)

signals from the sea surface by the ice particles in the upper regions of the storm, which may or may not be related to precipitation. A thick cirro-stratus cloud may cause significant scattering at 85 GHz with no precipitation occurring below. Heavy graupel and hail can cause much scattering in the upper regions, therefore often the S85 and similar algorithms are indicative of strong convection. Figure 4.2d shows the infrared brightness temperature as seen by a geostationary satellite, the GMS, over the Eastern Pacific in 1992, when these data were collected. The microwave algorithms used in Fig. 4.2 are from Petty (1994). Passive microwave satellite data were quickly adopted to aid in applying the classical Dvorak (1984) technique for estimating TC central pressures and intensity (e.g. Edson and Lander, 2002).

The lowest frequency on SMMR was not used well for SST in the beginning due to radio-interference, while for SSM/I the lower two frequencies were missing. However, an algorithm based on the 37 GHz signal on both instruments could be used for wind estimates away from precipitation. Global passive microwave data on wind speed have been collected by the SSM/I, which for some time operated simultaneously on three satellites. In the section on scatterometry below we discuss the merging of the wind speed data from microwave radiometers with the vector wind data derived from scatterometers.

4.2.2 TRMM Rain Radar and Radiometers

The advent of a rain radar in space on TRMM, even though it had only a 215 km wide swath, allowed calibration of the estimates of precipitation over the ocean obtained from the microwave radiometers. The TRMM rain radar was calibrated against coastal radars, which in turn have been calibrated with arrays of rain gauges in their vicinity, so this has been a great step forward. TRMM is also in a lower non-sun-synchronous Earth orbit only viewing 45° on either side of the equator, thereby crossing the swaths of the microwave radiometers. It also carries its own overlapping radiometer, the TRMM Microwave Imager (TMI). Figure 4.6 below (in the section on SAR), illustrates the results of the precipitation algorithm from TRMM data in conjunction with a SAR image. These high quality precipitation estimates are valuable for interpretation of signals from other spaceborne sensors, such as the SAR.

A wonderful application of TMI estimates of precipitation due to TC's in all ocean basins was provided by Lonfat et al. (2004). They produced a climatology of the rainfall in the path of tropical cyclones at various stages of development (categories 1–5 on the Saffir-Simpson scale) and for various radii of maximum wind. The quantitative aspect of precipitation derived from the TRMM satellite's rain algorithm allowed this climatology to be developed. It can be used to predict the amount of rainfall due to land-falling hurricanes and be an aid in forecasting flooding.

4.2.3 Scatterometry and TCs

A scatterometer is an active system that measures wind vectors by the varied return from the rough sea surface, when viewing the same pixel on the sea surface at different incidence angles (see Robinson, 2004, for details). Upwind, crosswind and downwind directions give different diffuse backscatter from the gravity-capillary waves. SEASAT carried a 3-stick Ku-band radar looking off to one side, a scatterometer named SASS. Many years were needed to fully develop algorithms and resolve the ambiguities that are due to noise in both the surface wave field and the electronics. It took 15 years until the European Space Agency (ESA) launched a C-band scatterometer of similar 3-stick design on the European Remote Sensing Satellite 1 (ERS-1), in August 1991. C-band penetrates clouds and precipitation better, but is less sensitive at low wind speeds than the Ku-band. The ERS-1 was quite successful and was followed by ERS-2 in 1996.

NASA launched NSCAT, the NASA scatterometer on the Advanced Earth Observation Satellite (ADEOS) a Japanese satellite in 1995. This satellite and a later version, ADEOS 2 (2002) were short-lived. In 1999, the US launched a single instrument SEAWINDS on the QuikSCAT satellite to compensate for the loss of

ADEOS. The SEAWINDS was of a new design developed for use on the ADEOS 2. The SEAWINDS is also at Ku-band, but observes in a 1,800 m wide conical swath. It has survived more than 10 years and become well used for hurricane research. In the early days of these wind retrievals, the author and several colleagues had a chance to look at the data for their usefulness in observing TCs, or as we choose to do, looking for early signs of surface wind circulations as tropical depressions developed off the African coast.

In our case we could look for the origin and early stages of known hurricanes (Katsaros et al., 2001). Later, direct calculation of vorticity using SEAWINDS vectors was reported by Sharp et al. (2002). Figure 4.3 shows hurricane Floyd in its early stages in the eastern Atlantic (Liu, 2001).

Scatterometry is now used regularly for tropical cyclone forecasting needs and the data are assimilated from the Advanced Scatterometer (ASCAT), on the European Meteorological Operational satellite (METOP) at the European Centre for Medium-range Weather Forecasts (ECMWF). It follows the ERS pattern of C-band and stick-radar, but this one has swaths on both side of the satellite nadir, so it can cover about 1,000 km of sea surface as it advances in its polar orbit. Developing methods for assimilation of these vector data into models and eliminating areas of too much precipitation have taken much effort over the past three decades.

Currently, active and passive wind data are merged and used in many contexts. In France, the *Institut Francais de Recherche et de l'Exploitation de la Mer* (IFREMER) and its *Centre d'Archivage et de Traitement* (CERSAT) distribute data and research has been performed there on the methods of merging data obtained at

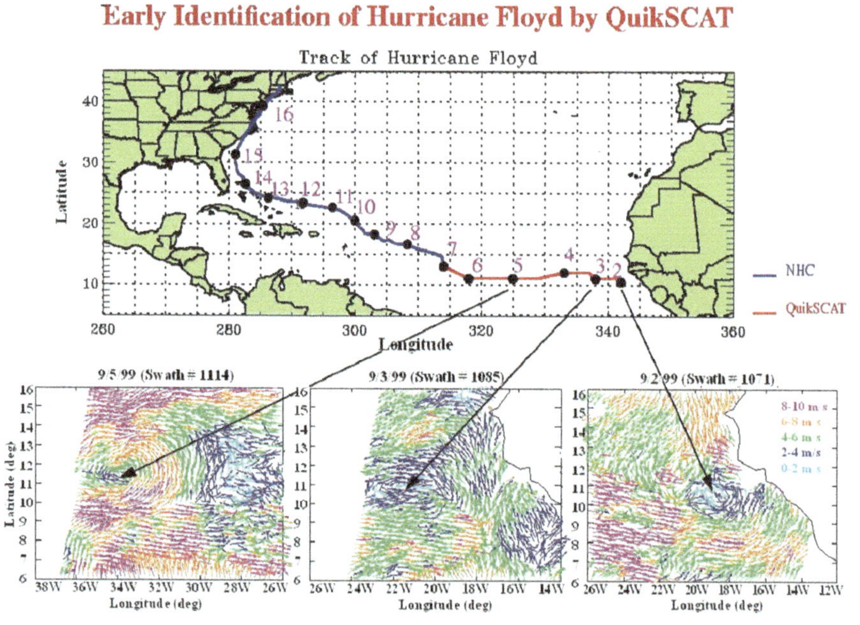

Fig. 4.3 Identification of circulation in the surface wind field of tropical depressions that would become hurricane Floyd (from Liu, 2001)

different times and in different orbit swaths (e.g. Bentamy et al., 2003). This work includes estimates of turbulent fluxes of latent and sensible heat.

4.2.4 Altimetry and TCs

An altimeter measures the distance between the sea surface and the satellite in a narrow swath at nadir (about 7 km × 7 km). SEASAT carried an altimeter and several more have followed on ERS 1 and ERS 2 and on the dedicated satellite, launched in 1992, the Topography Experiment TOPEX/Poseidon (where Poseidon was an experimental additional altimeter operated by the French Space Agency). With a determination of the Earth's average gravitational geoid over time, deviations or Sea Height Anomalies (SHAs) can be determined. These allow interpretation in terms of ocean currents and the heat content of upper ocean eddies. This latter aspect has been developed for aid in determining the Tropical Cyclone Heat Potential, TCHP, of warm eddies in the path of TCs.

Exceptionally high values of heat content have been linked to rapid intensification of a passing hurricane or typhoon. The early estimates of this possibility was for hurricane Opal in the Caribbean sea (Shay et al., 2000). A notable example of intensification while passing over an eddy is that of hurricane *Katrina* in 2005, as illustrated in Figs. 4.4 and 4.5.

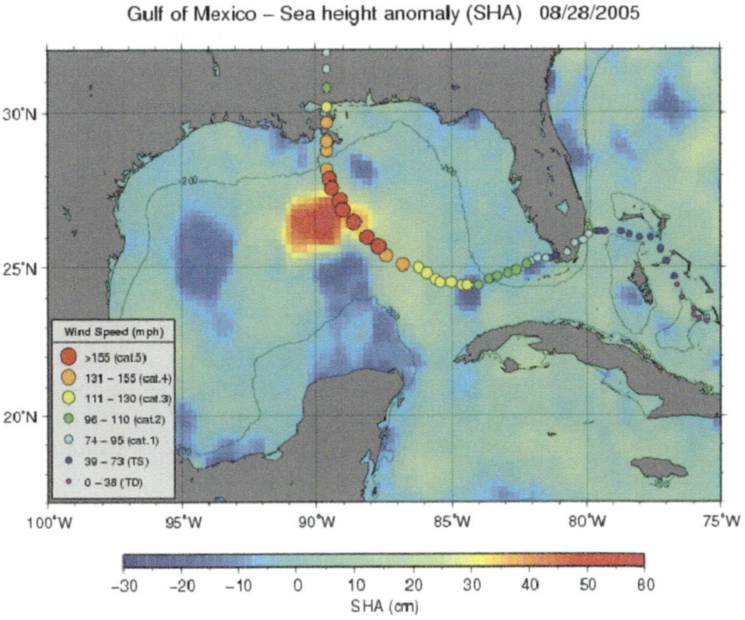

Fig. 4.4 Illustration of the Sea Height Anomaly present as hurricane *Katrina* passed it on its approach to New Orleans, Louisiana, USA in August of 2005. The colours and size of the *position circles* of the storm indicate the category rating of the storm at the time. *Katrina* intensified as she passed over the warm eddy

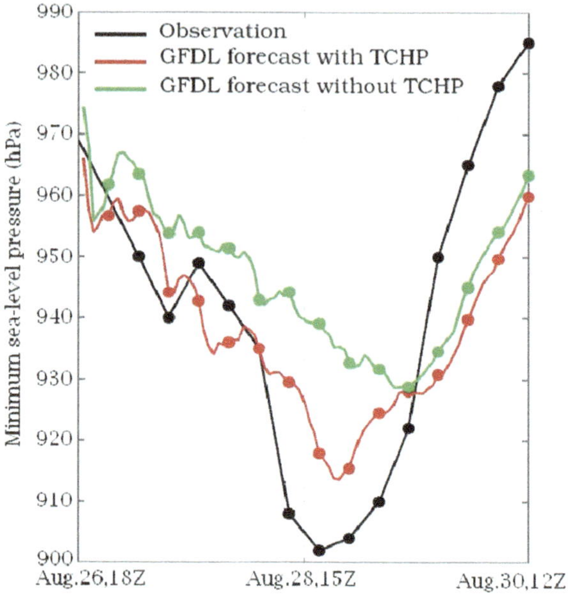

Fig. 4.5 Minimum atmospheric pressure at sea level during the passage of hurricane *Katrina* in the Gulf of Mexico in 2005; showing the observations (*black*), and the reduction of error in the GFDL model with (*red*) and without (*green*) using the TCHP in the coupled atmosphere-ocean TC model. Provided by G. Goni

The concept has been further developed and improved with regular information on the structure in the upper ocean provided by airborne extended bathythermographs, AXBTs, dropped into the sea ahead of storm passage, or more recently by the data from drifting ARGO buoys. The depth of the 26°C isotherm, is used as a piece of input data in calculating THCP. The 26°C value is chosen because this is the temperature associated with TC genesis. Operational estimation of TCHP based on altimeter data is now available via the web.[2] The work has been extended to the warm eddies off Taiwan and reported by Lin et al. (2009).

4.2.5 SAR Observes TCs

The Synthetic Aperture Radar, SAR, on SEASAT was very experimental and much practical and theoretical work was required to interpret the data from a moving sea surface. ERS 1 and 2 also carried a SAR from whose data sea surface wave spectra could be calculated. The RADARSAT 1 SAR launched by the Canadian Space Agency, CSA, in 1998, had several modes of operation. One of them was in "wide swath mode" or SCAN-SAR at about 250 km swath width with 50–100 m resolution. Such data were only collected occasionally and had to be ordered weeks in advance.

However, as a few images were collected, which showed very interesting features, the CSA developed a program called "Hurricane Watch" to plan for data

[2]See http://www.aoml.noaa.gov/phod/cyclone/data/

collection on short notice. Extreme waves and wind structures at so-called mesoscale have been observed with these data and similar data are now available from the European environmental satellite, ENVISAT (e.g. Lehner et al., 1998, 2000).

In the first year as RADARSAT crossed the category five hurricane *Mitch*, an exceptional image was collected which showed streaks parallel to the wind on the sea surface. These were speculated to be surface signatures in the wind field of roll vortices, having the correct orientation and separation (e.g. Brown, 1980). Helical rolls have been observed in aircraft studies and from space (e.g. LeMone, 1973; Alpers and Brümmer, 1994). The image of hurricane *Mitch* in Fig. 4.6 shows these features (Katsaros et al., 2000, 2002). The right hand side of the picture shows the associated precipitation in rain bands derived from the TRMM Microwave Radiometer, TMI, with support from the rain radar data. Clearly the roll-features occur between rain bands. Recent work with aircraft penetration has shown rolls in the wind speed and turbulence measurements of momentum and water vapor flux at aircraft elevations (French et al., 2007; Drennan et al., 2007; Zhang et al., 2008). Theory of roll formation in tropical storm situations is presented by Foster (2005).

New types of observations of the eyes in TCs were provided by the RADARSAT-SAR as illustrated in Fig. 4.7 (e.g. Vachon and Katsaros, 1999; Du and Vachon, 2003). The full interpretation is difficult, since these wide-field signals may react to other effects such as the rain in the eye wall or the rain induced waves on the sea surface, plus the effects of foam and tilting of the roughness elements by large waves.

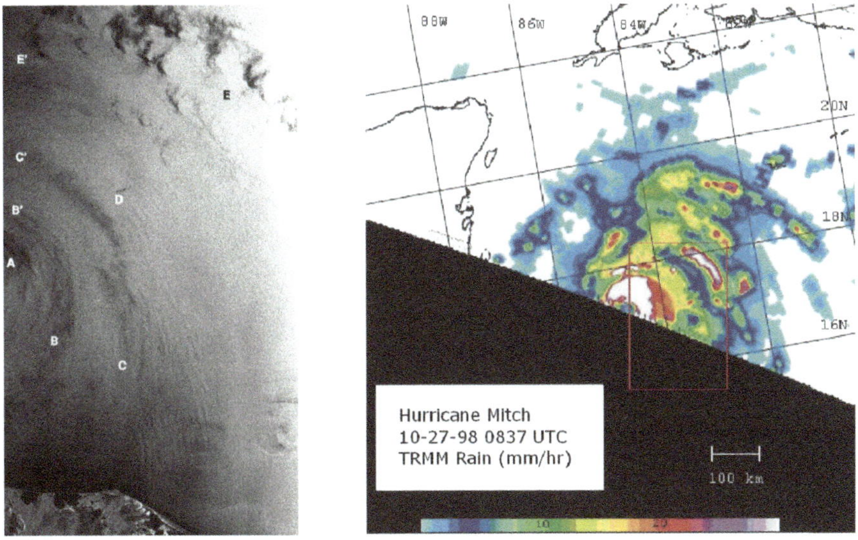

Fig. 4.6 RADARSAT-1 SAR (*left*) (copyright CSA) and TRMM (*right*) images from Hurricane Mitch on 27 October 1998. The SAR image at 1,133 UT covers 184 km × 322 km. The TRMM image at 0837 UT has the same orientation as the SAR image and is 1,060 km × 1,100 km. The *red box outlines* the SAR swath. The *color scale* is TRMM rainfall in mm/h. In the *right panel*: *Green* = 10 mm/h; *red* = 20 mm/h (from Katsaros et al., 2000)

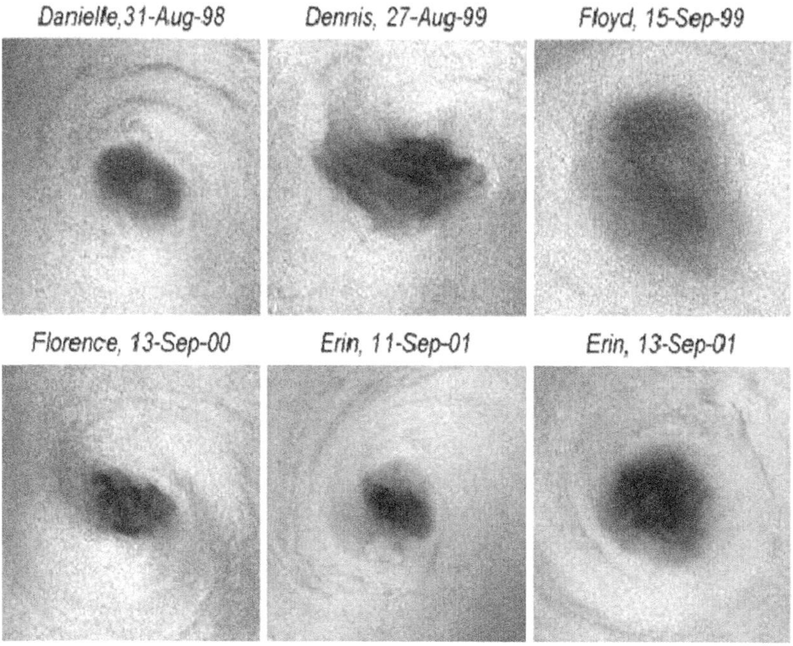

Fig. 4.7 Hurricane eyes observed with RADARSAT-1 SAR in 1998–2001. After Du and Vachon (2003). Copyright CSA

Nonetheless, the wavy aspects of the edges of the eyes are reminiscent of the numerical and laboratory studies of breakdown of vortices (Kossin and Schubert, 2001, 2004). Meso-vortices were also suspected in regions of enhanced destruction in hurricane Andrew 1992, where the eyewall entered the land (Black, 1998, personal communication, based on ground examinations after hurricane Andrew).

4.3 Future Developments

Katsaros et al. (2010) discuss various data types and the importance of coordinated sampling, calibration and algorithm development, as well as long-term archiving with metadata and convenient distribution.

The relevant "virtual constellations" for hurricanes include efforts to coordinate altimeter missions in the Ocean Surface Topography, OST, constellation (Wilson et al., 2009; Fu et al., 2009) and scatterometer missions (Liu et al., 2008). In effect, we already have had a microwave radiometer and rain radar constellation for precipitation estimates operating since TRMM was launched in 1997 due to the many microwave radiometers already in service. Plans for continuation beyond the eventual demise of TRMM exists with the Global Precipitation Measuring (GPM) mission, which is planned for the 2013 time frame (Committee on the Future of

Rainfall Measuring Missions, 2007). A review paper by Stephens and Kummerow (2007) discuss the benefits of measuring many aspects of the cloud and precipitation simultaneously.

4.4 Conclusion

Microwave remote sensing of tropical cyclones has contributed tremendously to insight about the structure and life-cycle of TCs and has allowed improved diagnostics and forecasting of these storms. In this review we highlighted the value of microwaves in allowing us to look inside cloud and storm systems with radiometers and rain radars, and to look through non-raining clouds for surface wind speed with scatterometers, observing circulation in low pressure systems early in their development. Altimeters allow determination of the oceanic heat content for potential development of cyclones. SAR sensors, which are not yet fully observing the globe, nonetheless allow high resolution looks at wind fields and surface waves among other features yet to be fully understood.

The three decades since the pioneering SEASAT satellite was launched and the first Oceans from Space Conference in 1980 have provided a bonanza of new discoveries about tropical cyclones. This short review has only highlighted a few of the many exciting developments, but the reader is encouraged to pursue the literature and contribute new research in this field, where every small advance can lead to safer living in the path of tropical cyclones from improved understanding and forecasting.

Acknowledgements This work was supported over many years by the authors' institutions of work and by NASA and NOAA in the USA and ESA in Europe. In particular, the writing of this review occurred under support from the NSF ATM 0631685. Thanks go to the following individuals for their figures: W. Timothy Liu for Fig. 4.3, Gustavo Goni for Figs. 4.4, 4.5; P. Vachon for the RADARSAT-SAR images (Figs. 4.6 and 4.7) and the use is courteously allowed by the copyright holder, Canadian Space Agency. To S. Öhrvik many thanks for assistance with manuscript production. I acknowledge with much appreciation the teaching about hurricanes by my colleagues at the Hurricane Research Division/AOML/NOAA and many others.

References

Alishouse JC, Snider JB, Westwater ER, Swift CT, Ruf CS, Snyder AS, Vongsathorn J, Ferraro RR (1990a) Determination of cloud liquid water content using the SSM/I. IEEE Trans Geosci Rem Sens 28:817–822

Alishouse JC, Snyder SA, Vongsathorn J, Ferraro RR (1990b) Determination of oceanic total precipitable water from the SSM/I. IEEE Trans Geosci Rem Sens 28:811–816

Alpers W, Brümmer B (1994) Atmospheric boundary layer rolls observed by the synthetic aperture radar aboard the ERS-1 satellite. J Geophys Res 99:12613–12621

Bentamy A, Katsaros KB, Mestas-Nuñez AM, Drennan WM, Forde EB, Roquet H (2003) Satellite estimates of wind speed and latent heat flux over the global oceans. J Clim 16(4):636–656

Brown RA (1980) Longitudinal instabilities and secondary flows in the planetary boundary layer: a review. Rev Geophys Space Phys 18:683–697

Committee on the Future of Rainfall Measuring Missions (2007) NOAA's Role in space-based Global Precipitation estimation and application. Published by the Board on Atmospheric Sciences and Climate, Division of Earth and Life Studies. Report available from National Academies Press, Washington, DC 20055, p. 131. International Standard Book Number 13: 978-0-309-10298-8; and Number 10: 0-309-10298-7

Drennan WM, Zhang JA, French JR, McCormick C, Black PG (2007) Turbulent fluxes in the hurricane boundary layer. Part II: Latent heat fluxes. J Atmos Sci 64:1103–1115

Du Y, Vachon PW (2003) Characterization of hurricane eyes in RADARSAT-1 images with wavelet analysis. Can J Remote Sens 29(4):491–498

Dvorak VF (1984) Tropical cyclone intensity analysis using satellite data. NOAA Tech Rep NESDIS 11, p. 47

Edson RT, Lander MA (2002) Evaluation of microwave imagery in the life cycle of tropical cyclones. Proceedings of 25th Conference on Hurricanes and Tropical Meteorology, American Meteorological Society, San Diego, CA

Foster RC (2005) Why rolls are prevalent in the hurricane boundary layer. J Atmos Sci 62: 2647–2661

French JR, Drennan WM, Zhang JA, Black PG (2007) Turbulent fluxes in the hurricane boundary layer. Part I: Momentum flux. J Atmos Sci 63:1089–1102

Fu LL, Alsdorf D, Rodriguez E, Morrow R, Mognard N, Lambin J, Vaze P, Lafon T (2009) The SWOT (Surface Water and Ocean Topography) Mission. White paper for ocean obs, http://www.oceanobs09.net/

Katsaros KB, Bentamy A, Bourassa M, Ebuchi N, Gower J, Liu WT, Vignudelli S (2010) Climate data issues from an oceanographaphic remote sensing perspective. Proceedings of PORSEC 2008, Science Press/Springer 2010 (in press)

Katsaros KB, Brown RA (1991) Legacy of the Seasat mission for studies of the atmosphere and air-sea interaction. Bull Am Meteorol Soc 72:967–981

Katsaros KB, Forde EB, Chang P, Liu WT (2001) QuikSCAT facilitates early detection of tropical depressions in 1999 hurricane season. Geophys Res Lett 28(6):1043–1046

Katsaros KB, Vachon PW, Black PG, Dodge PP, Uhlhorn EW (2000) Wind fields from SAR: could they improve our understanding of storm dynamics? Johns Hopkins APL Technical Digest 21(1):86–93

Katsaros KB, Vachon PW, Liu WT, Black PG (2002) Microwave remote sensing of tropical cyclones from space. J Oceanogr 58(1–2):137–151

Kossin JP, Schubert WH (2001) Mesovortices, polygonal flow patterns, and rapid pressure falls in hurricane-like vortices. J Atmos Sci 58:2196–2209

Kossin JP, Schubert WH (2004) Mesovortices in Hurricane Isabel. Bull Amer Meteor Soc 85: 151–153

Kummerow C, Simpson J, Thiele O, Barnes W, Chang ATC, Stocker E, Adler RF, Hou A, Kakar R, Wentz F, Ashcroft P, Kozu T, Hong Y, Okamoto K, Iguchi T, Kuroiwa H, Im E, Haddad Z, Huffman G, Krishnamurti T, Ferrier B, Olson WS, Zipser W, Smith, EA, Wilheit TT, North G, Nakamura K (2000) The status of the tropical rainfall measuring mission (TRMM) after two years in orbit. J Appl Meteorol 39:1965–1982

Lehner S, Horstmann J, Koch W, Rosenthal W (1998) Mesoscale wind using recalibrated ERS SAR images. J Geophys Res 103:7847–7856

Lehner S, Schulz-Stellenfleth J, Schättler B, Breit H, Horstmann J (2000) Wind and wave measurements using complex ERS-2 wave mode data. IEEE Trans Geosci Rem Sens 38:2246–2257

LeMone MA (1973) The structure and dynamics of horizontal roll vortices in the planetary boundary layer. J Atmos Sci 20:1077–1091

Lin I-I, Chen CH, Pun IF, Liu WT, Wu CC (2009) Warm ocean anomaly, air sea fluxes, and the rapid intensification of tropical cyclone. Geophys Res Lett 36:L03817, doi:10.1029/2008GL035815

Liu WT (2001) Wind over troubled water. Backscatter 12(2):14

Liu WT, Tang W, Xie X, Navalgund RR, Xu K (2008) Power density of ocean surface wind from international scatterometer tandem missions. Int J Remote Sens 29:6109–6116

Lonfat M, Marks FD, Chen SS (2004) Precipitation distribution in tropical cyclones using the Tropical Rainfall Measuring Mission (TRMM) microwave imager: a global perspective. Monthly Wea Rev 132:1645–1660

Petty GW (1994) Physical retrievals of over-ocean rain rate from multichannel microwave imagery. Part I: Theoretical characteristics of normalized polarization and scattering indices. Meteorol Atmos Phys 54:79–100

Robinson IS (2004) Measuring the Oceans from Space: Theoretical Principles and Methods of Satellite Oceanography. Springer Verlag, Berlin, Heidelberg, New York, p. 646

Sharp RJ, Bourassa MA, O'Brien JJ (2002) Early detection of tropical cyclones using Seawinds vorticity. Bull Amer Meteorol Soc 83:879–889

Shay LK, Goni GJ, Black PG (2000) Effects of a warm oceanic feature on Hurricane Opal Mon Weather Rev 128:1369–1383

Simpson J, Adler RF, North GR (1988) Proposed tropical rainfall measuring mission (TRMM) satellite. Bull Amer Meteor Soc 69:278–295

Stephens GL, Kummerow CD (2007) The remote sensing of clouds and precipitation from space: a review. J Atmos Sci 64:3742–3765

Vachon PW, Katsaros KB (1999) Atmospheric cyclones from spaceborne SAR. Backscatter 10(4):14–19

Willoughby HE, Clos JA, Shoribah MG (1982) Concentric eyewalls, secondary wind maxima, and the evolution of the hurricane vortex. J Atmos Sci 39:395–411

Wilson S, Parisot F, Escudier P, Fellous J, Benveniste J, Bonekamp H, Drinkwater M, Fu L, Jacobs G, Lin M, Lindstrom E, Miller L, Sharma R, Thouvenot E (2009) Ocean surface topography constellation: the Next 15 years in satellite altimetry. White Paper of OceanObs' Conference, Venice, Italy, 21–24 September, 2009. See also http://www.oceanobs09.net/

Zhang JA, Katsaros KB, Black PG, Lehner S, French JR, Drennan WM (2008) Effects of roll vortices on turbulent fluxes in the hurricane boundary layer. Boundary-Layer Meteorol 128:173–189

Zhao H (1994) Analysis of Tropical Cyclones Using Microwave Data from the Special Sensor Microwave/Imager, Master's Thesis, Department of Atmospheric Sciences, University of Washington, p. 82. Available on request from katsaros@whidbey.net

Chapter 5
Direct Surface Current Field Imaging from Space by Along-Track InSAR and Conventional SAR

Roland Romeiser, Johnny Johannessen, Bertrand Chapron, Fabrice Collard, Vladimir Kudryavtsev, Hartmut Runge, and Steffen Suchandt

5.1 Introduction

Since the SEASAT mission in 1978 (Fu and Holt, 1982), spaceborne synthetic aperture radars (SARs) have acquired millions of high-resolution images of ocean scenes, which have been used for applications such as wave and wind retrievals, oil pollution monitoring, ship detection, sea ice monitoring, and the interpretation of signatures of surface current gradients over oceanic fronts, internal waves, and shallow-water bathymetry. Unfortunately, despite the fact that a SAR is a Doppler radar, conventional SAR images do not provide direct information on target velocities, since the Doppler information in the raw data is normally utilised to obtain the highest possible spatial resolution in flight (azimuth) direction. In a process called aperture synthesis, targets are mapped to azimuthal locations in the image where their contribution to the spectrum of the received signal during the SAR overpass appears at a Doppler frequency of 0. This implies the assumption that targets have a radial (line-of-sight) velocity of 0. Targets with a nonzero radial velocity will appear shifted in azimuth direction, and it is sometimes possible to retrieve their velocity from the visible displacement (e.g. between train and track or between ship and wake), but this is not possible for distributed targets such as the ocean surface.

Within the last decade, considerable progress has been made in the development of two techniques that do permit a direct retrieval of line-of-sight surface current fields from SAR data. One technique, called along-track interferometry (ATI), requires a second antenna. The other technique is based on Doppler centroid estimates from conventional SAR raw data at a reduced spatial resolution. Both techniques have been demonstrated in several experiments, and they are available for immediate use with existing satellites. The direct imaging of surface currents at relatively high spatial resolutions is particularly attractive for applications for which radar altimetry (e.g. Wunsch and Stammer, 1998) does not work and

R. Romeiser (✉)
Rosenstiel School of Marine and Atmospheric Science, University of Miami, Miami, FL 33149-1031, USA
e-mail: rromeiser@rsmas.miami.edu

V. Barale et al. (eds.), *Oceanography from Space*,
DOI 10.1007/978-90-481-8681-5_5,

ground- or ship-based remote sensing systems (e.g. Palmer, 1991; Essen et al., 2000; Plant et al., 2005) cannot be deployed easily.

5.2 How to Measure Currents by SAR

Although the idea of exploiting Doppler centroid anomalies of conventional SAR raw data for current retrievals was formulated more than 30 years ago (Shuchman, 1979), it did not get much attention until an impressive demonstration with ENVISAT ASAR data was published in 2005 (Chapron et al., 2005). In the meantime, the ATI technique had been proposed by Goldstein and Zebker (1987) and demonstrated in several experiments. Initially, the ATI technique promised current measurements at full SAR resolution, while a Doppler centroid anomaly analysis seemed to reduce the SAR to a coarse-resolution real aperture radar, but as a result of suboptimal system parameters of available spaceborne ATI systems on the one hand and the development of optimised methods for Doppler centroid estimates on the other hand, differences between actual results of the two techniques are much smaller than one might expect. Let us have a brief look at the theoretical background.

5.2.1 Along-Track Interferometry

The ATI technique is based on interferometric combination of two complex SAR images of the same scene, which are acquired with a short time lag on the order of milliseconds. Phase differences between pixels of the two images are proportional to Doppler shifts of the backscattered signal. To obtain two interferometric SAR (InSAR) images with a short time lag from a moving platform, one needs two antennas separated by a corresponding distance in flight direction. Accordingly, the technique is called along-track interferometry, not to be confused with cross-track interferometry (XTI) for topographic mapping. As mentioned above, the ATI concept was first proposed by Goldstein and Zebker (1987). First airborne ATI results were shown by Goldstein et al. (1989). Thompson and Jensen (1993) presented results of another experiment and demonstrated the importance of correcting ATI-derived velocity fields for contributions of wave motions. They were able to estimate required corrections theoretically. Further airborne ATI experiments were carried out, for example, by Ainsworth et al. (1995), Graber et al. (1996), Siegmund et al. (2004), Bjerklie et al. (2005), Romeiser (2005), and Toporkov et al. (2005).

The time lag τ between the two ATI images is determined by the along-track antenna separation L and platform velocity V. Depending on the transmit/receive sequence of the antennas, one obtains $\tau = L/V$ or $\tau = L/2\,V$, where L or $L/2$, respectively, is called effective baseline. For current measurements, τ needs to be sufficiently long to obtain significant phase signatures from current variations of interest and sufficiently short to avoid phase ambiguities and a decorrelation of the backscattered signal. The decorrelation time depends on radar frequency and wind/wave conditions. According to model results of Romeiser and Thompson (2000), decorrelation times at X band (10 GHz) and L band (1 GHz) are on the order of 5–15 ms and 50–150 ms, respectively, where the lower (higher) values are for high (low) wind speeds. For a satellite ($V \approx 7$ km/s) this translates into maximal along-track antenna distances between about 35 and 2,100 m for different system parameters and wind conditions.

While temporal decorrelation at long time lags can lead to a useless quasi-uniform phase difference distribution with no recoverable velocity information, a low signal-to-instrument-noise ratio at short time lags can be reduced by averaging over a sufficient number of independent full-resolution pixel values, since the instrument-related phase noise is a zero-mean contribution. The number of phase samples that need to be averaged to obtain velocity estimates with a given accuracy is a good measure of the data quality of an ATI system, since it describes the relation between measuring accuracy and effective spatial resolution. Two diagrams showing the theoretical behaviour of this parameter as function of effective ATI baseline and instrument noise level for a spaceborne ATI system ($V = 7{,}000$ m/s) at X band, VV (vertical transmit and receive) polarisation, an incidence angle of 30°, and wind speeds of 5 and 15 m/s are shown in Fig. 5.1. Ideal baselines for the given parameters are in the range of about 20–40 m. Black dots indicate that the parameters of the X band section of the radar system used for the Shuttle Radar Topography Mission (SRTM) and of TerraSAR-X are clearly suboptimal. As will be shown later, the current fields obtained from these two systems have an effective spatial resolution on the order of 1 km, which is consistent with these diagrams.

ATI images are affected by the same velocity-related SAR mapping artefacts as conventional SAR images, and detected velocities need to be corrected for contributions of sub-resolution-scale wave motions, which may vary within an image due to wave-current interaction (Thompson and Jensen, 1993; Romeiser and Thompson, 2000). An iterative correction on the basis of numerical simulations was demonstrated by Romeiser (2005).

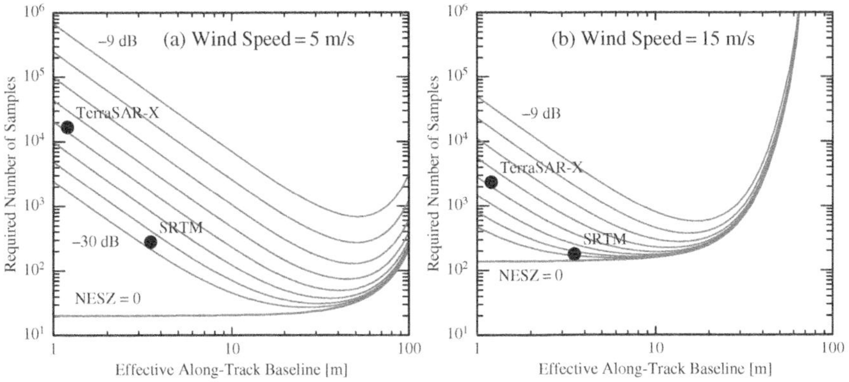

Fig. 5.1 Theoretical number of ATI phase samples to be averaged to obtain current estimates with an rms error of 0.1 m/s vs. effective along-track baseline, for instrument noise levels (NESZ) of 0 and −30 to −9 dB in steps of 3 dB and for wind speeds of 5 and 15 m/s. Radar frequency = 9.65 GHz, polarisation = VV, incidence angle = 30°. *Black dots* indicate properties of SRTM and TerraSAR-X

5.2.2 Doppler Centroid Anomaly Analysis

Direct instantaneous frequency determination from the phase history analysis of single antenna returns is a standard methodology to process SAR images (e.g. Madsen, 1989). Single-antenna Doppler estimates can indeed be directly obtained from the measured return signal spectral peak frequencies, i.e. Doppler centroids. Such estimates are commonly used to focus the SAR image. However, when compared to geometrically predicted Doppler frequencies (considering the relative motion between satellite and rotating earth), systematic differences or Doppler shift anomalies were reported for conventional SAR ocean scenes (Chapron et al., 2002). Analysis of global Wave Mode data from ENVISAT ASAR (one 10 km × 6 km "imagette" every 100 km) proved that these Doppler anomalies originated from geophysical conditions. Their analysis works best for homogeneous scenes, exhibiting small image intensity variations, and yields estimates with a spatial resolution of about 8 km × 4 km for ENVISAT ASAR Wide Swath Mode images (swath width ≈ 400 km) and 1 km × 1 km for ERS and ENVISAT Image Mode products (swath width ≈ 100 km). Note that the azimuthal resolution can be significantly better than the real aperture resolution of the radar antenna on the order of 5 km, since the Doppler centroid analysis can be combined with some amount of SAR processing. Like in ATI data processing, there is a tradeoff between the effective spatial resolution and the relative rms error (compared to the local expectation value) of retrieved radial velocity maps. From ASAR Wave Mode data, one velocity estimate is obtained every 100 km along the track.

As interpreted (Chapron et al., 2005; Johannessen et al., 2008), the Doppler anomaly is associated with an overall bulk velocity including the mean velocity of the radar detected surface scatters and the desired ocean surface current. The Doppler anomaly has been found to depend on radar frequency, incidence angle, polarisation, and environmental conditions, mostly wind speed and direction (Mouche et al., 2008). The mean velocity of the radar detected scatterers is generally larger for HH than VV polarisation and decreases with the radar wavelength and for incidence angles greater than 30°. Although no direct comparison has been performed so far, the physical mechanisms that lead to differences between actual surface currents and uncorrected Doppler velocities obtained from Doppler centroid anomalies seem to be the same as the ones that affect ATI data (Thompson and Jensen, 1993; Romeiser and Thompson, 2000; Romeiser, 2005).

The partitioning of the Doppler anomaly to the different contributions is a challenging problem, but the technique is robust and has the potential to meet high spatial resolution requisites. In all cases, the measurement of Doppler anomalies can complement the generation of conventional SAR images, so that geometrical and dynamical properties of the ocean scene can be derived together. Under favourable and well known environmental conditions, it is highly feasible to clearly identify mesoscale and submesoscale features and to infer absolute surface velocities along the radar line-of-sight direction, as pointed out by Johannessen et al. (2008).

5.3 Experimental Results

While the single-antenna Doppler anomaly analysis can be applied to thousands of existing SAR images from the last 2 decades, the archive of existing spaceborne ATI images of ocean scenes is limited to a few images from the 11-day SRTM mission in February 2000 and a few test images that have been acquired with TerraSAR-X since spring 2008. Here is a summary of key results we have obtained so far.

5.3.1 Along-Track InSAR Results

The single-pass XTI system installed on space shuttle *Endeavour* for the SRTM mission in February 2000 had an along-track antenna distance of 7 m (in addition to the cross-track distance of 60 m, which is not relevant for current measurements), which permitted current retrievals at a few test sites of opportunity, while the primary mission objective was topographic mapping over land (Rabus et al., 2003). TerraSAR-X, launched in June 2007, has a programmable phased-array antenna panel that can be divided into two parts with a theoretical phase center distance of 2.4 m for receiving. In both cases, the effective baseline is half the given distance, and the effective ATI time lags are clearly suboptimal (see Fig. 5.1), but the data have been good enough for a demonstration of current measurements from space by ATI. Figure 5.2 shows artistic views of SRTM and TerraSAR-X.

Figure 5.3a shows a line-of-sight current field retrieved from an interferometric X band image of the Dutch Wadden Sea from SRTM (from Romeiser et al., 2005). Due to the required pixel value averaging for noise reduction (see Fig. 5.1), the

Fig. 5.2 Artistic illustrations of SRTM (*left*, courtesy NASA/JPL-Caltech) and TerraSAR-X (*right*, Siemens press picture) in space. The SAR antennas of SRTM are in the cargo bay and at the end of the 60-m long boom; the SAR antenna of TerraSAR-X is the *grey panel* at the *bottom*

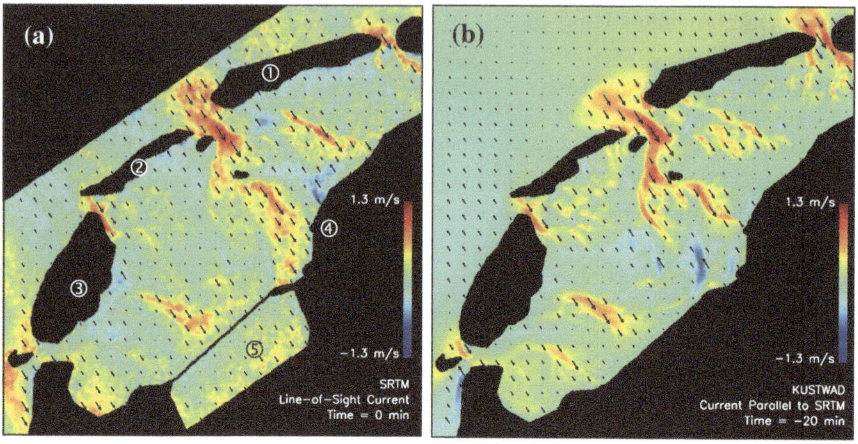

Fig. 5.3 Current field in the Dutch Wadden Sea from (**a**) SRTM and (**b**) a numerical circulation model. Area size = 70 km × 70 km, grid resolution = 100 m × 100 m. Map legend: 1 Terschelling, 2 Vlieland, 3 Texel, 4 Harlingen, 5 Lake IJssel

effective spatial resolution is on the order of 1 km, and the rms difference between SRTM-derived currents and reference currents from the numerical circulation model KUSTWAD (ten Cate et al., 2000) was found to be better than 0.1 m/s (Fig. 5.3b).

Romeiser et al. (2007) used another SRTM image to retrieve currents in the Elbe river (Germany). Results are shown in Fig. 5.4. Assuming that the dominant flow is

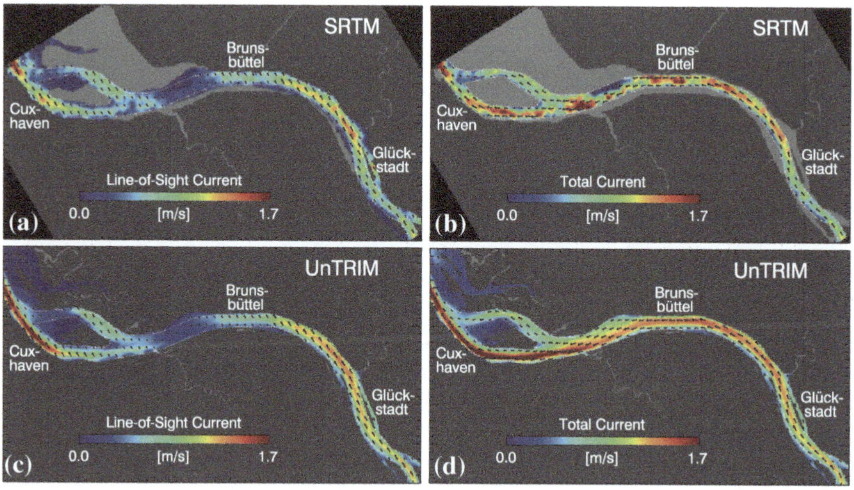

Fig. 5.4 Current field in the Elbe river (Germany); (**a**) SRTM-derived line-of-sight currents, (**b**) SRTM-derived quasi-2-D total surface currents, (**c**) model-derived component parallel to the look direction of SRTM, (**d**) model-derived total 2-D current field. Area size = 55 km × 30 km, grid resolution = 100 m × 100 m

parallel to the river bed, it was possible to construct a fully 2-D surface current field (Fig. 5.4b) from the ATI-derived component (Fig. 5.4a). Again, the SRTM-derived currents were found to be consistent with a numerical model or the river, UnTRIM (Casulli and Walters, 2000).

The divided antenna of TerraSAR-X has an even shorter ATI time lag than SRTM, and the instrument noise level is higher. However, according to Romeiser and Runge (2007), the smaller pixel size of TerraSAR-X permits more averaging of original pixel values at the same efffective spatial resolution (see also Fig. 5.1), which actually overcompensates the phase sensitivity and instrument noise handicap. The effective spatial resolution of current fields from TerraSAR-X in stripmap mode (swath width = 30 km, nominal pixel resolution = 3 m) was expected to be better than 1 km at an rms error of current estimates of 0.1 m/s. A major advantage of TerraSAR-X compared to SRTM is the pure ATI geometry of the divided antenna, which facilitates absolute current measurements. With a pure ATI system, a phase difference of 0 corresponds to a line-of sight velocity of 0, while phase differences from combined ATI/XTI systems include a topographic contribution that is often not well known.

ATI data acquisitions with TerraSAR-X are possible in various modes of operation, all of which are still in an experimental stage of development. In spring and summer 2008, a first series of ATI images was acquired in the so-called Aperture Switching (AS) mode, which uses a single receiver for both antenna halves in an alternating way at a doubled pulse repetition frequency. This is less desirable, but easier to implement than the full Dual Receive Antenna (DRA) mode, which uses two receivers in parallel. In AS mode, the swath width of stripmap images is reduced to about 16 km, the noise level is a little higher than in DRA mode, and ambiguities in the SAR processing can produce ghost images of bright targets on land over water. Nevertheless, Romeiser et al. (2010) were able to process and analyse six AS-mode images of the Elbe river quite successfully. Again, UnTRIM model results were used as reference. Example results for three of the six cases are shown in Fig. 5.5. The data quality of TerraSAR-X AS-mode data seems to be consistent with the theoretical predictions, and the retrieval of absolute currents from the pure ATI data of TerraSAR-X (in contrast to relative current variations within the scene from SRTM data) seems to work, but a final quantitative evaluation has not yet been done due to the preliminary state of the existing data processing routines.

5.3.2 Doppler Anomaly Analysis Results

At first order, the Doppler anomaly is mostly wind dependent, as revealed when collocated monthly wind fields from the European Centre for Medium Range Weather Forecasts (ECMWF) were projected along the radial direction of ENVISAT ASAR Wave Mode data. Based on the collocated data set obtained this way, a neural network model, called CDOP, was created for an incidence angles of 23° and 33°, where inputs are wind speed and relative wind direction with respect to azimuth;

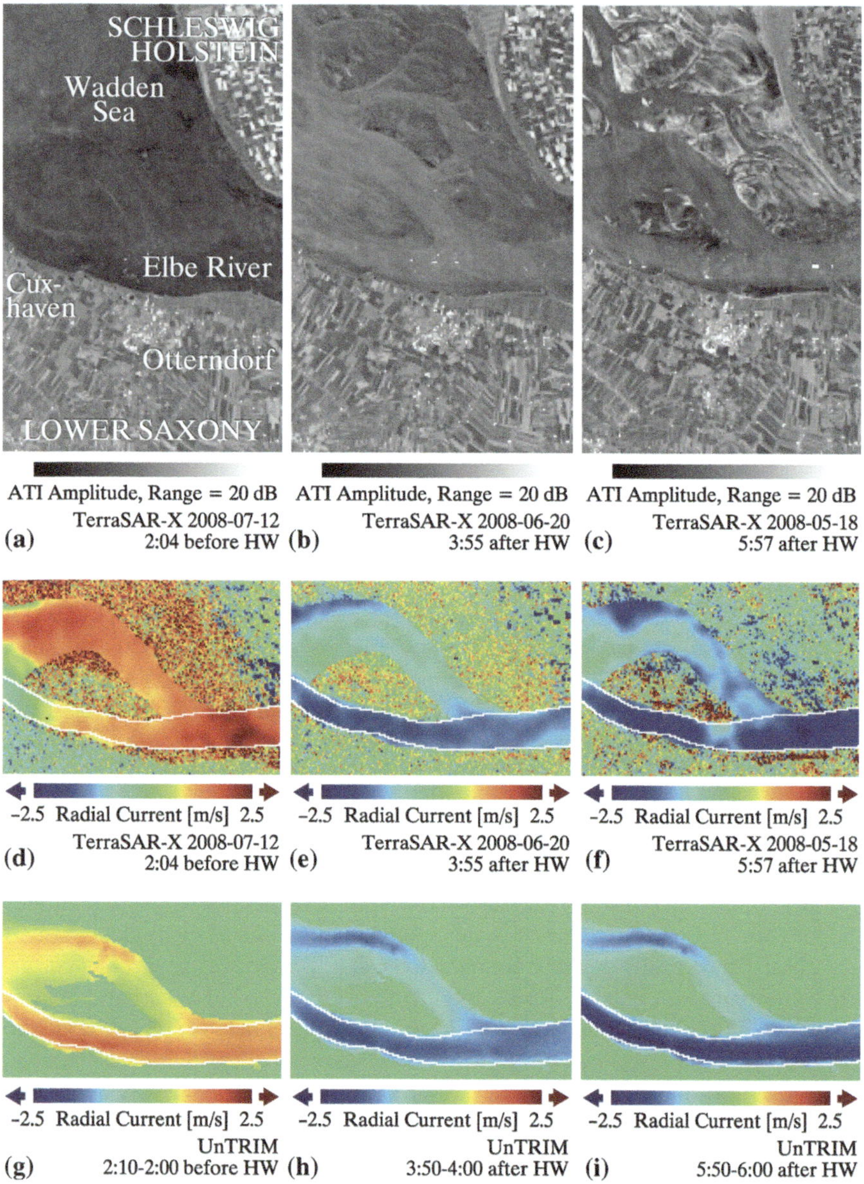

Fig. 5.5 First TerraSAR-X ATI results for the Elbe river; (**a**–**c**) three amplitude images acquired on different days at different tidal phases, (**d**–**f**) corresponding ATI-derived line-of-sight current fields, (**g**–**i**) corresponding line-of-sight current fields according to numerical model UnTRIM. Area size (amplitude images) = 16 km × 25 km, grid resolution = 100 m × 100 m, look direction = *left to right*, incidence angle = 31.0° (near range) to 32.5° (far range)

output is the wind-dependent contribution to Doppler velocities. Later, using the Doppler grid now available for each Wide Swath product, the CDOP model was extended to incidence angles from 17° to 42° (Collard et al., 2008).

Taking benefit of the large number of ASAR imagette observations and the development of CDOP, the detection capability of the Pacific equatorial current regime was examined. A monthly averaged residual radial current field, obtained after the removal of the wind effect, is presented in Fig. 5.6a. It exhibits a band with significant easterly (negative) directed radial velocities around 7°N latitude and two bands of westerly (positive directed) values on either sides centered at 2°N and 13°N latitude. This latitudinal variation of the line-of-sight surface current is in agreement with the expected positions of the equatorial current and counter current.

The zonal flow field at three selected transects was compared to zonal surface currents from the numerical global ocean circulation model MERCATOR (http://www.mercator-ocean.fr/), the drifter-derived climatology of global near-surface currents produced by the National Oceanic and Atmospheric Administration (NOAA) in the framework of the Global Drifter Program, and surface currents derived from altimetry data and wind field analysis from NOAA through the Ocean Surface Current Analyses – Real Time (OSCAR) project (Bonjean and Lagerloef, 2002). As shown in Fig. 5.6b–d, the overall agreement is noteworthy. The location of

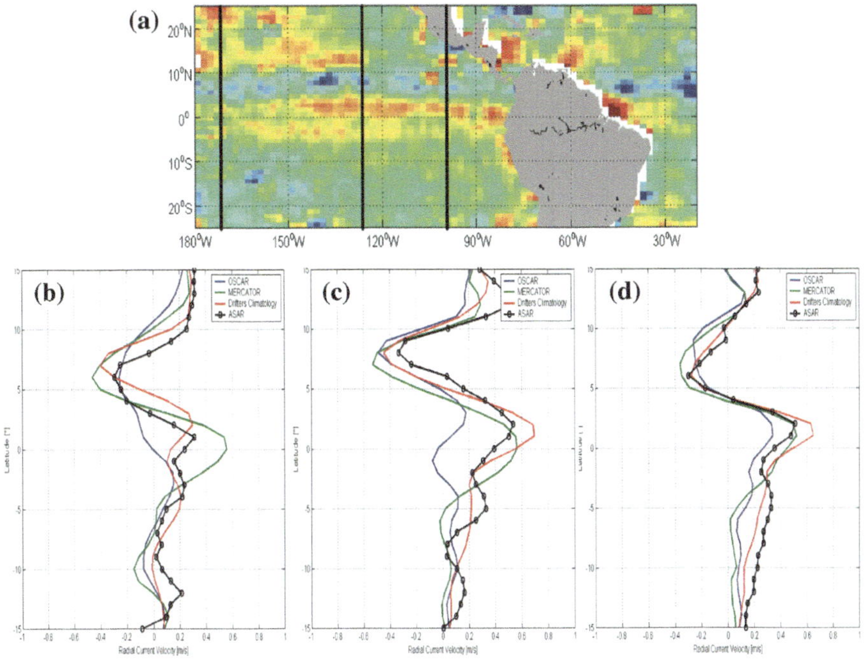

Fig. 5.6 Monthly mean residual radial surface velocity in November 2006 at 2° × 1° resolution from Doppler anomaly analysis (**a**) and comparison of radial velocities from MERCATOR, OSCAR, drifters, and ASAR at (**b**) 170°W, (**c**) 128°W, (**d**) 100°W and latitudes from 10°S to 10°N (*black lines* in (**a**))

the equatorial currents and counter currents are predicted and measured at the same latitudes, while the range of SAR-derived radial currents agrees with the surface current predictions and the independent measurements.

Using the ASAR wide swath mode, synoptic imaging of intense current regimes is also possible. After removal of the wind contribution to the Doppler signal over the Agulhas Current, it is evident that this residual surface velocity is connected with the strength and pathway of the greater Agulhas Current (Gründlingh, 1983), as demonstrated by Johannessen et al. (2008). Residual line-of-sight velocities reach 2 m/s in the Agulhas Current and nearly 1.5 m/s in the return current, as shown in Fig. 5.7. Currents of this magnitude were found to be consistent with

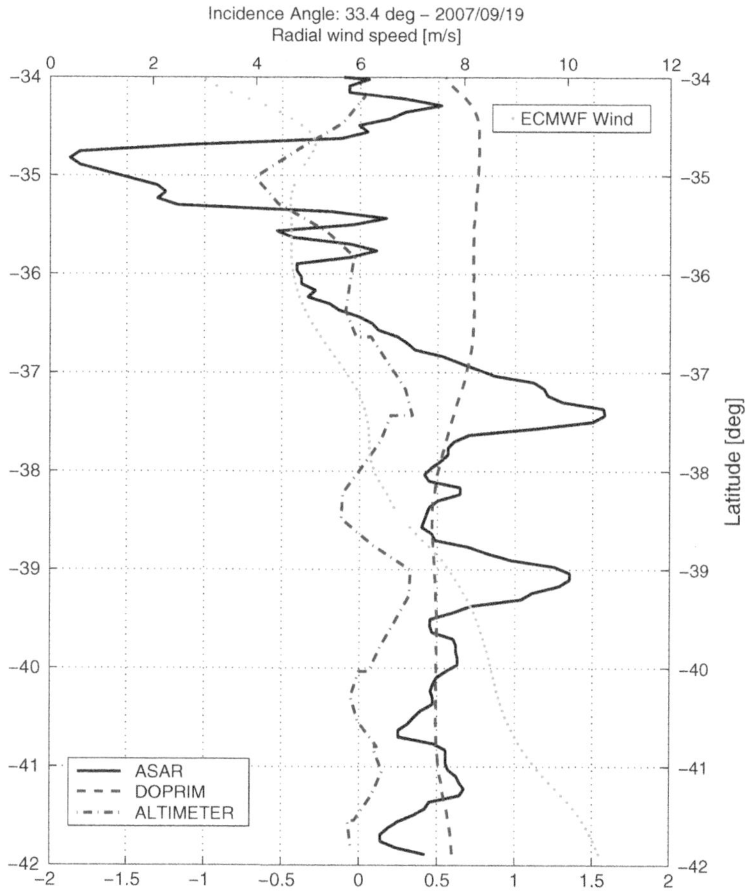

Fig. 5.7 Range-directed velocities along a transect in azimuth through the ASAR image of Fig. 5.8b: Residual velocity from Doppler anomaly analysis (*solid black*), geostrophic current according to altimeter data (*blue dash-dot*), wind-induced Doppler velocity contribution according to numerical model DopRIM (*red dashed*). *Light-blue dotted line* shows radial wind speeds from ECMWF

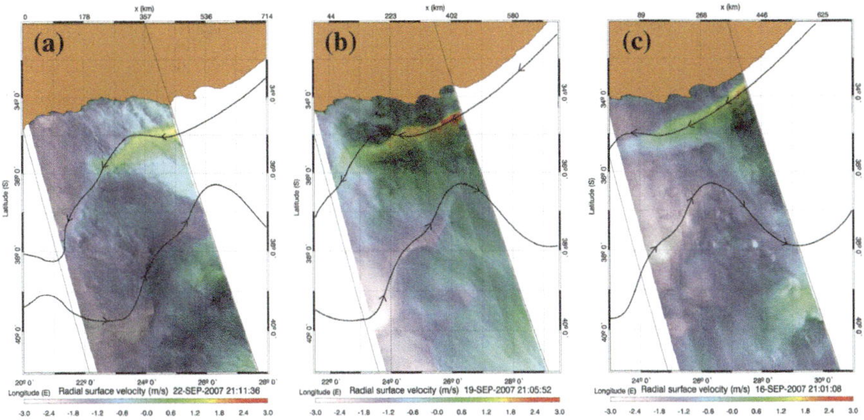

Fig. 5.8 Time series of range-directed Doppler velocity from ascending ASAR wide swath (420 km) images on (**a**) 22, (**b**) 19, (**c**) 16 September 2007 covering the greater Agulhas Current region. *Colour bar marks* radial velocities from −3 to +3 m/s. Positive speed is directed towards the SAR look direction (i.e. perpendicular to the swath, to the *left*). *Black curve* marks position of maximum geostrophic current derived from altimetry 7-day mean

drifter data, while altimeter data indicate a maximum geostrophic current of only about 0.7 m/s.

From this promising result, further evidence of the persistent range-directed surface velocity magnitude and pattern associated with the greater Agulhas Current can be assessed by the time series shown in Fig. 5.8 for acquisitions on 22, 19, and 16 September 2007. The full validation of such measurements remains a challenge as coincident direct sea surface current measurements are non-existent.

5.4 Outlook

Doppler centroid estimates are now included in ASAR image products from ESA. TerraSAR-X ATI acquisitions over selectable test areas should be available to registered users at the time of publication of this paper. Furthermore, a second TerraSAR-X type satellite called TanDEM-X (Moreira et al., 2004) should have been launched for formation flight with TerraSAR-X. While its main purpose is a high-resolution topographic mapping of land surfaces, the combination of TerraSAR-X and TanDEM-X will also enhance the ATI capabilities. In certain latitude bands, the along-track distance between the two satellites will permit ATI with a longer baseline and a corresponding improved accuracy and spatial resolution of surface current measurements. Furthermore, it may be possible to rotate TerraSAR-X and TanDEM-X into slightly different look directions and operate them both in split-antenna mode for two-dimensional vector current measurements, as proposed by Schulz-Stellenfleth et al. (2006; see also Frasier and Camps, 2001; Toporkov et al., 2005). ESA is currently considering the possibility to fly a passive bi-static SAR antenna in ATI formation with SENTINEL-1. On the long term, technically optimised ATI satellites for ocean applications could permit vector current measurements at an effective spatial resolution of 100 m and with improved spatial and temporal sampling characteristics. The technology for such instruments is readily available. Further development in this field must be driven by user demand and should include dedicated validation campaigns. In the following we discuss promising applications. Some text and figures have been adapted from Romeiser and Runge (2008).

5.4.1 Global Oceanography

Chapron et al. (2005) have demonstrated the generation of global surface current maps at a spatial resolution on the order of 100 km from ENVISAT ASAR Wave Mode data. Johannessen et al. (2008) have shown that ASAR Wide Swath data can reveal large surface currents of the Agulhas Current that are not detected by altimeters.

Both capabilities can be very useful for global circulation studies as well as for studies on the dynamics of strong mesoscale current features in certain areas. The potential spatial resolution improvement of the ATI technique by more than one order of magnitude can be useful for studies on small-scale current features that have a strong larger-scale effect on the upper-ocean circulation as well as on chemical and biological processes.

For example, McGillicuddy et al. (2007) describe a stimulation of mid-ocean plankton blooms by eddy/wind interactions. Some of the eddies of interest can be seen in altimeter data, but an analysis of high-resolution current fields from single-antenna Doppler anomaly or ATI measurements, ideally in combination with water colour and temperature data from other sensors, would be highly desirable for more detailed investigations and improved model developments in this field.

5.4.2 Coastal Oceanography

The general circulation patterns in coastal seas are often well known and reproducible by numerical circulation models. Furthermore, many HF radars have been installed within the last decade for a continuous monitoring of currents, waves, and ship traffic in coastal regions (e.g. Gurgel and Schlick, 2007). However, quasi-geostrophic dynamic processes on spatial scales of 1–10 km are still not entirely understood, and they may be of interest for scientists and other users who are unable to install coast-based remote sensing systems or to perform extensive in-situ measurements. Such processes include the formation of mesoscale eddies, fronts, internal waves, the response of upper layer dynamics to rapid changes in the wind field, or effects of changes in bottom topography, river discharges, or other boundary conditions that are not well known and difficult to predict with purely theoretical approaches. Despite the coarse temporal sampling, repeated high-resolution current measurements from space can be valuable for basic research and routine monitoring in such regions.

5.4.3 Bathymetric Mapping and Monitoring

The mapping and monitoring of underwater bathymetry in coastal waters with strong tidal currents (for example, the Wadden Sea off the coast of the Netherlands, Germany, and Denmark) on the basis of conventional SAR intensity imagery was demonstrated by Calkoen et al. (2001) and has been available as an operational commercial service for several years. This technique exploits that bathymetric features become visible in SAR images due to a modulation of the tidal flow by the spatially varying water depth and a corresponding surface roughness modulation via wave-current interaction.

Since the imaging mechanism is quite indirect and includes several nonlinearities and dependencies on parameters that are not well known, the inversion is done through an iterative optimisation scheme for water depths and model parameters, which are modified until best possible agreement between observed and simulated radar image intensity variations is obtained with correct depths at known calibration points. This method is sufficient for the identification of major bathymetric changes, which can then be examined in more detail by conventional echosoundings. This way, the use of SAR data improves the cost efficiency of ship operations of the responsible monitoring agencies.

Even more improvement can be expected from the use of ATI data instead of conventional SAR images, since the relation between water depths and surface currents is clearly more direct than the one between water depths and SAR image intensities. Romeiser et al. (2002) demonstrated a bathymetric mapping on the basis of airborne ATI data using a very simple approach, exploiting just linearised surface current – water depth relations derived from reference data at a few locations in the test area by a regression analysis. An example result from an experiment at the German island of Sylt is shown in Fig. 5.9. In combination with a full physics-based flow model, such as the one used by Calkoen et al. (2001), bathymetry retrievals from ATI data should be more accurate and reliable than the conventional SAR-based approach, and the method should be applicable to more complex scenarios since the data interpretation is less ambiguous.

5.4.4 Coastal and Offshore Engineering

Coastal and offshore engineers are interested in effects of currents, waves, and winds on shore and water based structures, as well as effects of modifications of such structures. For example, British and German engineers have recently formulated requirements for the site selection of electric power generators in tidal waters and for further investigations on variations in the 3-D flow around such generators (European Commission, 2005). While it is obvious that the turbines should be placed in areas with long periods of strong and uniform currents, data from a British prototype system indicate that currents acting on the rotor exhibit unexpected strong variations on short time scales, which affect efficiency and wear of the device quite strongly. To study and optimise the relevant dynamic processes systematically,

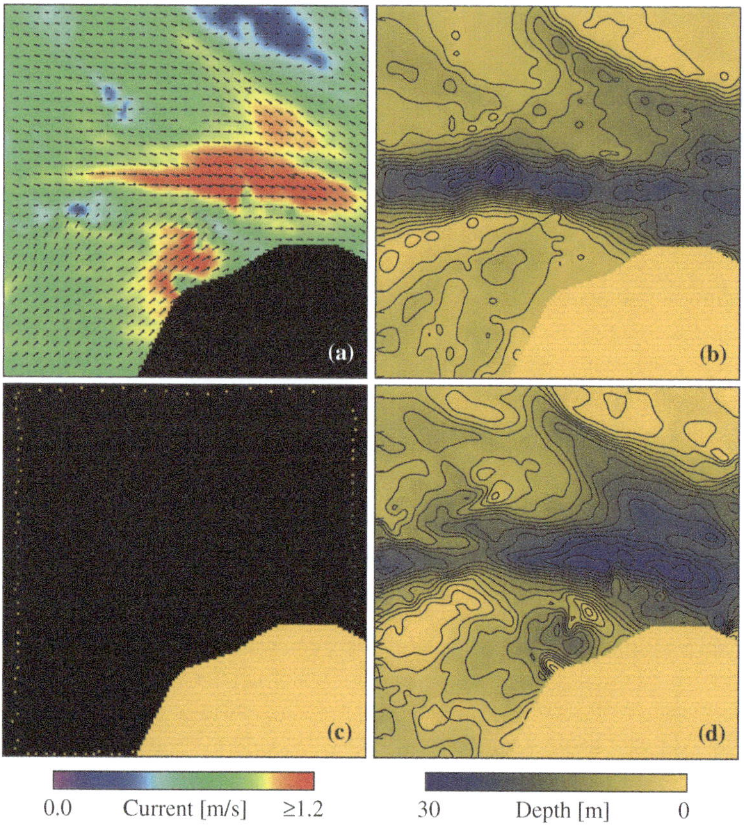

Fig. 5.9 Bathymetry retrieval example; (**a**) airborne ATI-derived current field north of the German island of Sylt; area size = 3.5 km × 3.5 km, grid resolution = 25 m × 25 m; (**b**) depth map from echosoundings with an effective resolution of 200 m; (**c**) 78 selected reference depth points near the boundaries; (**d**) depth map derived from the reference depths in combination with the ATI-derived current field

high-resolution and consistent wind, wave, and surface current measurements in combination with theoretical investigations are required. SAR/ATI-based measurements can make a significant contribution to this specific research as well as to similar offshore engineering tasks.

5.4.5 River Runoff Monitoring

Another promising field of application is the monitoring of river runoff, which is important for coastal oceanography, hydrology, and climate research. Furthermore, the redistribution of water due to climate changes and changes in population, industrialisation, and land use can have major effects on the earth system and on economical and political developments. At present, many rivers are monitored locally, and data from stations throughout the world are collected and archived at the Global

Runoff Data Centre (GRDC) in Koblenz, Germany. However, in some regions measurements are practically impossible for various reasons, and many countries do not publish existing runoff data. The development of a satellite-based monitoring system is highly desirable (Alsdorf et al., 2003).

British scientists have already demonstrated the monitoring of water levels in rivers on the basis of reprocessed conventional radar altimeter data (Berry, 2002). Concepts for more specific high-resolution altimeter missions for river applications has been proposed to ESA and NASA. The use of stationary microwave Doppler scatterometers for current measurements in rivers has been demonstrated by Plant et al. (2005).

Results of current retrievals in the Elbe river by SRTM and TerraSAR-X ATI have been discussed in Section 5.1. Scientists at the University of Hamburg are currently studying concepts for a data synthesis system for optimal river runoff assessments on the basis of spaceborne ATI and altimeter data, other available data, and numerical model computations.

An example of a velocity field in the Amazon river from ENVISAT ASAR, obtained using the Doppler anomaly technique, is shown in Fig. 5.10. The outflow direction is well aligned with the radar look direction. The surface velocity is quite

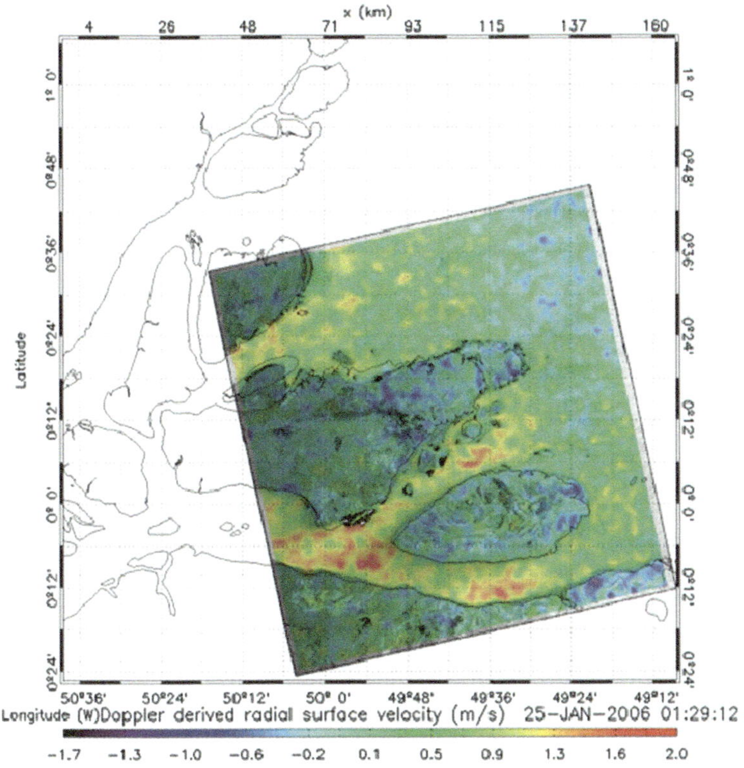

Fig. 5.10 Doppler velocity field in the Amazon river mouth, as obtained from ENVISAT ASAR data using the Doppler anomaly analysis technique

inhomogeneous and variable, as can be expected in the river mouth of this strong river, which possibly passes a range of different water depths that would trigger bands of acceleration and deceleration.

5.5 Summary and Conclusions

We have given an overview of the state of the art in the field of high-resolution surface current measurements from space by conventional and interferometric SAR, including a description of the two existing techniques, a review of experimental results from ENVISAT ASAR and TerraSAR-X, and a discussion of possible and desirable further developments and promising applications. While the basic concepts of Doppler anomaly analysis of conventional SAR data and of along-track interferometry were developed earlier, it was in the last decade that first successful current measurements from space by both methods could be demonstrated, using conventional SAR data from ERS and ENVISAT and interferometric SAR data from SRTM. With the development of robust data processing and correction methods, the addition of Doppler information to standard ASAR data products from ESA, and the implementation of experimental ATI capabilities on TerraSAR-X, conventional SAR- and ATI-based current measurements can now be made available for a variety of applications, and further development in this field will depend on user demand and financial support rather than on the solution of major technological problems.

The Doppler centroid anomaly analysis of conventional SAR data works with all SAR images (including archived ones) and provides maps of line-of-sight currents with spatial resolutions in the kilometer range within swaths of up to 400–500 km. The effective spatial resolution of ATI-derived line-of-sight current fields is on the order of 1 km within a swath of up to 100 km, but it could be improved by about one order of magnitude with improved system parameters (longer along-track baseline, reduced instrument noise) on future satellites. Furthermore, a spaceborne dual-beam along-track InSAR could measure two current components during a single overpass to obtain fully two-dimensional vector current maps.

So far we have demonstrated the retrieval of coarse-resolution global current maps from ENVISAT ASAR Wave Mode data and the observation of regional current features, such as the Agulhas Current, with Image Mode and Wide Swath data, as well as the imaging of current fields in coastal waters and rivers with ATI data from SRTM and TerraSAR-X. We see a variety of promising applications in the open ocean, coastal waters, and river estuaries, such as scientific studies on turbulence structures, shallow-water bathymetry monitoring, studies related to coastal and offshore engineering projects, and river runoff monitoring. Altogether, current measurements by spaceborne SAR and ATI seem to have the potential to become a widely used remote sensing technique, with a level of maturity comparable to radar altimetry, wind scatterometry, and SAR-based oil spill detection or wave spectra retrievals, within the coming 10 years.

Acknowledgements We thank H. Breit, M. Eineder, and U. Steinbrecher (DLR, Oberpfaffenhofen, Germany), M. Gade, S. Grünler, and J. Sprenger (University of Hamburg, Germany), N. Winkel, A. Sohrmann, and H. Weilbeer (BAW, Hamburg), W. Sent (BSH, Hamburg), and K. de Jong and J. Vogelzang (RWS, Netherlands) for important contributions to the research presented in this chapter. This work has been partly funded by the U.S. Office of Naval Research under grant N00014-09-1-0366.

References

Ainsworth TL, Chubb SR, Fusina RA, Goldstein RM, Jansen RW, Lee JS, Valenzuela GR (1995) INSAR imagery of surface currents, wave fields, and fronts. IEEE Trans Geosci Rem Sens 33:1117–1123

Alsdorf D, Lettenmaier D, Vörösmarty C, NASA Surface Water Working Group (2003) The need for global, satellite-based observations of terrestrial surface waters. EOS Trans AGU, 84(269):275–276

Berry PAM (2002) A new technique for global river and lake height monitoring using satellite altimeter data. Int J Hydropower Dams 9(6):52–54

Bjerklie DM, Moller D, Smith LC, Dingman SL (2005) Estimating discharge in rivers using remotely sensed hydraulic information. J Hydrol 309:191–209

Bonjean F, Lagerloef GSE (2002) Diagnostic model and analysis of the surface currents in the tropical Pacific Ocean. J Phys Oceanogr 32:2938–2954

Calkoen CJ, Hesselmans GHFM, Wensink GJ, Vogelzang J (2001) The bathymetry assessment system: efficient depth mapping in shallow seas using radar images. Int J Rem Sens 22: 2973–2998

Casulli V, Walters RA (2000) An unstructured grid, three-dimensional model based on the shallow water equations. Int J Numer Methods Fluids 32:331–348

Chapron B, Collard F, Ardhuin F (2005) Direct measurements of ocean surface velocity from space: interpretation and validation. J Geophys Res 110:C07008, doi:10.1029/2004JC002809, 17pp

Chapron B et al. (2002) Ocean geophysical results from ASAR. Presentation at ASAR Validation Review 2002, available at http://envisat.esa.int/workshops/validation_12_02/asar/wv-asar-validation-review_bchapron_12-dec-2002.ppt

Collard F, Mouche A, Chapron B, Danilo C, Johannessen JA (2008) Routine high resolution observation of selected major surface currents from space. Proceedings of the SeaSAR 2008, ESA Communication Production Office, ESTEC, Noordwijk, Netherlands, 8pp

Essen HH, Gurgel KW, Schlick T (2000) On the accuracy of current measurements by means of HF radar. IEEE J Oceanic Eng 25:472–480

European Commission (2005) SEAFLOW pilot project for the exploitation of marine currents. Final Report JOR3-CT98-0202, European Commission, Brussels

Frasier SJ, Camps AJ (2001) Dual-beam interferometry for ocean surface current vector mapping. IEEE Trans Geosci Rem Sens 39:401–414

Fu LL, Holt B (1982) Seasat Views Oceans and Sea Ice with Synthetic-Aperture Radar, JPL Publisher, Jet Propulsion Lab, Pasadena, CA, USA, pp. 81–120, 200

Goldstein RM, Barnett TP, Zebker HA (1989) Remote sensing of ocean currents. Science 246:1282–1285

Goldstein RM, Zebker HA (1987) Interferometric radar measurement of ocean surface currents. Nature 328:707–709

Graber HC, Thompson DR, Carande RE (1996) Ocean surface features and currents measured with synthetic aperture radar interferometry and HF radar. J Geophys Res 101:25813–25832

Gründlingh ML (1983) On the course of the Agulhas Current. S Afr Geogr J 65:49–57

Gurgel KW, Schlick T (2007) Land-based over-the-horizon radar techniques for monitoring the NE Atlantic coastal zone. In: Barale V, Gade M (eds.) Remote Sensing of the

European Seas, Springer Science and Business Media, Berlin, Heidelberg, New York, pp. 447–460

Johannessen JA, Chapron B, Collard F, Kudryavtsev K, Mouche A, Akimov D, Dagestad KF (2008) Direct ocean surface velocity measurements from space: improved quantitative interpretation of Envisat ASAR observations. Geophys Res Lett 35:L22608, doi:10.1029/2008GL035709, 6pp

Madsen SN (1989) Estimating the doppler centroid of SAR data. IEEE Trans Aerosp Electron Syst AES 25:134–140

McGillicuddy DJ et al. (2007) Eddy/wind interactions stimulate extraordinary mid-ocean plankton blooms. Science 316:1021–1026

Moreira A, Krieger G, Hajnsek I, Hounam D, Werner M, Riegger S, Settelmeyer E (2004) TanDEM-X: a TerraSAR-X add-on satellite for single-pass SAR interferometry. Proceedings of the IGARSS 2004, IEEE, Piscataway, NJ, 4pp

Mouche AA, Chapron B, Reul N, Collard F (2008) Predicted doppler shifts induced by ocean surface wave displacements using asymptotic electromagnetic wave scattering theories. Waves Random Complex Media 18:185–196

Palmer AJ (1991) Surface current mapping performance of bistatic and monostatic Δk-radars. IEEE Trans Geosci Rem Sens 29:1014–1016

Plant WJ, Keller WC, Hayes K, Spicer K (2005) Streamflow properties from time series of surface velocity and stage. J Hydr Eng 131:657–664

Rabus B, Eineder M, Roth A, Bamler R (2003) The shuttle radar topography mission – a new class of digital elevation models acquired by spaceborne radar. ISPRS J Photogramm Rem Sens 57:241–262

Romeiser R (2005) Current measurements by airborne along-track InSAR: measuring technique and experimental results. IEEE J Ocean Eng 30:552–569

Romeiser R, Breit H, Eineder M, Runge H, Flament P, de Jong K, Vogelzang J (2005) Current measurements by SAR along-track interferometry from a space shuttle. IEEE Trans Geosci Rem Sens 43:2315–2324

Romeiser R, Runge H (2007) Detailed analysis of ocean current measuring capabilities of TerraSAR-X in several possible along-track InSAR modes on the basis of numerical simulations. IEEE Trans Geosci Rem Sens 45:21–35

Romeiser R, Runge H (2008) Current measurements in coastal waters and rivers by along-track InSAR. In: Barale V, Gade M (eds.) Remote Sensing of the European Seas, Springer Science and Business Media, Berlin, Heidelberg, New York, pp. 411–422

Romeiser R, Runge H, Suchandt S, Sprenger J, Weilbeer H, Sohrmann A, Stammer D (2007) Current measurements in rivers by spaceborne along-track InSAR. IEEE Trans Geosci Rem Sens 45:4019–4030

Romeiser R, Seibt-Winckler A, Heineke M, Eppel D (2002) Validation of current and bathymetry measurements in the German Bight by airborne along-track interferometric SAR. Proceedings of the IGARSS 2002, IEEE, Piscataway, NJ, 3pp

Romeiser R, Suchandt S, Runge H, Steinbrecher U, Grünler S (2010) First analysis of TerraSAR-X along-track InSAR-derived current fields. IEEE Trans Geosci Rem Sens 48:820–829

Romeiser R, Thompson DR (2000) Numerical study on the along-track interferometric radar imaging mechanism of oceanic surface currents. IEEE Trans Geosci Rem Sens 38-II:446–458

Schulz-Stellenfleth J, Hajnsek I, Lehner S (2006) Use of TanDEM-X in a squinted split antenna mode configuration to retrieve 2-D current and ocean wave information. Proceedings of the IGARSS 2006, IEEE, Piscataway, NJ, USA, 4pp

Shuchman R (1979) The feasibility of measurement of ocean current detection using SAR data. Proceedings of the 13th International Symposium on Remote Sensing of Environment, Ann Arbor, MI, USA, pp. 93–103

Siegmund R, Bao M, Lehner S, Mayerle R (2004) First demonstration of surface currents imaged by hybrid along- and cross-track interferometric SAR. IEEE Trans Geosci Rem Sens 42: 511–519 ten Cate HH, Hummel S, Roest MRT (2000) An open model system for 2d/3d hydrodynamic simulations. Proceedings of the Hydroinformatics 2000, Int Assoc Hydraulic Engineering and Research, Madrid, Spain

Thompson DR, Jensen JR (1993) Synthetic aperture radar interferometry applied to ship-generated waves in the 1989 Loch Linnhe experiment. J Geophys Res 98:10259–10269

Toporkov JV, Perkovic D, Farquharson G, Sletten MA, Frasier SJ (2005) Sea surface velocity vector retrieval using dual-beam interferometry: first demonstration. IEEE Trans Geosci Rem Sens 43:2494–2502

Wunsch C, Stammer D (1998) Satellite altimetry, the marine geoid, and the oceanic general circulation. Ann Rev Earth Planet Sci 26:129–253

Chapter 6
Scatterometer's Unique Capability in Measuring Ocean Surface Stress

W. Timothy Liu, Xiaosu Xie, and Wenqing Tang

6.1 Introduction

Ocean surface stress, the turbulent transfer of momentum between the ocean and the atmosphere, is a vector quantity with a magnitude and a direction. It is closely related to wind but not solely driven by wind, as it depends also on ocean parameters such as surface current and temperature. While the general public appreciates wind as air in motion, very few people know what is stress. Even for oceanographers, the concept of stress distribution is largely derived from that of wind, because there was no large-scale measurement of stress over the ocean until the launch of the first scatterometer. A scatterometer measures the ocean surface roughness that is supposed to be in equilibrium with stress, and therefore it has the unique capability of measuring stress over the global ocean.

Although scatterometer measurements were related to stress, in prelaunch studies of the first scatterometer on Seasat (e.g. Wentz, 1978; Jones and Schroeder, 1978), and were validated against stress measurements (Liu and Large, 1981), scatterometers have been promoted as wind measuring instruments for the past 3 decades. The geophysical product of the scatterometer is the equivalent neutral wind (U_N). U_N, by definition, has an unambiguous relation with surface stress, provided that ocean surface current is negligible, while the relation between actual wind and surface stress depends also on atmospheric density stratification (see Section 6.2). Over most of the ocean, the atmosphere is generally believed to be near neutral, current speed is much smaller than wind speed, and U_N is assumed to be the actual wind, particularly in operational weather prediction.

Wind over ocean is much needed for marine weather forecast and to avoid shipping hazard. The significance of wind measurement is clearly felt, for example, when a hurricane suddenly intensifies and changes course or when the unexpected delay of monsoon brings drought. Surface wind convergence brings moisture and latent heat that drives deep convection and fuels atmospheric circulation. Detailed

W.T. Liu (✉)
Jet Propulsion Laboratory, California Institute of Technology, Pasadena, CA 91109-8099, USA
e-mail: w.t.liu@jpl.nasa.gov

V. Barale et al. (eds.), *Oceanography from Space*,
DOI 10.1007/978-90-481-8681-5_6,

distribution of wind power is also needed for the optimal deployment of floating wind farms in open sea that are enabled by new technology (Liu et al., 2008a). Just a few decades ago, almost all ocean wind measurements came from merchant ships. However, the quality and geographical distribution of these wind reports were uneven. Today, operational numerical weather prediction (NWP) also gives us wind information, but NWP depends on models, which are limited by our knowledge of the physical processes and the availability of data. The scatterometers have provided observations for important science and operational applications, in the past decades, as reviewed by Liu (2002) and Liu and Xie (2006). At this tenth anniversary of QuikSCAT launch, we will go back to the basics of scatterometry and turbulence transfer to demonstrate the uniqueness of scatterometer stress measurements that may enable new scientific applications from new perspectives.

For oceanographers, it is stress more than wind that directly drives ocean circulation. The two-dimensional stress field is needed to compute the divergence and curl (vorticity) that control the ocean vertical mixing. The mixing brings short-term momentum and heat trapped in the surface mixed layer into the deep ocean, where they are stored over time. It also brings nutrients and carbon stored in the deep ocean to the surface, where there is sufficient light for photosynthesis. Horizontal currents, driven in part by stress, distribute the stored heat and carbon in the ocean. The magnitude of stress affects the turbulent transfer between ocean and atmosphere of heat, moisture and gases that are critical for climate changes.

The relation between wind and stress, affecting the interpretation of scatterometer measurements, will be described in Section 6.2. Talking about measuring stress with a scatterometer, but using U_N as the actual wind, and explaining the variation of scatterometer observations from wind theories, would lead to a misinterpretation of physical processes. Hence, the difference between wind and stress is tackled in Section 6.3. Using a neutral drag coefficient to derive stress from the U_N provided by the scatterometer has inherent deficiency, so a new geophysical model function (GMF) to retrieve stress is discussed in Section 6.4. With the potential benefit of direct retrieval of stress, which is also driven by smaller-scale ocean surface parameters such as current and temperature, a re-thinking of the feedback processes is explored in Section 6.5. A constellation of scatterometers to meet operational and research needs is presented in Section 6.6.

6.2 Turbulence Parametrization and Scatterometer Geo-Physical Product

Ocean surface stress (τ) is the turbulent transfer of momentum between the ocean and the atmosphere. The turbulence is generated by atmospheric instability caused both by wind shear (difference between wind and current) and buoyancy (vertical density stratification resulting from temperature and humidity gradients). In the past, direct τ measurements, by the so-called eddy-correlation method, have only been done in a few field campaigns (Smith, 1980). In practice, knowledge of τ is derived

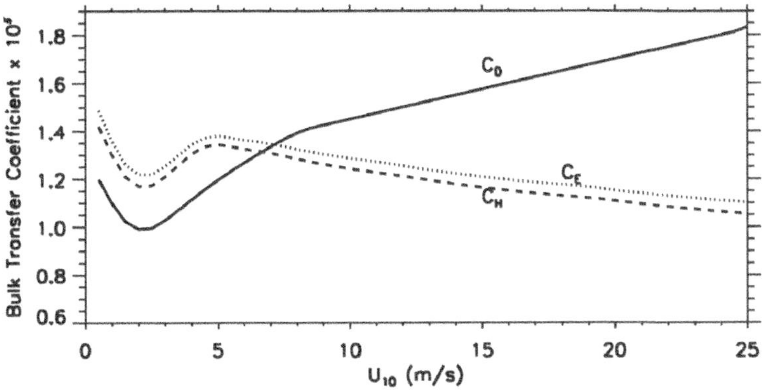

Fig. 6.1 Variation of the bulk transfer coefficients of momentum (drag coefficient), heat, and moisture with wind speed by Liu et al. (1979)

from winds (U) at a reference height, through a drag coefficient C_D, as defined by

$$\tau = \rho C_D (U - U_S)^2 \tag{6.1}$$

where U_s is the surface current and ρ is the air density. C_D has been derived largely from field measurements (Kondo, 1975; Smith, 1980; Large and Pond, 1981). Figure 6.1 (Liu et al., 1979) illustrates the variation of C_D with wind speed, at neutral stability, compared with the transfer coefficient for heat (C_H) and water vapor (C_E). At low wind speed ($U < 3$ m/s), the flow is smooth and C_D increases with decreasing wind speed. At moderate wind ($3 < U < 25$ m/s), C_D is an increasing function of wind speed for a rough sea with open fetch.

The drag coefficient is, of course, only a simple approximation to relate what we want – stress, to the measurement that is available – wind. We imbed our insufficient knowledge of turbulence transfer in this coefficient. Secondary factors, such as sea states, swell, and spray from breaking waves (e.g. Donelan et al., 1997; Bourassa et al., 1999), whose data are not generally available, are not included in this parameterization schemes and should be part of the errors. Although we include surface current in the formulation (Equation 6.1), it is generally ignored because no current measurement is readily available. These factors, together with the stability effects may contribute to the uncertainties of the drag coefficient.

Liu et al. (1979) first proposed a parameterization method of stress, which is equivalent to a C_D including the stability effects and molecular constraints at the interface, by solving the similarity equation (non-dimensional flux-profile relation) in the surface layer, where the vertical gradient of stress is negligible:

$$\frac{U - U_S}{U_*} = 2.5\left(\ln\frac{z}{z_0} - \psi_u\right) = \frac{1}{\sqrt{C_D}}, \tag{6.2}$$

where $U_* = (\tau/\rho)^{1/2}$ is the friction velocity, z_0 is the roughness length, and ψ_u is a function of the stability parameter, which is the ratio of buoyancy to shear production of turbulence. Equation (6.2) is solved simultaneously with similarity equations for temperature and humidity. An alternative to using the neutral C_D is to express z_0 as a function of U_*. For example, Liu and Tang (1996) incorporated such a relation in solving the similarity equation. They combined a smooth flow (Nikuradse, 1933) and rough flow (Charnock, 1955) relations to give

$$Z_0 = 0.11\frac{\nu}{U_*} + 0.011\frac{U_*^2}{g} \tag{6.3}$$

where ν is the kinematic viscosity and g is the acceleration due to gravity.

Typical wind profiles at various stabilities are shown in Fig. 6.2 as illustration. Neglecting U_s and ψ_u in Equation (6.2), U becomes U_N and it is uniquely related to U_* (or τ). Although the atmosphere is believed to be near neutral over most ocean area, exact neutral stability ($\psi_u = 0$) is rare, and to compute U_N from conventional wind measurements of U (point A in Fig. 6.2), the stability effect has to be removed. First, U_* and z_0 are computed based on the parameterization scheme of Liu et al. (1979) (the computer codes and subsequent modifications were presented by Liu and Blanc, 1984; Liu and Tang, 1996), as the slope and intercept at the surface of the curve in Fig. 6.2. The neutral relation (straight line) defined by U_* and z_0 will then give U_N (point B). This method has been used in development and calibration of all the GMF of the NASA scatterometer.

Liu et al. (1979) first postulated that, in a rough sea, under a moderate range of winds, C_H and C_E do not increase with wind speed because molecular constraint

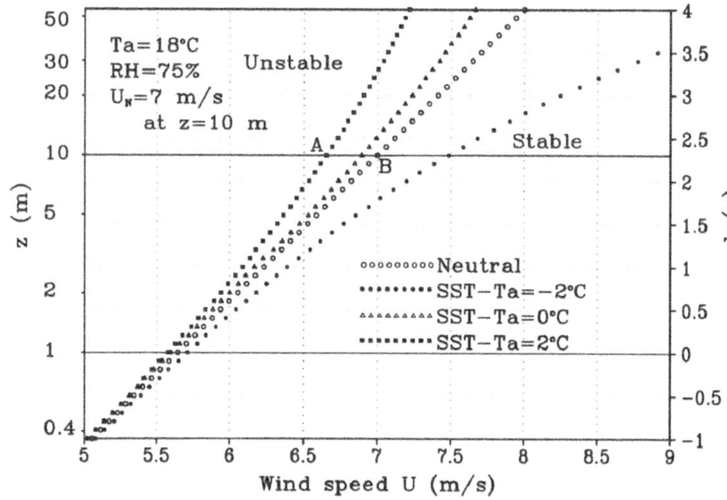

Fig. 6.2 Typical wind profiles at various stability conditions derived from the flux-profile relation by Liu et al. (1979). B is the equivalent neutral wind corresponding to the actual wind measurement at A

at the interface, while C_D may still increase since momentum is transported by form-drag. The heat and moisture transfer coefficients are defined by

$$H = \rho c_P C_H (T - T_s)(U - U_S) \tag{6.4}$$

$$E = \rho C_E (Q - Q_s)(U - U_S) \tag{6.5}$$

where c_P is the isobaric specific heat, H is the heat flux, E is moisture flux (evaporation). T is the potential temperature, T_s the sea surface temperature, and Q is the specific humidity at a reference level and Q_s is the specific humidity at the interface. Liu's hypothesis on C_H and C_E, up to 20 m/s wind speed, as illustrated in Fig. 6.1, was subsequently supported by measurements in field experiments (e.g. DeCosmo et al., 1996).

Emanuel (1995) argued, from theoretical and numerical model results, that the scenario of Liu et al. (1979) could not be extrapolated to the strong wind regime of a hurricane. To attain the wind strength of a hurricane, the energy dissipated by drag could not keep increasing while the energy fed by sensible and latent heat does not increase with wind speed. His results, showing that the maximum wind speed in mature storm is sensitive to ratio of C_H and C_E to C_D, and that the ratio could not exceed a very small range, put limit on the increase of C_D as a function of wind speed.

Under strong winds, flow separation occurs, and wind is detached from roughness growth. The postulation of the level-off of the increase of C_D with wind speed at hurricane scale winds was supported by the results of the laboratory studies of Donelan et al. (2004), and the aircraft experiments by Powell et al. (2003) at wind speed above 30 m/s, as illustrated in Fig. 6.3. The result of Large and Pond (1981) derived for the range of moderate wind speeds is extrapolated to the range of strong wind speeds for comparison in the figure. Such flow separation may explain the high wind saturation of the scatterometer discussed in Section 6.3.2.

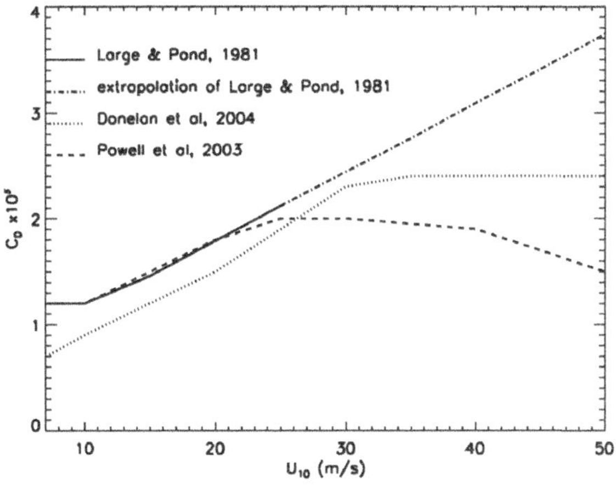

Fig. 6.3 Variation of the drag coefficients in strong winds

6.3 Characterization of Stress Versus Wind Distribution

6.3.1 Dependence on Surface Temperature and Current

When the first Quick Scatterometer (QuikSCAT) data came back in 1999, the science team was surprised to see that the scatterometer signal in the equatorial Pacific propagates westward with the ocean temperature front of the tropical instability waves, in the area where we expected to see steady trade winds (e.g. Liu et al., 2000; Chelton et al., 2001). Such coincident propagation was previously observed by Xie et al. (1998) in European Research Satellite (ERS) data. Since then, the spatial coherence between scatterometer measurements and sea surface temperature (T_S) has been observed over many locations and under various atmospheric conditions, e.g. over the Kuroshio Extension (e.g. Nonaka and Xie, 2003), over the circumpolar current (e.g. O'Neill et al., 2003), Indian Ocean (Vecchi et al., 2004), in the East China Sea during winter cold air outbreak (Xie et al., 2002), over the Gulf Stream Ring (Park and Cornillon, 2002), and over typhoon wake (Lin et al., 2003). Following traditional paths to study atmospheric boundary layer processes, many scientists were quick to postulate explanations of the wind and T_S correlation, based on boundary layer height change, pressure gradient force, secondary flow, cloud entrainment, and organized convection, but none of these is generally applicable to the ubiquitous correlation, as pointed out in the review by Small et al. (2008).

The first explanation by oceanographers, when they saw the results of Liu et al. (2007) showing that QuikSCAT measurements deviate from the mean winds with rotation in opposite direction to the underlying surface current of the Agulhas Extension meanders, was that either the drifter (current) or the scatterometer measurements were erroneous. Their reasoning was that the strong current meanders should impart its rotation on the prevailing westerly wind through drag, and the wind anomalies should show the same rotation as the current. Misinterpreting scatterometer stress as wind is the cause of confusion. Stress must be spatially coherent with T_S and ocean current, which create buoyancy and wind shear. As pointed out by Liu et al. (2007) and Liu and Xie (2008), at small turbulence scales at the surface, factors that affect atmospheric boundary layer dynamics (wind), such as Coriolis force, pressure gradient force, baroclinicity, cloud entrainment, etc., are not important. That is why the spatial coherence is ubiquitous, under all kinds of atmospheric circulations. Stress is the vector difference between wind and current. For a uniform wind blowing over a rotating current, the vector differences will have opposite rotation to the current (Park et al., 2006). The ocean signals, of course, will affect winds aloft through stress. The dynamic factors will then become important.

Liu et al. (2007) observed T_S signatures in cloud and atmospheric temperature high in the atmosphere over the Agulhas Extension. Even stronger penetrating signals have been found, not only in temperature profiles but also in precipitation profiles over the Kuroshio Extension. Present numerical models of atmospheric circulation in the mid-latitudes do not propagate the surface stress signal vertically much beyond the atmospheric boundary layer. The observations posted a challenge to understanding the transition from random turbulence to organized convection in the atmosphere.

The difference between winds and stress over oceanfronts is well documented by Liu et al. (2007) for the Agulhas Extension and by Liu and Xie (2008) for the Kuroshio Extension. Figure 6.4 shows the results of a conceptual experiment over the Kuroshio Extension by Liu and Xie (2008). A uniform wind field (average U_N speed blowing from west to east) over the area covered by the current meanders is assumed, with high and low T_s anomaly centers measured by the Advanced Microwave Scanning System – Earth Observing System (AMSR-E) marked as ° and * respectively.

U_N is computed with the stability-dependent bulk parameterization scheme of Liu et al. (1979), using air temperature from the National Center for Environmental Prediction (NCEP). The coherence between U_N and T_s is obvious even for a uniform wind field (Fig. 6.4a). The U_N computed from a uniform wind field shows the same

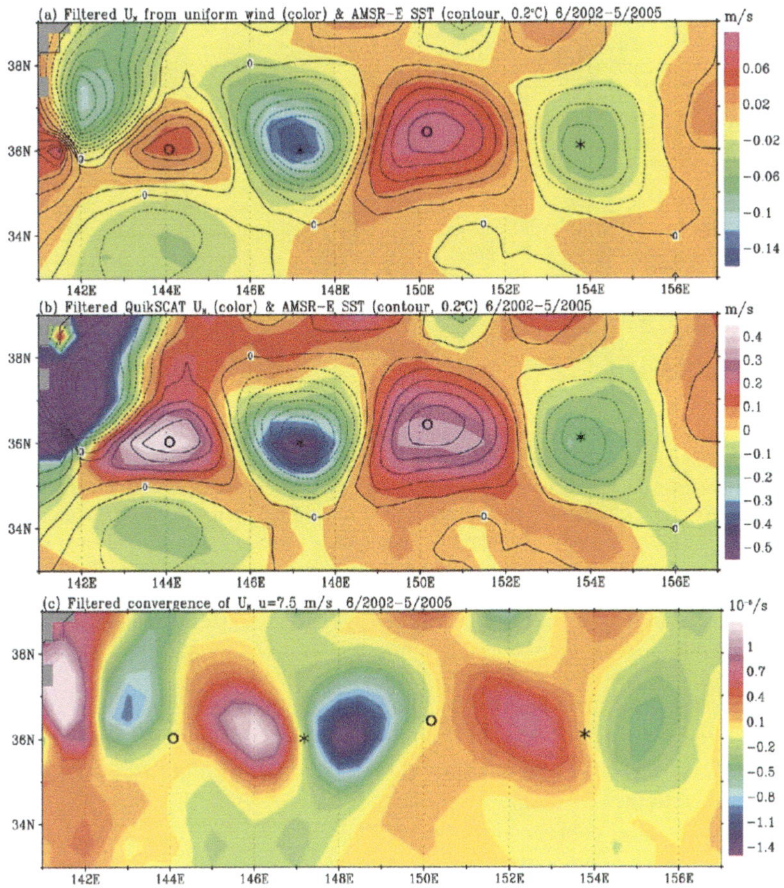

Fig. 6.4 (continued)

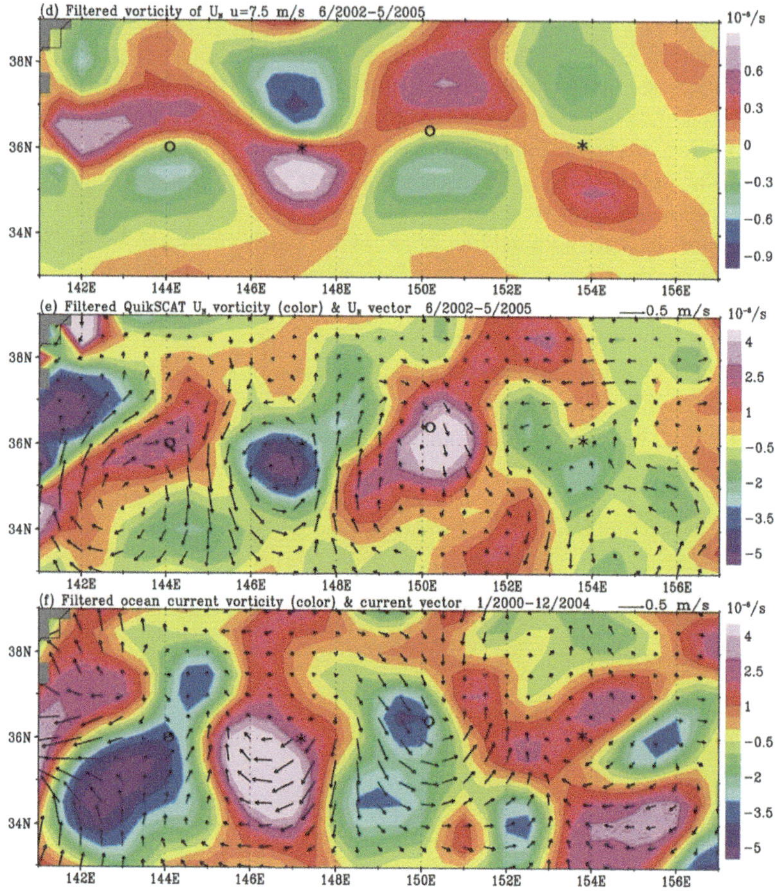

Fig. 6.4 (**a**) Isotherms of filtered *Ts* measured by AMSR-E (0.2°C interval) superimposed on filtered U_N computed from a uniform wind field of $u = 7.5$ m/s (color, m/s). *Solid* and *broken lines* represent positive and negative values, respectively. (**b**) Same as (**a**), except for filtered magnitude of QuikSCAT U_N (*color*, m/s). (**c**) Convergence of filtered U_N computed from the uniform wind (unit is 10^{-6}/s). (**d**) Vorticity of filtered U_N computed from the uniform wind (unit is 10^{-6}/s). (**e**) Filtered vector (*black arrows*) superimposed on vorticity (color, 10^{-6}/s) of U_N observed by QuikSCAT. (**f**) Filtered vector (*black arrows*) superimposed on vorticity (color, 10^{-6}/s) of the surface current measured by Lagrangian drifters. The large-scale gradients are removed by a two-dimensional filter

coherent pattern as that measured by QuikSCAT (Fig. 6.4b). Figure 6.4c shows that the convergence of U_N computed from the uniform wind field is in quadrature (90° phase difference) with T_s in the downwind direction, implying that the convergence is in-phase with the downwind T_s gradient.

A similar spatial coherence is observed between T_s and QuikSCAT U_N convergence. Figure 6.4d shows that the vorticity of the computed U_N is in quadrature with T_s in the crosswind direction; vorticity is in-phase with the crosswind T_s gradient. Positive vorticity is found to the south and negative anomalies to the north of the cold eddy, and vice versa for the warm eddy. QuikSCAT should observe the vorticity distribution, if there is no ocean current (e.g. O'Neill et al., 2003), but its observations do not show such crosswind relation (Fig. 6.4e). Positive vorticity is collocated (in-phase) with warm water and negative vorticity is collocated with cold water. The vorticity of the observed U_N is opposite to the vorticity of the surface current measured by drifters (Fig. 6.4f), confirming scatterometers measure the vector difference between wind and current, as associated with stress. Liu et al. (2007) found similar results over the Agulhas Extension.

The stability effect on the difference between wind and stress, in the form of the difference between U_N and U, is governed by ψ_u, as

$$\delta U = U_N - U = 2.5 U_* \psi_u \tag{6.6}$$

Using U_N provided by QuikSCAT, sea surface temperature from AMSR-E, air temperature and air humidity from the reanalysis of the European Center for Medium-Range Weather Forecasts (ECMWF), Liu et al. (2010) computed δU at 10 m averaged over a 3 years period, for January and July (as shown in Fig. 6.5). This distribution of the stability effect on wind speed follows closely the distribution of sea-air temperature difference shown in Fig. 6.6. Because atmospheric temperature

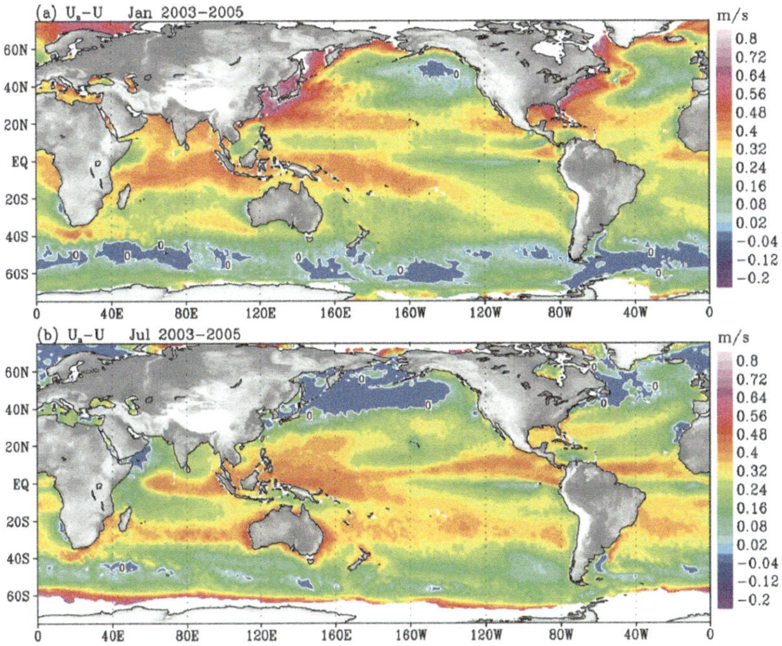

Fig. 6.5 Difference between equivalent neutral wind and actual wind at 10 m for (**a**) January and (**b**) July

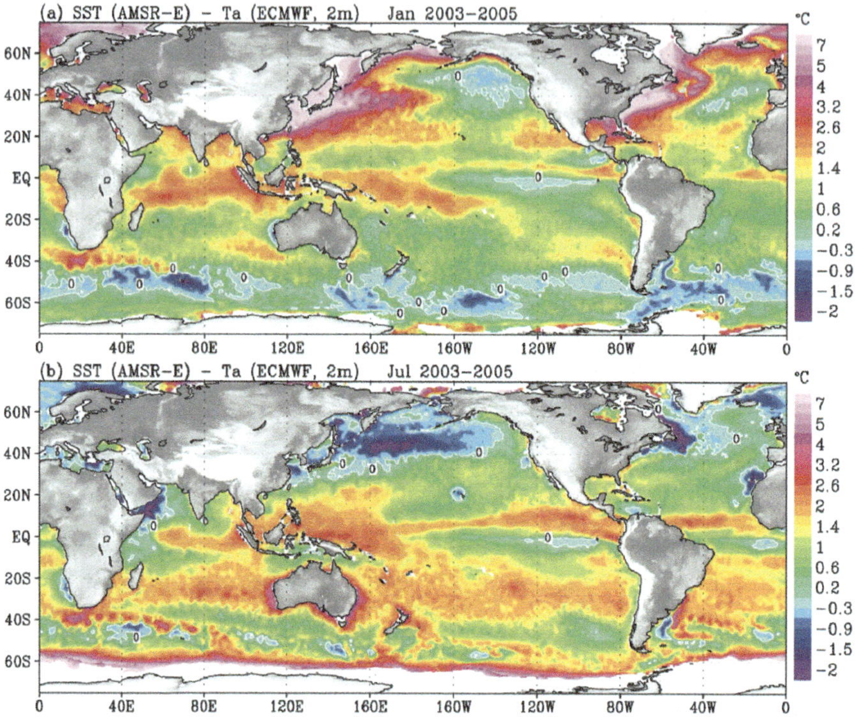

Fig. 6.6 Difference between sea surface temperature and air temperature (2 m) for (**a**) January and (**b**) July

variations have larger scale than oceanic temperature variations, the buoyancy-generated turbulence should have strong spatial correlation with T_s at small scales.

The formulation of ψ_u was largely based on experiment data on land, validated only with small amount of measurements over ocean. Although there have been many investigations to improve flux parameterization, (e.g. Fairall et al., 1996) in the past few decades, there is no significant change in the formulation of ψ_u. As shown by Liu et al. (2007) and Liu and Xie (2008), and discernable in Fig. 6.4, the dynamic range of U_N over T_s anomalies computed from uniform wind fields is smaller than the QuikSCAT measurement. One of the plausible reasons is that the ψ_u in use now underestimates the stability effect on vertical wind speed changes, and needs to be investigated. Another reason is a positive atmospheric feedback through convection (see Liu et al., 2007; Liu and Xie, 2008).

Although both NCEP and ECMWF have assimilated QuikSCAT data since 2002, the data were assimilated as actual wind vector and not stress. The wave-number spectra constructed from ECMWF 10 m winds collocated with one swath of QuikSCAT data on August 25, 2005, over the Atlantic (Fig. 6.7) shows that QuikSCAT U_N has higher power than the ECMWF winds at high wave numbers.

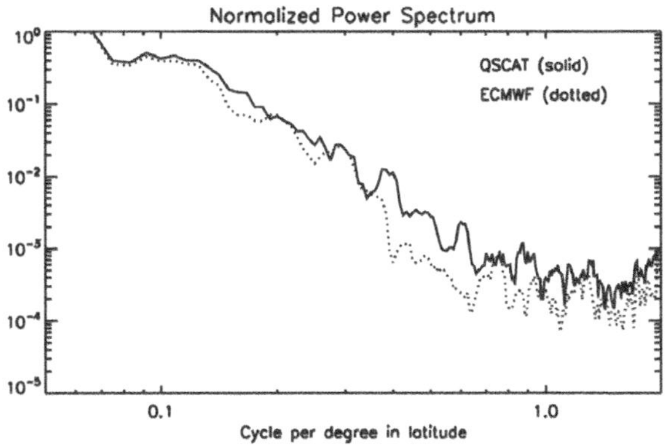

Fig. 6.7 Power spectrum of UN measured by QuikSCAT and 10 m wind from ECMWF

It implies that the stress measured by QuikSCAT has more small-scale information than the winds provided by ECMWF, unless the comparison reflects the deficiency in the models.

6.3.2 High Wind Saturation

Engineers have long been puzzled by the high wind saturation of scatterometer measurements causing an underestimation of the strength of winds in tropical cyclones and extratropical storms by scatterometers (see review in Liu and Xie, 2006). Retrieving strong winds from scatterometer observations is known to be difficult because of the lack of in-situ standards for calibration. The problem is clearly illustrated in Fig. 6.8, where QuikSCAT measurements at Ku-band are compared with collocated HWind speed operationally produced by the Hurricane Research Division at the Atlantic Oceanographic and Meteorological Laboratory, at 1 m/s bin size (Liu, 2010). Data for the 12 hurricanes of the North Atlantic in the 2005 seasons, excluding those with over 10% chances of rain, were examined. Figure 6.8 shows that, in moderate winds ($U < 35$ m/s), the logarithm of the backscatter (in term of the normalized radar cross section σ_0 in dB) increases linearly with the logarithm of wind speed, at both polarizations.

The error bar (one standard deviation of the 1 m/s wind speed bin) is mainly due to the dependence on azimuth angle (see Section 6.4). At strong winds ($U > 35$ m/s), however, σ_0 increases at a much slower rate with increasing wind speed. Errors increase because there is a small number of data in the high speed bins. Similar saturation is found in the C-band European Advanced Scatterometer (ASCAT). Strong wind saturation has been postulated (e.g. Donelan and Pierson, 1987) and observed from aircraft in hurricane experiments (Donnelly et al., 1999; Yueh et al., 2000).

When the model function developed over the moderate wind range is applied to the strong winds, an underestimation of wind speed results. Efforts have been made to adjust the model function (slope in Fig. 6.8) in strong winds and to find a sensor channel (combination of polarization, frequency, incident angle) for future

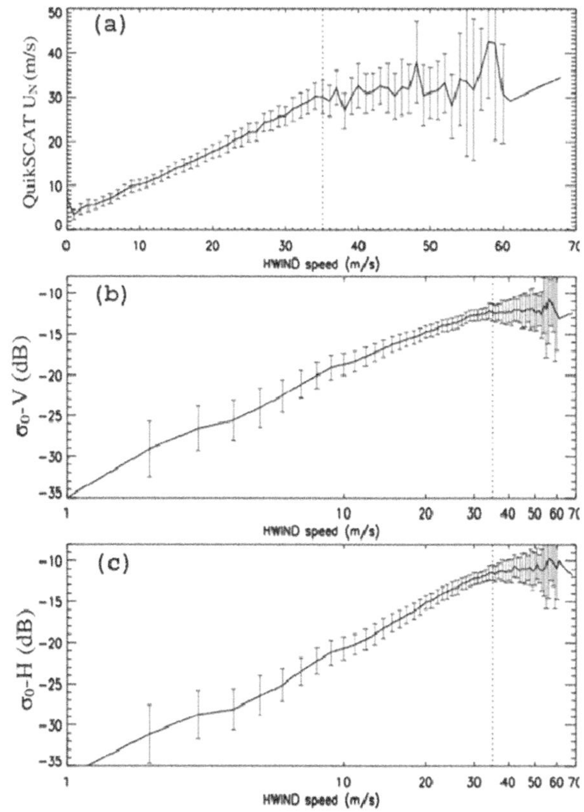

Fig. 6.8 Bin-average of (**a**) U_N, (**b**) normalized radar cross section at vertical, and (**c**) horizontal polarization measured by QuikSCAT for 12 hurricanes as a function of co-located Hwind. Error bars represent 1 standard deviation of data in each bin

scatterometers that would be sensitive to hurricane-scale winds (Esteban-Fernandez, 2006). As seen in Section 6.2, flow separation occurs at high winds and conventional formulations of drag coefficient fail. Although retrieving the very strong winds of a hurricane may not be an important and practical way to use the scatterometer, its measurements are significant not only in revealing the relation between wind and stress, but also in studying oceanic feed back to the dynamic forcing that is critical in hurricane maintenance and intensification.

6.4 Geophysical Model Function for Stress Retrieval

6.4.1 Equivalent Neutral Wind Retrieval

At present, QuikSCAT geophysical retrieval is through GMF-W, which relates the scatterometer normalized radar cross-section σ_0 with U_N, as

$$\sigma_0 = f\,(U_N, \chi, \vartheta, p) \tag{6.7}$$

where χ is the relative azimuthal angle between the plane of incidence of the radar beam and the wind direction, θ is the incidence angle (relative to nadir) and p represents the polarization. At fixed χ, σ_0 (in dB) increases approximately linearly with log (U_N). The azimuthal variation of σ_0 can be characterized as harmonics with "upwind-downwind asymmetry" and "up- wind-crosswind modulation". It has been demonstrated that the azimuthal angle dependence can be separated from the incidence angle and U_N functions using the 3-term Fourier series (Wentz et al., 1984; Freilich, 1996):

$$\sigma_0 = A_0\,(U_N,\theta) + A_1\,(U_N,\theta)\cos(\chi) + A_2\,(U_N,\theta)\cos(2\chi) \tag{6.8}$$

The core of GMF-W is consisted of the A coefficients as tabulated empirical data. The forward GMF-W accepts U_N vector as input and gives σ_0 as output. The inverse GMF, however, is not unique. A single measurement of σ_0 generates a range of potential wind vectors, all of which would have given rise to the observed backscatter. To solve the inverse problem, σ_0 at multiple azimuth angles are used. At least three collocated observations of σ_0 differing in "look", i.e. geometry (χ, θ), allows the determination of a unique wind vector. Theoretically, the solution could be found from data without noise. Noise complicates the solution and a maximum likelihood estimator (MLE) has to be used (Pierson, 1984). A common practice is to keep several ambiguous solutions at each wind vector cell. In order to select the proper ambiguity, we assume that the wind is unlikely to shift radically from one cell to the next and a median filter technique has been used (Shaffer et al., 1991). The median filter technique is an iteration procedure generally initialized by the NWP field in the so called "nudging".

6.4.2 Stress Retrieval

There are many reasons for a GMF-S to retrieve stress (or U_*) directly rather than the present GMF-W to retrieve U_N. A first reason lies in the present GMF-W, which should be developed and calibrated with U_N computed from research-quality in-situ wind measurements, using methods based on the similarity relations of Liu et al. (1979), as in Section 6.2. Indeed, such computation of U_N was performed before credible ocean surface wind products became available from operational NWP centers. Most of the tuning of the revised GMF after Seasat was based on NWP products (e.g. Wentz and Smith, 1999) that are not U_N (not corrected for stability dependence). The resulting errors are not reversible and difficult to gauge.

Ideally, stress could be derived from U_N retrieved from scatterometer, using a neutral drag coefficient. However, if the drag coefficient is not the same as that used to derive U_N for development of the GMF, an error will be introduced through the uncertainty of the drag coefficient. This is the second reason for a GMF-S. Weissman and Graber (1999) provide an example of the very few attempts to tabulate stress instead of U_N in the A coefficients of Equation (6.8). Two additional reasons are related to the directional difference between wind and stress. The procedure to "select" the stress from multi-solutions with ambiguous direction needs modification in two steps. The first is testing various "nudging" fields that are more relevant to stress than wind. Our feasibility studies suggest that such changes in the nudging field is significant in the region where strong ocean current exists and stress should point to the direction of the vector difference between wind and current. A more flexible median filter should also be designed to accommodate the small spatial scale of stress as compared with winds.

The development of GMF-S could follow closely those for GMF-W, and our feasibility study using a small sample of stress vectors and "nudging" data that include ocean current are successful. One of the reason usually given for promoting scatterometer as a wind sensor instead of stress sensor is that there are more wind than stress measurements to develop and calibrate the GMF. Such explanation runs in contrast to the rationale behind using U_N as the geophysical product because of its unambiguous relation to stress. To provide each U_N for development or calibration the GMF from measured wind, stress or U_* has to be computed first as discussed in Section 6.2. There are as many stress available as U_N.

The stress prepared from wind in such way is not ideal because it addresses only the stability problem but does not include current information. Such deficiency may be somewhat alleviated through the ambiguity removal process by using more appropriate filter size and nudging with the vector difference between wind and optimal surface current information that is available. Ocean surface current measurements are very sparse. The current velocity has been derived from Argos satellite collections of the displacements of drifters with drogues centered at 15 m-depth (Niiler, 2001). Ocean surface currents are also provided by the Ocean Surface Currents Analyses – Realtime program, using a combination of scatterometer and altimeter data (Lagerloef et al., 1999). Global, high frequency current data are only available from numerical models, such as, Estimating the Circulation and Climate of the Ocean (Fukumori, 2002) and Regional Ocean Modeling System (Chao et al., 2009).

6.5 Potential Oceanic Feedback

Although Pacanowski (1987) showed the importance of the feedback of ocean current on stress in a numerical experiment more than two decades ago, many ocean scientists are still forcing their ocean models using stress that is entirely determined from the wind field, independent of local changes caused by ocean. With the availability of QuikSCAT data, Polito et al. (2001) discussed the surface current and temperature feedback to the stress forcing, and Pezzi et al. (2004) demonstrated negative feedback with ocean general circulation model, at the tropical instability waves. With the recent demonstration of spatial coherence between scatterometer measurement and T_S over extensive ocean regions, as discussed in Section 6.3, many numerical studies of T_S feedback to stress forcing were performed with numerical models (e.g. Seo et al., 2007; Spall, 2007; Hogg et al., 2009). The measurement of stress field that depends not only on the fast and large-scale atmospheric circulation but also on the small scale and slow ocean processes as reflected in surface current and temperature will enable new studies from new perspective in the near future.

6.6 Satellite Constellations

One polar-orbiting scatterometer at a low altitude (e.g. 800 km) orbit, can sample at a location on Earth not more than two times a day. The temporal sampling may not be sufficient to monitor wind/stress with high frequencies. Meso-scale weather system, such as hurricane may be missed through orbit gaps. Additional instrument flying in tandem will allow the description of higher temporal variability and the reduction of the aliasing (bias introduced by sub-sampling) of the mean wind/stress, as described by Lee and Liu (2005) in their study on the impact on ocean mixed layer depth.

Besides QuikSCAT, there are two more scatterometers in operation. One is the C-band and fan-beam ASCAT on board the Meteorological Operational Satellite (METOP), launched in October 2006 by the European Space Agency. It has dual swath separated by a broad nadir gap. The other is a Ku-band, pencil beam, conically scanning scatterometer on Oceansat-2, launched in September 2009 by the Indian Space Agency.

The local time of ascending node for the three scatterometers is 5:54 AM, 9:30 PM, and noon, respectively. Figure 6.9 shows that the zonal average of revisit interval decreases from the equator to the poles because the orbits come closer at

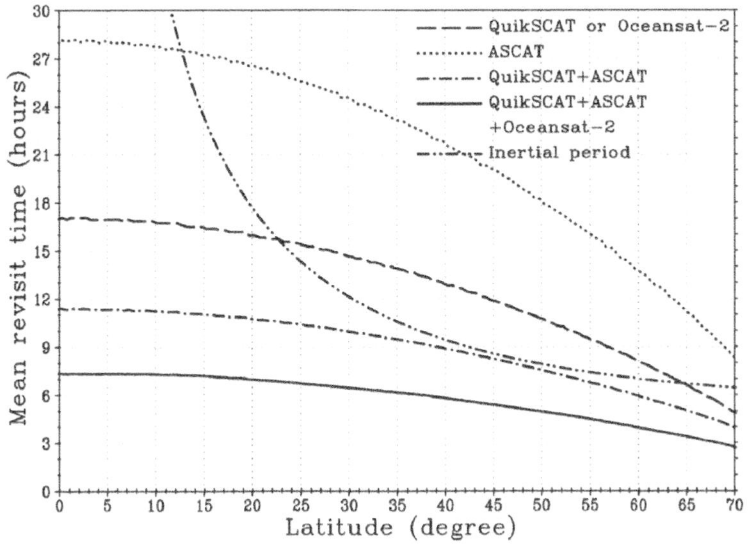

Fig. 6.9 The latitudinal variation of zonally averaged revisit interval for various tandem missions

high latitudes. Liu et al. (2008b) demonstrated that the combination of the three sensors brings the revisit interval close to 6 h from equator to the pole, meeting the 6 hourly operational NWP requirement and the inertial frequency required by the oceanographers. Because the scatterometer on Oceansat-2 has similar design as QuikSCAT, the transportability of GMF-S developed for QuikSCAT to a well-calibrated Oceansat-2 scatterometer could be tested. China is planning to launch a Ku-band scatterometer similar to QuikSCAT on the Haiyang-2 mission in 2011. Its sampling characteristic is expected to be very similar to QuikSCAT (see Liu et al., 2008b).

6.7 Discussions

It is insufficient to view the distribution of ocean surface stress by deriving it only from wind field through traditional drag coefficient. The overall large-scale difference between directly retrieved stress and stress derived from U_N may not be large, but the regional differences are significant. Our preliminary studies show the feasibility of developing a prototype GMF-S to retrieve stress from QuikSCAT observations, which we intended to proceed expeditiously when support becomes available. Processing of QuikSCAT observations for a decade long of continuous and consistent ocean surface stress field would enable the science community for new applications from a new perspective.

QuikSCAT is operating far beyond its expected life span and may fail at anytime. The transportability of such model function to a similar sensor is important. With a follow-on mission to QuikSCAT not yet confirmed, the Indian and Chinese components of a scatterometer constellation become all the more important, not only to improve the sampling of ocean surface wind and stress, but also to meet the contingency of filling any data gap.

Not all spacebased ocean surface wind and stress measurements are comparable in quality. Standardizing the technology requirements for observation accuracies of different research and operational applications and international cooperation are very desirable. Many scientific reports have affirmed the need of high quality, continuous, and consistent long time-series of ocean surface wind and stress vectors.

Acknowledgements This study was performed at the Jet Propulsion Laboratory, California Institute of Technology, under contract with the US National Aeronautics and Space Administration (NASA). It was jointly supported by the Ocean Vector Winds and the Physical Oceanography Programs of NASA.

References

Bourassa MA, Vincent DG, Wood WL (1999) A flux parameterization including the effects of capillary waves and sea state. J Atmos Sci 56:1123–1139

Chao Y, Li Z, Farrara J, McWilliams JC, Bellingham J, Capet X, Chavez F, Choi JK, Davis R, Doyle J, Frantaoni D, Li PP, Marchesiello P, Moline MA, Paduan J, Ramp S (2009)

Development, implementation and evaluation of a data-assimilative ocean forecasting system off the central California coast. Deep Sea Res II, 56:100–126
Charnock H (1955) Wind stress on a water surface. Quart J Roy Meteor Soc 81:639–640
Chelton DB, Esbensen SK, Schlax MG, Thum N, Freilich MH, Wentz FJ, Gentemann CL, McPhaden MJ, Schoff PS (2001) Observations of coupling between surface wind stress and sea surface temperature in the eastern tropical Pacific. J Clim 14:1479–1498
DeCosmo J, Katsaors KB, Smith SD, Anderson RJ, Oost WA, Bumke K, Chadwick H (1996) Air-sea exchange of water vapor and sensible heat: the humidity exchange over the sea (HEXOS) results. J Geophys Res 101:12001–12016
Donelan MA, Drenan WM, Katsaros KB (1997) The air-sea momentum flux in conditions of wind sea and swell. J Phys Oceanogr 27:2087–2099
Donelan MA, Haus BK, Reul N, Plant WJ, Stiassnie M, Graber HC, Brown OB, Saltzman ES (2004) On the limiting aerodynamic roughness of the ocean in very strong winds. Geophys Res Lett 31:L18306, doi:10.1029/2004GL019460
Donelan MA, Pierson WJ (1987) Radar scattering and equilibrium ranges in wind-generated waves with application to scatterometry. J Geophys Res 92:4971–5029
Donnelly WJ, Carswell JR, McIntosh RE, Chang PS, Wilkerson J, Marks F, Black PG (1999) Revised ocean backscatter models at C and Ku band under high-wind conditions. J Geophys Res 104:11485–11497
Emanuel K (1995) Sensitivity of tropical cyclones to surface exchange coefficients and a revised steady state model incorporating eye dynamics. J Atmos Sci 52:3969–3976
Esteban-Fernandez D, Carswell JR, Frasier S, Chang PS, Black PG, Marks FD (2006) Dual-polarized C- and Ku-band ocean backscatter response to hurricane-force winds. J Geophys Res 111:C08013, doi:10.1029/2005JC003048
Fairall CW, Bradley EF, Rogers DP, Edson JB, Young GS (1996) Bulk parameterization of air-sea fluxes in TOGA COARE. J Geophys Res 101:3747–3767
Freilich MH (1996) SeaWinds Algorithm Theoretical Basis Document. NASA ATBD-SWS-01
Fukumori I (2002) A partitioned Kalman filter and smoother. Mon Wea Rev 130:1370–1383
Hogg AMC, Dewar WK, Berloff P, Kravstov S, Hutchinson DK (2009) The effects of mesoscale ocean-atmosphere coupling on the large-scale ocean circulation. J Clim 22:4066–4082
Jones WL, Schroeder LC (1978) Radar backscatter from the ocean: dependence on surface friction velocity. Bound Layer 13:133–149
Kondo J (1975) Air-sea bulk transfer coefficients in diabetic conditions. Bound-Layer Meteor 9:91–112
Lagerloef GSE, Mitchum G, Lukas R, Niiler P (1999) Tropical Pacific near surface currents estimated from altimeter, wind and drifter data. J Geophys Res 104:23313–23326
Large WG, Pond S (1981) Open ocean momentum flux measurements in moderate to strong winds. J Phys Oceanogr 11:324–336
Lee T, Liu WT (2005) Effects of high-frequency wind sampling on simulated mixed layer depth and upper ocean temperature. J Geophys Res 110(C5):C05002, doi:10.1029/2004JC002746
Lin I-I, Liu WT, Wu CC, Chiang JC, Sui CH (2003) Satellite observations of modulation of surface winds by typhoon-induced upper ocean cooling. Geophys Res Lett 30:1131, doi:10.1029/2002GL015674
Liu WT (2002) Progress in scatterometer application. J Oceanogr 58:121–136
Liu WT (2010) Sea surface wind/stress vector. Encyclopedia of Remote Sensing, Springer, Heidelberg (in press)
Liu WT, Blanc TV (1984) The Liu, Katsaros and Businger (1979) Bulk Atmospheric Flux Computational Iteration Program in FORTRAN and BASIC. NRL Memo. Rep. 5291, Naval Research Laboratory, Washington DC, 16pp
Liu WT, Katsaros KB, Businger JA (1979) Bulk parameterization of air-sea exchanges in heat and water vapor including the molecular constraints at the interface. J Atmos Sci 36:1722–1735
Liu WT, Large WG (1981) Determination of surface stress by Seasat-SASS: a case study with JASIN Data. J Phys Oceanogr 11:1603–1611

Liu WT, Tang W (1996) Equivalent Neutral Wind, JPL Publication 96-17, Jet Propulsion Laboratory, Pasadena, California, 16pp

Liu WT, Tang W, Xie X (2008a) Wind power distribution over global ocean. Geophys Res Lett 35:L13808, doi:10.1029/2008GL034172

Liu WT, Tang W, Xie X (2010) Wind power at sea as observed from Space. In: Muyeen SM (ed.) Wind Power, Intech, Vienna (in press)

Liu WT, Tang W, Xie X, Navalgund R, Xu K (2008b) Power density of ocean surface wind-stress from international scatterometer tandem missions. Int J Remote Sens 29(21):6109–6116

Liu WT, Xie X (2006) Measuring ocean surface wind from space. In: Gower J (ed.) Remote Sensing of the Marine Environment, Manual of Remote Sensing, 3rd Edition, Vol. 6, American Society for Photogrammetry and Remote Sensing, Bethesda MD, USA, Chapter 5, pp. 149–178

Liu WT, Xie X (2008) Ocean-atmosphere momentum coupling in the Kuroshio Extension observed from Space. J Oceanogr 64:631–637

Liu WT, Xie X, Niiler PP (2007) Ocean-atmosphere interaction over Agulhas Extension Meanders. J Clim 20(23):5784–5797

Liu WT, Xie X, Polito PS, Xie S, Hashizume H (2000) Atmosphere manifestation of tropical instability waves observed by QuikSCAT and Tropical Rain Measuring Missions. Geophys Res Lett 27:2545–2548

Niiler PP (2001) Ocean circulation and climate-observing and modeling the global ocean. In: Chruch J, Siedler G, Gould J (eds.) The World Ocean Circulation, Academic Press, San Diego, pp. 193–204

Nikuradse J (1933) Stromungsgesetze in rauben Rohren. V.D.I. Forschungsheft 361:22

Nonaka M, Xie SP (2003) Covariations of sea surface temperature and wind over the Kuroshio and its extension: evidence for ocean-to-atmosphere feedback. J Clim 16:1404–1413

O'Neill LW, Chelton DB, Esbensen S (2003) Observations of SST-induced perturbations of the wind stress field over the Southern Ocean on seasonal timescales. J Clim 16: 2340–2353

Pacanowski RC (1987) Effect of equatorial currents on surface stress. J Phys Oceanogr 17:833–838

Park KA, Cornillon P (2002) Stability-induced modification of sea surface winds over Gulf Stream rings. Geophys Res Lett 29(24):2211, doi:10.1029/2001GL014236i

Park KA, Cornillon P, Codiga DL (2006) Modification of surface winds near ocean fronts: effects of the Gulf Stream rings on scatterometer (QuikSCAT, NSCAT) wind observations. J Geophys Res 111:C03021, doi:10/1029/2005JC003016

Pezzi LP, Vialard J, Richard KJ, Menkes C, Anderson D (2004) Influence of ocean-atmosphere coupling on the properties of tropical instability waves. Geophys Res Lett 31:L16306, doi:10/1029/2004GO019995

Pierson WJ (1984) A Monte-Carlo comparison of the recovery of winds near upwind and downwind from the SASS-1 model function by means of the sum of squares algorithm and a maximum likelyhood estimator. NASA Rep. 3939

Polito PS, Ryan JP, Liu WT, Chavez FP (2001) Oceanic and atmospheric anomalies of tropical instability waves. Geophys Res Lett 28:2233–2236

Powell MD, Vickery PJ, Reinhold T (2003) Reduced drag coefficient for high wind speeds in tropical cyclones. Nature 422:279–283

Seo H, Miller AJ, Roads JO (2007) The scripps coupled ocean-atmosphere regional (SCOAR) Model, with applications in the eastern Pacific sector. J Clim 20:381–402

Shaffer SJ, Dunbar RS, Hsiao SV, Long DG (1991) A median-filter-based ambiguity removal algorithm for NSCAT. IEEE Trans Geosci Rem Sens 29:167–174

Small RJ, de Szoeke S, Xie SP, O'Neill L, Seo H, Song Q, Cornillon P, Spall M, Minobe S (2008) Air–sea interaction over ocean fronts and eddies. Dyn Ocean Atmos 45:274–319

Smith SD (1980) Wind stress and heat flux over the ocean in gale force winds. J Phys Oceanogr 10:709–726

Spall MA (2007) Effect of sea surface temperature-wind stress coupling on baroclinic instability in the ocean. J Phys Oceanogr 37:1092–1097

Vecchi GA, Xie SP, Fischer AS (2004) Ocean-atmosphere covariability in the western Arabian Sea. J Clim 17:1213–1224
Weissman DE, Graber HC (1999) Satellite scatterometer studies of ocean surface stress and drag coefficient direct model. J Geophys Res 104:11329–11335
Wentz FJ (1978) Estimation of the sea surface's two-scale backscatter parameters. NASA Contractor Rep. 145255
Wentz FJ, Peteherych S, Thomas LA (1984) A model function for ocean radar cross sections at 14.6 GHz. J Geophys Res 89:3689–3704
Wentz FJ, Smith DK (1999) A model function for the ocean-normalized radar cross section at 14 GHz derived from NSCAT observations. J Geophys Res 104:11499–11514
Xie SP, Hafner J, Tanimoto Y, Liu WT, Tokinaga H, Xu H (2002) Bathymetric effect on the winter climate through the sea surface temperature in the yellow and East China seas. Geophys Res Lett 29(24):2288, doi:10.1029/2002GL015884
Xie SP, Ishiwatari M, Hashizume H, Takeuchi K (1998) Coupled ocean-atmospheric waves on the equatorial front. Geophys Res Lett 25:3863–3866
Yueh S, West R, Li F, Tsai W, Lay R (2000) Dual-polarized Ku-band backscatter signatures of hurricane ocean winds. IEEE Trans Geosci Rem Sens 38:73–88

Chapter 7
Interpretation of SAR Signatures of the Sea Surface: A Multi-sensor Approach

Leonid M. Mitnik and Vyacheslav A. Dubina

7.1 Introduction

During the last decade, satellite microwave sensors such as multichannel scanning radiometers, scatterometers and altimeters have provided extensive time series of upper ocean data. When combined with longer time series from visible and infrared (IR) sensors, the picture of the complexity of upper ocean and coastal zone processes, and of air-sea-ice interaction, is broadened significantly. Large-scale ocean processes are becoming better understood, leading researchers and modellers to probe further into mesoscale processes (up to 200 km) where significant energy exchanges and air-sea-ice interactions are occurring. Meandering currents are a dominant feature of many open ocean and coastal waters and often appear in patterns of Sea Surface Temperature (SST) and Chlorophyll a (chl-*a*) concentration in visible and thermal images acquired by satellite sensors. Furthermore, these currents often develop sharp fronts and eddies that affect wind-wave-current interactions, leading to both wave refraction and small scale surface roughness anomalies. These in turn, provide distinct expressions in both high resolution visible and Synthetic Aperture Radar (SAR) images (Johannessen et al., 2006). SAR backscatter signals come from the sea surface roughness with wavelengths which are approximately similar to the SAR wavelength (between a few to a few of tens centimeter). The roughness is controlled by local wind, wave-current interaction, as well as by the presence of surface active films or grease ice.

In the early 2000s, the unique and detailed ocean surface information in SAR imagery from ERS-2, RADARSAT-1, Envisat ASAR and recently launched ALOS, TerraSAR-X, COSMO-SkyMed, and RADARSAT-2 has been used successfully with other ocean sensors, buoy and ship data, and regional wind/wave models to improve our understanding of coastal and open ocean processes on these scales. Combining SAR images with those collected by the MODIS spectroradiometer,

L.M. Mitnik (✉)
Satellite Oceanography Department, V.I. Il'ichev Pacific Oceanological Institute, Far Eastern Branch of the Russian Academy of Sciences, Vladivostok 690041, Russia
e-mail: mitnik@poi.dvo.ru

V. Barale et al. (eds.), *Oceanography from Space*,
DOI 10.1007/978-90-481-8681-5_7, © Springer Science+Business Media B.V. 2010

SeaWiFS and Landsat ETM+, together with in-situ quasi-simultaneous measurements of ocean colour and SST, has opened a new era for observations of upper ocean circulation, air-sea-ice interaction, biological productivity, and coastal environmental monitoring. In the following, the role and importance of integration of SAR, visible, IR and passive microwave data is considered in the context of detecting and interpretating oceanic processes.

The examples shown here were selected mostly from the archives of SAR images and ancillary remote and/or in-situ observations of the northwestern Pacific Ocean and the Indonesian Seas. They cover various oceanic processes, for which the application of orbital SAR is well established. The high spatial resolution of the SAR datasets (25–150 m) is significantly better than that provided by MODIS and AVHRR (visible and IR images with resolution ranging from approximately 250 m to 1 km), AMSR-E (fields of the brightness temperatures T_{B}s at frequencies in the range of 6.9–89.0 GHz, with resolution of 5–20 km) and the scatterometer QuikSCAT (fields of the sea surface wind with resolution of 12.5–25 km). Further, in contrast to SAR observations, visible observations of the sea surface are limited by the availability of sunlight, while both visible and IR observations are further limited by the presence of clouds.

In the following, the mesoscale variability addressed will focus mostly on that of eddies of various scales, including those in the marginal ice zone, frontal features, and internal waves (IWs) in the Japan Sea (Section 7.2), Okhotsk Sea and Kuroshio-Oyashion frontal zone (Section 7.3) and Indonesian Seas (Section 7.4). These phenomena are detected by SAR and thermal IR sensors, due to the high SST gradients along the current/eddies boundaries, and visible sensors, due to the sea surface roughness anomalies induced by short-wave-current interaction (Johannessen et al., 1996; Ufermann et al., 2002; Gagliardini and Clemente Colón, 2004). Finally, Section 7.5 will provide a summary of the cases presented as well as a future outlook.

7.2 Oceanic Phenomena in the Japan Sea

Test cases to study SAR oceanic signatures in the Japan Sea demonstrate the effectiveness of combining SAR images with information obtained by other satellite sensors. The Japan Sea may be thought of as a miniature ocean, since it "possesses a western boundary currents as the East Korean Warm Current, a mid-ocean jet as one of two branches of the Tsushima Current, a polar front as the northern boundary of the Tsushima Current" (Ichie, 1984). Figure 7.1 shows a map of the area and the major circulation features of the Japan Sea.

This sea is characterized by great variability in the upper 200-m layer where water properties are altered by lateral exchanges through the shallow straits and vertical exchanges with the atmosphere. At the scales of 10–500 km, the upper column is known to be a combination of warm and cold currents, eddies and upwelling zones. At still lesser scales, a significant contribution to the variability gives the narrow streamers of the warmer and colder water, IWs, river plumes, etc.

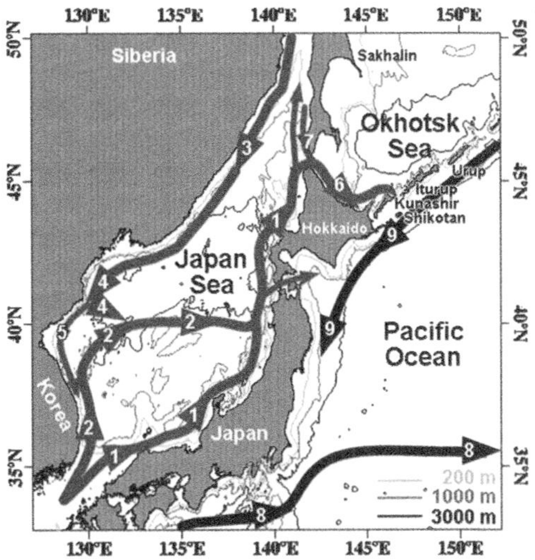

Fig. 7.1 Schematic of surface circulation in the Japan and Southern Okhotsk Seas and in the Pacific Ocean east of Japan: (1) Tsushima Current, (2) East-Korean Current, (3) and (4) Primorye Current, (5) North-Korean Current, (6) Soya Warm Current, (7) West-Sakhalin Current, (8) Kuroshio and (9) Oyashio

During the cold season, outbreaks of cold air from Siberia are accompanied by increased air/sea interaction, so that intensive stirring and cooling of the upper layer waters are typical. The subarctic (polar) front, dividing the warmer waters of the Tsushima Current and the colder waters in the northern portion of the sea, is located along $\approx 38-40°N$. The coldest waters with SST $\leq 0°C$ are surrounded by the numerous eddy structures of different scales with SST $\approx 2-3°C$. The velocity of coastal currents and currents on the eddy boundaries can reach 40–70 cm/s. Plankton blooms are observed during spring. The area with increased chl-*a* concentration shifts northward with the increase of SST. Water stratification forming in a warm season is favorable for internal wave generation and propagation, both in the coastal and open areas of the Japan Sea. Complicated structures of oceanic and atmospheric phenomena manifest themselves in the fields of various physical parameters.

7.2.1 Subarctic Front in the Japan Sea

The interaction of the East Korean Warm Current (EKWC) and the Primorye Current is quite interesting. Both currents are actually oppositely-directed coastal currents flowing along Primorye/Korean coasts (Fig. 7.1). The southward Primorye Current encounters the northward EKWC at about 38–40°N. Their interaction creates a very complicated frontal zone, the exact location of which varies with the seasons, as shown by the analysis of SST time series.

The complex structure of the subarctic frontal zone (i.e. of its northwestern part) at the end of the cold season is shown in Fig. 7.2a (Mitnik and Dubina, 2005). The most impressive details are 3 mesoscale eddies the size of 60–70 km (labeled A, B, C), seen also in a concurrent AVHRR thermal image (Fig. 7.2b). The highest

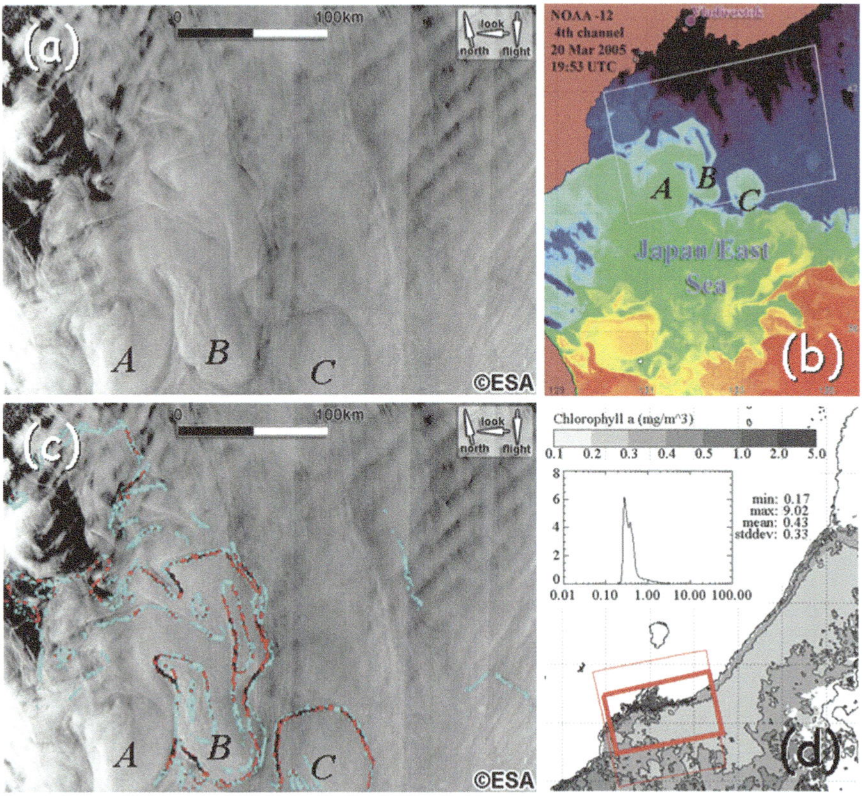

Fig. 7.2 Mesoscale eddies in the subarctic frontal zone of the Japan Sea on 20 March 2005: (**a**) ASAR image the size of 230 km × 290 km taken at 12:50 UTC, (**b**) AVHRR infrared image at 19:53 UTC, (**c**) SST gradients superimposed on the same ASAR image and (**d**) map of chl-*a* concentration for 14–21 March. A *white rectangle* in (**b**) and the *bold red rectangle* in (**d**) mark the boundaries of the ASAR image

radar contrasts are observed at the eddies boundaries, where both surface current and current shift reach maximum values (see a comparison with thermal gradients in Fig. 7.2c). Eddies are known to influence the dynamics of the entire Japan Sea. However, their formation, structure, and time evolution are not well understood.

7.2.2 Convergence Zone

Sometimes the boundary between cold and warm water masses shifts to the north, to about 42°N, from the usual average position near 40–41°N. Such a situation was observed in November 2003 (Fig. 7.3). Eddy-like instabilities of different space and time scales were observed in the frontal zone. They manifested themselves both in the SST field (Fig. 7.3b) and in radar reflectivity field (Fig. 7.3a). The size of

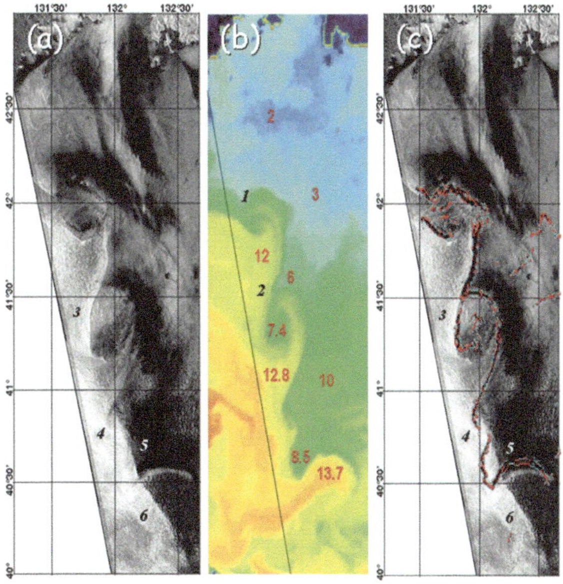

Fig. 7.3 Eddy-like variability in the area dividing warm and cold waters in the Japan Sea, to the south of Vladivostok, in (**a**) Envisat ASAR image at 12:53 UTC; and in (**b**) NOAA AVHRR infrared image at 07:55 UTC; on 13 November 2003. In (**c**), same ASAR with superimposed SST gradients. The *dark slanting line* in (**b**) marks the western boundary of the ASAR image; red digits are SST values in °C and dark digits are the signatures discussed in a paper

cyclonic eddies *1* and *2* (as indicated in Fig. 7.3b) is about 40 km, while the distance between their centers is 65 km (Mitnik and Dubina, 2005).

The highest contrast of the radar cross-section σ° against the uniform background occurs in the northern part of the convergence zone (Fig. 7.3a). Sections across the zone show positive σ° increment for its western (warm water) side and negative increment for the eastern (cold water) side, as well as the decrease of the average σ° level of the eastern side relative to the western one. Such σ° profiles correspond to combination of the surface current features (convergence and shift) with SST front and near surface wind direction (Kudryavtsev et al., 2005).

The western part of the ASAR image is characterized by higher brightness in comparison with its eastern part, a fact that can be explained by the change of the atmosphere from an unstable state (water is warmer then air) to stable one. An angle between SAR look direction and a convergence zone is another important factor determining the change of the radar brightness across the zone. This is evident from the comparison of legs *3–4* with a leg *5* directed at right angle to each other (Fig. 7.3a, c). A leg *6* appearance on SAR image is caused by the surface current and wind features since the sharp SST gradients are absent in warm waters where this leg is located (Fig. 7.3b).

During the ASAR sensing, the wind speed varied over a range of 2–5 m/s, as derived by the QuikSCAT scatterometer. At weak winds, the σ° values (i.e. the SAR image brightness) will depend on surface current direction relative to wind direction. Southward currents in the western sides of cyclonic eddies *1* and *2* are the reverse to the northward currents in their eastern sides (Fig. 7.3b). Wind and current directions coincide in the eddies eastern sides and the wind speed relative to the sea surface decreases till 2–2.5 m/s when backscatter is absent. As a result, the corresponding regions have a dark tone on SAR image. Further east, outside of eddies, the brightness of the image increases, but it is lower than to the west of the front where SST is higher.

7.2.3 Eddies in the Coastal Zone

Coastal zones are of crucial importance to society. These areas are characterized by interaction of complex and coupled physical and bio-geo-chemical processes in the upper ocean and atmospheric boundary layer, at various spatial and temporal scales. The same areas are strongly influenced also by terrestrial processes, especially by run-off. A quantitative understanding of the processes impacting the coastal region is required to determine how wind, waves, current and river discharge variability, as well as coastal orography, will affect coastal systems. The study of dynamic phenomena in this zone is difficult, given the wide spectrum of temporal and spatial variability of physical processes occurring, which requires the development of synergic approaches through the combined use of remote sensing and in-situ data, together with modeling.

Mesoscale and small-scale features of surface circulation have often the form of eddies and vortex pairs which are visualized in the ocean color, SST, sea surface roughness or sea ice fields. Figure 7.4 shows eddies of various scales and warm water flows in ASAR and AVHRR thermal images of the Japan Sea to the east of the Korean coastline (Mitnik and Dubina, 2005).

7.3 Okhotsk Sea and Oyashio-Kuroshio Frontal Zone

7.3.1 Eddies in the Soya Warm Current Area

The Soya Warm Current (SWC) flows southeastward along the coast of Hokkaido, in area with depth less than 150 m (Fig. 7.4). It is formed by Japan Sea waters that enter the Okhotsk Sea through the Soya (LaPerouse) Strait. The typical width of the SWC is approximately 50 km. Seasonal variations of monthly-mean velocity of the SWC are clear pronounced. The velocity reaches a maximum of approximately 100 cm/s in summer (August and September) and became weak in winter (January and February) (Ebuchi et al., 2009). SWC waters are characterized by high temperature and salinity, and by a sharp front separating them from the lower-temperature and lower-salinity offshore waters. The distance of the front from the coast is about 35–50 km.

Frequently, radar signatures in the form of eddies, eddy streets and waves are observed in the SWC area to the southeast of Cape Krilion, north of Hokkaido. Waves are produced after the flow separates from the cape and propagates downstream. In the region of flow separation, strong horizontal shear induces barotropic instability, in which the waves are generated (Ohshima and Wakatsuchi, 1990). On

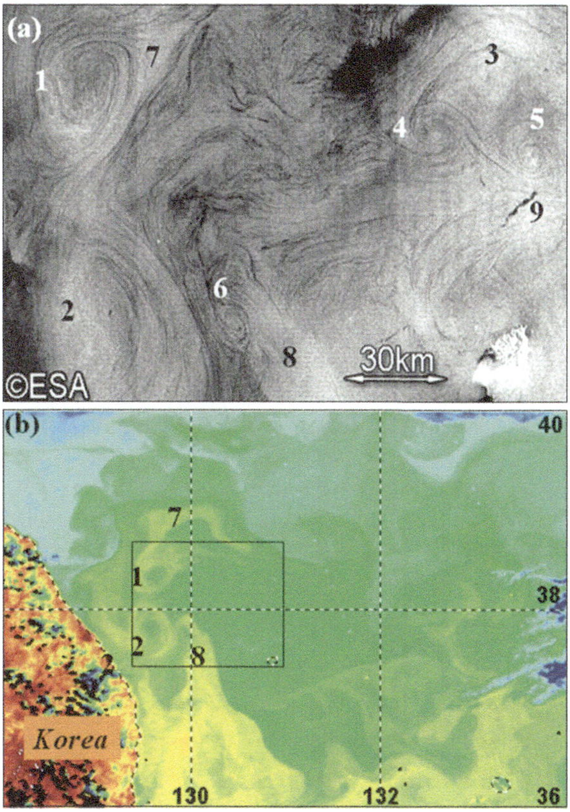

Fig. 7.4 Cyclonic eddies (1–6), warm water flows (7 and 8) and oil spills (9) in the Japan Sea, east of Korea, on 14 April 2004: (**a**) Envisat ASAR image, at 01:28 UTC, and (**b**) NOAA AVHRR infrared image, at 15:20 UTC. The *dark rectangle* in (**b**) marks the boundaries of ASAR image

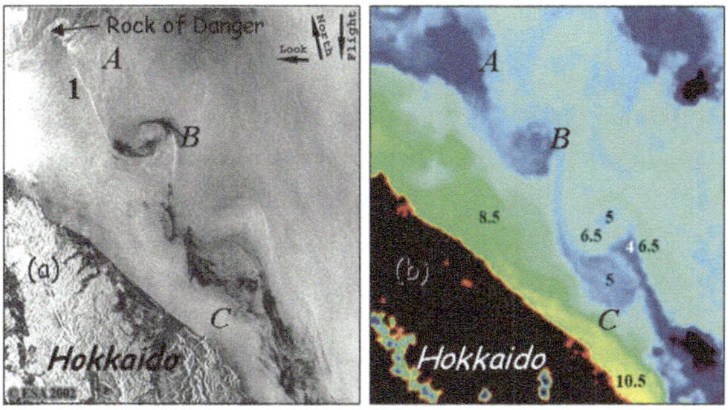

Fig. 7.5 ERS-2 SAR image the size of 100 km × 100 km taken at 01:18 UTC (**a**) and NOAA AVHRR infrared image at 03:46 UTC (**b**) for 13 May 2002 showing eddies *A*, *B* and *C* in the Soya Warm Current area. Digits are AVHRR-derived SST values

a SAR image for 13 May (Fig. 7.5a), a narrow bright band *1* of ∼1 km width caused by current shift starts near Rock of Danger (45.8°N, 142.2°E) 13 km to the south-east of Cape Krilion. An extensive stony area of about 5 km in length is located around the small (150 m × 50 m) rock. Shallow areas between Krilion Cape and Rock of Danger and to the east of them, creates conditions for vertical mixing and homogenization of the oceanic characteristics in the area southeast of Krilion Cape (Danchenkov et al., 2003) where cold waters are located (Fig. 7.5b). The band *1* is close to the boundary dividing warm and cold waters (as the position of the current shift zone may be different from the thermal boundary). A first eddy-like structure *A* is located to the southeast of Rock of Danger. The eddy is better distinguished in the AVHRR thermal image. A second eddy *B*, consisting of a cold core, of about 8 km in size, and two spiral "tails" are apparent in both images. The dark features cover the eddy's core from the north/south and increase its radar contrast. They are very likely due to damping of the small-scale sea surface roughness by the surfactant films. The distance between "tails" is ~15 km and can be considered as the eddy's size. The third eddy *C* also has a cold central area with SST of about 5°C. The coldest waters (4°C) are in a band to the southeast of the eddy's centre. The cold waters in the third eddy's area are characterized by a decreased backscatter level.

The current shift zone is visible north of the eddy, where SST is around 6.5°C and rather uniform. The eddies are located at a distance of ~25 km from the coastline, while the distance between their centers is ~37 km, in close agreement with the results of model experiments (Ohshima and Wakatsuchi, 1990) and coastal radar observations (Ebuchi, 2006).

7.3.2 Ice Eddies

In the winter season, pack ice visualizes the features of surface circulation with weak winds and sea ice concentration of less than ~70%. An increased viscosity of the ice-water surface layer influences the interaction of an eddy with the surrounding waters. Figures 7.6a, c show a complex surface circulation pattern in the partly ice-covered Southern Okhotsk Sea, imaged by ASAR on 8 February 2003. The images cover the high ice concentration area, the Marginal Ice Zone with ice bands and eddies, and ice-free waters, with brightness variations caused by surface wind variations accompanying cellular convection in the atmospheric boundary layer (Fig. 7.6a) and high winds (area *1*). The most interesting objects are eddies *I* and *II*. Eddy *I* (size of ~90 km) is formed by two spirals consisting of 5–7 elliptical eddies (size of ~7–10 km). These small-scale eddies, at the periphery of the large ones, are manifestation of the increased viscosity of the ice-water surface layer influencing the interaction of an eddy with the surrounding waters.

The large, elliptical eddy *II*, located to the southeast from eddy *I*, was formed by two bands of grease ice (a soupy layer of frazil crystals clumped together, which makes the ocean surface resemble an oil slick). It looks dark on the images since grease ice damps the small-scale waves. The brightness variations in the ice-free area *1* surrounded by the ice bands are also caused by mesoscale convection in the atmosphere.

These ice eddies were observed again by ASAR in approximately 11 hours (Fig. 7.6c). The ice concentration in the central area of eddy *I* decreased sharply due to strong winds. Several small-scale spiral elliptical eddies in the western part

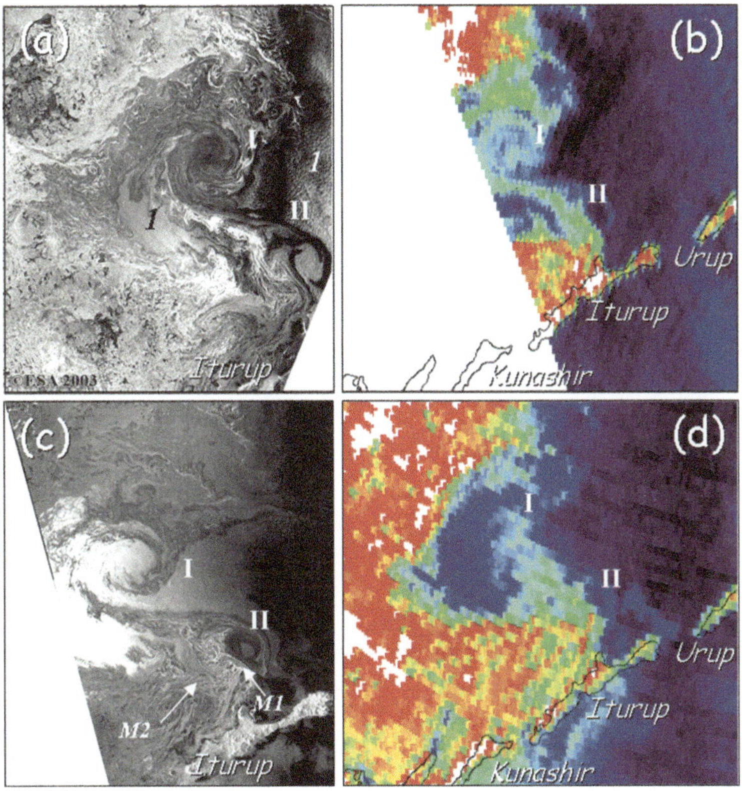

Fig. 7.6 Sea ice in the Southern Okhotsk Sea on 8 February 2003: Envisat ASAR images at VV polarization acquired at 00:30 UTC (**a**) and at 11:50 UTC (**c**) and Aqua AMSR-E brightness temperatures at 89 GHz with horizontal polarization at 02:35 UTC (**b**) and at 16:20 UTC (**d**)

of *I*, where the wind was higher, also melted. However they are easily recognized in the eastern part bordering the open sea. Ice formation continued in the eddy *II* area. Narrow curvilinear bands of the pancake ice appeared in the region where the grease ice was detected in the previous image. The two vortex pairs *M1* and *M2* are seen north of Iturup. Pair *M1*, especially its cap, looks brighter than *M2*, suggesting that the eddies consist of different types of sea ice (Fig. 7.6a, c). In the AMSR-E observations, carried out 3 h later than the ASAR imaging (Fig. 7.6b, d), the brightness temperatures T_Bs at 89.0 GHz for the open water area within the eddy *I* increased during the interval between overpasses, but ice formation continued in the area of eddy *II*, north of Iturup.

The backscatter characteristics of sea ice vary in the broad range that can be explained by the variation of ice types, concentration, conditions of formation, etc. Sea ice can look both brighter and darker against the open sea, the brightness of which depends on wind speed and direction. These features hinder the interpretation of SAR signatures and sea ice classification. Concurrent analysis of

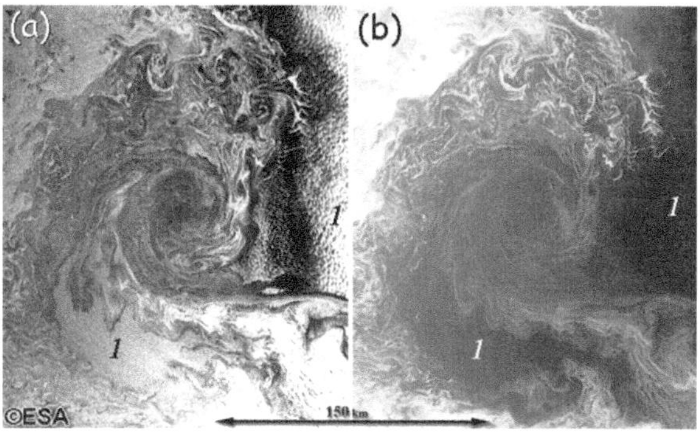

Fig. 7.7 Spiral ice eddy in the Southern Okhotsk Sea on Envisat ASAR image taken at 00:30 UTC (**a**) and on Terra MODIS visible image taken at 00:50 UTC (**b**) on 8 February 2003; 1 marks the ice-free area

SAR, visible and passive microwave data decreases the ambiguity of interpretation. This is evident from a comparison of the SAR images with the AMSR-E T_Bs and MODIS visible images. Both eddies have higher T_Bs values at 89 GHz in comparison with the open sea as seen in (Fig. 7.6b, d). The spatial resolution of T_Bs data is 4 km × 6 km. In the visible image, having a spatial resolution of 250 m × 250 m, the open sea has a dark tone even at high winds (Fig. 7.7b) as opposite to SAR images where the increase in wind speed causes the increase in radar backscatter (Fig. 7.7a). The brightness of sea ice depends on its thickness (type) and also on the presence of snow. So, grease ice and dark nilas (ice which is under 5 cm in thickness and is very dark in colour) have very low brightness contrast, as opposite for example to grey-white or white ice. The concurrent analysis of SAR and visible images (Fig. 7.7) gives a better idea of ice types and structure of ice eddies at different scales. Moreover Envisat ASAR, Terra MODIS and Aqua AMSR-E images form an extended time series, which allows to study over time the surface circulation in this dynamic area under influence of both oceanic and atmospheric factors.

7.3.3 Usage of Multiple SAR Sensors

The analysis of co-located ERS-2 SAR and Envisat ASAR images, acquired with time difference of about 30 min, allows the estimation of the velocity of the SAR signatures caused by both oceanic (currents, IWs, etc.) and atmospheric dynamic phenomena (Gade et al., 2003; Dubina and Mitnik, 2005; Liu and Hsu, 2009).

Two SAR images, consisting of several successive frames and covering the Northern Japan Sea and the Southern Okhotsk Sea, were used. The first was taken by ASAR at 12:11–12:14 UTC, in wide swath mode, and the second by ERS-2 SAR at 12:40 UTC, on 4 October 2003. To estimate the speed V, the original high-resolution ERS-2 SAR image was down-sampled to 75 m pixel size with the use of bi-cubic interpolation. The resampled image was warped onto the ASAR original image (75 m × 75 m pixel size) by triangulation technique. The resulting pixels of the warped ERS SAR image were derived using cubic convolution.

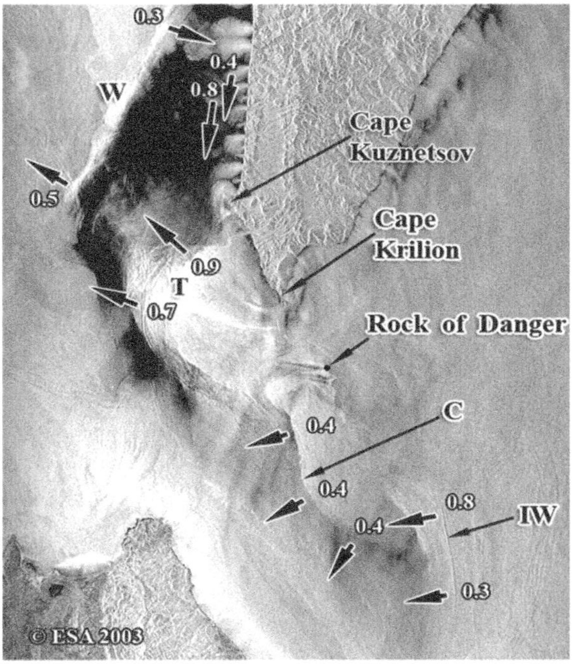

Fig. 7.8 ERS-2 SAR image (~ 100 km × 100 km) taken on 4 October 2003 at 12:40 UTC. The *arrows* show movement direction of the radar signatures and the numbers denote their speed in m/s estimated with the use of the co-located Envisat ASAR and ERS-2 SAR images

The highest *V* values (5–6 m/s) were derived for the atmospheric features (not shown). The velocities of the oceanic signatures (the Soya Warm Current front, packets of IWs, etc.) are significantly less and lie between 0 and 0.9 m/s (see Fig. 7.8). The arrows in Fig. 7.8 indicate direction of movement, while the numbers near the arrows are the respective velocities in m/s. The southward West Sakhalin Current is observed to the south of Cape Kuznetsov. The speed of this current changes rapidly, as the distance *D* from the coast increases: at $D = 3\,\text{km}$ $V = 0.4\,\text{m/s}$ and at $D = 5$ km $V = 0.8\,\text{m/s}$. The small-sized radar signatures, aligned from Cape Krilion and Rock of Danger to the west, indicate the presence of strong tidal outflow from the Okhotsk Sea. The signatures near Cape Krilion have an arched shape. The tidal outflow moves the radar signature *C* toward the Hokkaido coast at the speed of 0.4 m/s.

At the same time, the radar signature *T* moves westward at the speed of 0.7–0.9 m/s (Fig. 7.8). Close inspection of the derived movement shows that the bright radar signature *W*, corresponding to the convergence zone between the Tsushima Current and the West Sakhalin Current, undergoes cyclonic displacement. Its southern part moves to the northwest at the speed of 0.5 m/s and northern one moves toward the coast at the speed of 0.3 m/s. A narrow light-dark band in the lower right hand section of Fig. 7.8 is very likely an IW soliton. Its northern side moves westward at a rate of 0.8 m/s, while the southern one displaces slower (0.3 m/s), due to the Soya Warm Current influence. Several packets of IWs are visible to the east of the soliton. Their presence confirms that the water stratification and current velocities were favorable for IWs generation. Several factors influence the accuracy of the retrieved *V* values: time interval between two images and accuracy of their coordinate relation; speed and movement direction of a SAR signature; its contrast against the background; pixel size, etc. The mean error of the signature speed determination was approximately 0.15 m/s.

The pair of Envisat ASAR and ERS-2 SAR images is not only a good example, but also a good basis for the detailed study of surface currents. The derived estimates of speeds and directions of a number of selected SAR signatures may serve for the assessment of "first-guess flows", to retrieve current fields from radar data with the help of a radar imaging model. The proximity of the oceanic features to the coast improves georeference and gives a fine opportunity to the current field reconstruction.

7.3.4 Oyashio-Kuroshio Frontal Zone

Intense mesoscale eddies are observed in the subarctic frontal zone east of Japan, where the warm Kuroshio waters interact with cold Oyashio waters. The probability of clouds over this zone is high. SAR data, together with IR data, help to monitor eddy formations and trace their evolution. Radar contrasts of eddies against the surrounding waters are determined not only by the differences in SST, influencing the stability of the marine boundary layer in the atmosphere, but also by several other factors, such as surface current velocity and its direction relative to the local wind direction, surface wind speed, concentration of biogenic films.

Figure 7.9 shows Envisat ASAR and Terra MODIS thermal images of the subarctic frontal zone, southeast of Hokkaido and the southern Kuril Islands, acquired with a 54-min time interval. Dark areas in (b) correspond to clouds. Several eddies can be

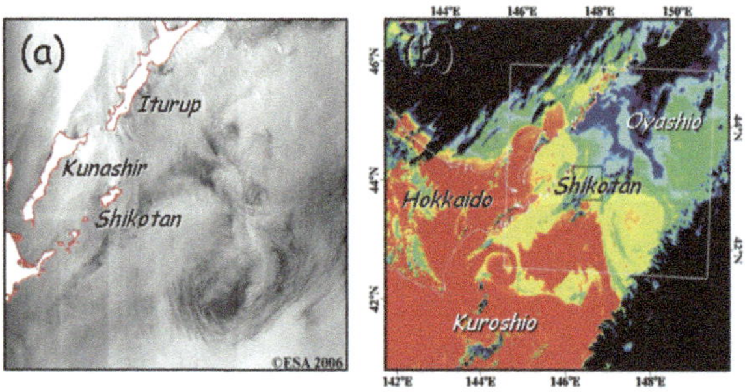

Fig. 7.9 Envisat ASAR image (**a**) and Terra MODIS thermal image (**b**) east of Hokkaido and Kuril Islands, taken on 24 September 2006 at 00:21 UTC (**a**) and 01:15 UTC (**b**). The *white square* in (**b**) shows the boundaries of the image in (**a**). The small *square* under the "Shikotan" name shows the boundaries of the fragments in Fig. 7.10

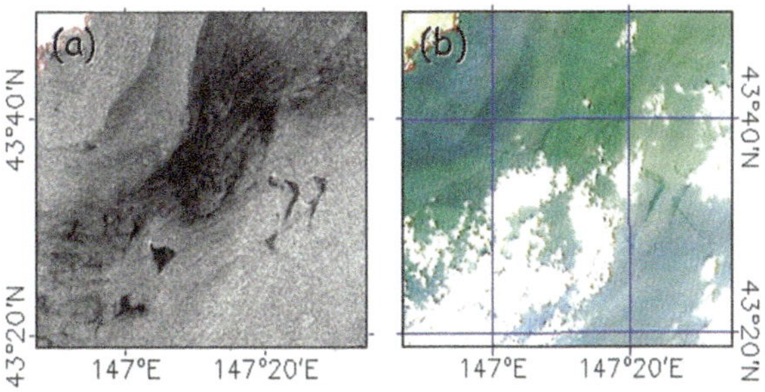

Fig. 7.10 Area of Pacific saury commercial fishing, as appearing in: an Envisat ASAR image (**a**) and an Terra MODIS visible image (**b**), taken on 24 September 2006 at 00:21 UTC (**a**) and at 01:15 UTC (**b**). Boundaries of the images are shown in Fig. 7.9. The *white area* in *upper left* is the southeast corner of Shikotan Island

seen on both images. Cold nutrient-rich Oyashio waters and a warm spiral anticyclonic synoptic eddy are characterized by weak backscattering due to the presence of biogenic films manifesting themselves as filaments. Rich catches of Pacific saury (*Cololabis saira*) are observed in the area when SST is around 15°C (the temperature preferred by migratory saury to aggregate intensively) and plankton abundance is high. Fat resulting from fish processing forms films that also damp the small-scale sea surface roughness.

The film-covered patches and bands starting near fish boats (light dots) look dark on the ASAR image (Fig. 7.10a). Several dark bands stand out against a background also on Terra MODIS visible image (Fig. 7.10b). A green-color area in Fig. 7.10b, caused by the increased chl-*a* (i.e. phytoplankton) concentration, corresponds to the darker area in Fig. 7.10a.

7.4 Internal Waves in the Indonesian Seas

The seas of Indonesia, Java, Banda, Sulu, Molucca, etc., as well as the South China Sea, Andaman Sea, Sulu Sea, etc., are ocean areas known for their large amplitude IWs.[1] IWs are observed throughout year since tropical waters are always stratified. Broad bands of increased sea surface roughness and foaming extending horizon-to-horizon are frequently observed from ship in tropical waters. IWs are generated as a result of the interaction of strong tidal currents between the islands with bottom elevations, sills and shelf edge. Impressive examples of the Iws radar signatures as seen in spaceborne SARs imagery, over the Indonesian Seas are reported in the literature (Mitnik and Alpers, 2000; Atlas of Internal Solitary-like Waves and their Properties, 2004; Mitnik, 2006; Jackson, 2007).

IWs become visible to SAR, as well as in images acquired in the visible range, because they are associated with variations of sea surface roughness. In the case of SAR, they can also be associated with a variable surface current which modulates the surface wave spectrum and thus the normalized radar cross section. In the past,

[1] For ERS1/2 SAR views of the tropical and subtropical ocean, see work by W. Alpers, L. Mitnik, Hock Lim and K.S. Chen in http://earth.esa.int/applications/ERS-SARtropical

spaceborne SAR imagery was used extensively in studies of IWs dynamics, but not so visible imagery. The conditions for IWs imaging by radar and optical sensors are quite different (Mitnik et al., 2000). While radar imaging is almost independent of weather conditions, optical imaging requires (at least partially) cloud free conditions and is strongly dependent on the viewing geometry. Usually, IWs are detected in sun glint area; however, their contrasts beyond sun glint area may be also high enough for their detection. When imaging IWs by optical sensors, one has to distinguish whether the sea surface manifestations of the IWs lie inside, outside the sun glint area, or in the transition zone. For a flat sea surface, sun glitter is seen when the sun elevation angle E (measured from the horizon) and the viewing (incidence) angle I (measured from the vertical) obeys the relationship $E + I = 90°$. If the sea is roughened by surface waves, each facet at the appropriate angle reflects the sun's rays to the sensor and a glitter pattern with diminishing brightness outward from its center results (Cox and Munk, 1954). In the sun glint area IWs become visible because the sun glitter radiance depends on the slope distribution of the surface waves, which is modulated by the surface current associated with the IWs. In the non-sun glint area IWs become visible because the diffuse reflection of sunlight depends also on the slope distribution of the surface waves. Wave breaking and appearance of foam patches can contribute significantly in optical contrast. This phenomenon is typical for intense fast-moving IWs.

When radar and optical images are used together, then the frequency of observation is increased significantly, and thus a better study of the internal wave dynamics becomes possible. In connection with this, consider multi-sensory satellite data on IWs in the Banda Sea.

The Banda Sea is about 1,000 km east to west, and about 500 km north to south. It occupies a total of 470,000 km^2 and opens to the Flores (west), Savu (southwest),

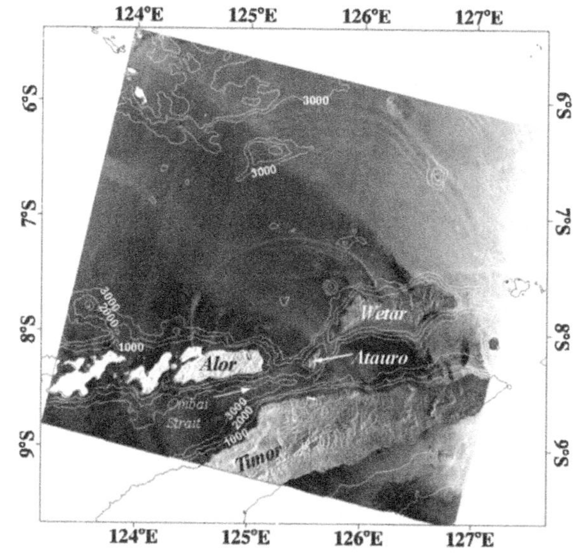

Fig. 7.11 Internal waves in the Banda Sea on Envisat ASAR image with vertical polarization acquired on 18 April 2003 at 01:18 UTC with superimposed bathymetric chart depth contours of 1,000, 2,000 and 3,000 m

Timor (south), Arafura (southeast), and Ceram and Molucca (north) seas. The Banda Sea is a part of the Indonesian Throughflow (ITF), which is an important pathway for the transfer of western tropical Pacific waters into the southeastern Indian Ocean through the Indonesian seas. Within the Banda Sea, Pacific waters are modified by the joint action of various physical processes and then exit into the Indian Ocean via the Timor Strait or the Ombai Strait. SAR and visible images allowed to conclude that IWs were very likely generated by a sill between Alor and Atauro Islands in the Ombai Strait, one of the major passages of ITF (Mitnik and Dubina, 2009.

Figure 7.11 shows two packets of IWs propagating northward into the Banda Sea. The IWs have the classic appearance of rank ordered non-linear waves that depress the pycnocline (Atlas of Internal Solitary-like Waves, 2004; Apel, 2004). IWs were clearly seen within the whole Envisat swath (405 km). The crests of solitons form almost correct semicircles, limited by the islands. The length of several

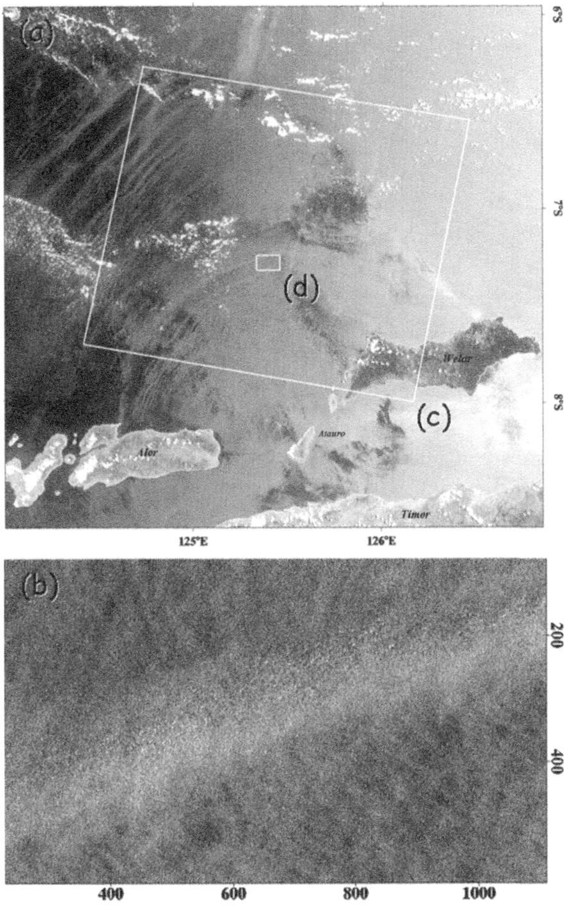

Fig. 7.12 Terra MODIS visible image taken at 02:00 UTC (**a**) and subsection of Landsat-7 ETM+ panchromatic image taken at 01:35 UTC on 30 September 2007 (**b**). The *white rectangle* (**c**) marks the boundaries of the whole Landsat-7 ETM+ image, while the small *white rectangle* (**d**) marks the boundaries of its subsection (**b**). *Vertical and horizontal scales* in (**b**) are given in pixels (15 m × 15 m)

crests exceeded 350 km. The leading soliton of the second packet intrudes the rear of the first packet that does not allow estimating the number of waves in the wave packet and its width. The wavelength of the solitons monotonically decreases from about 15 km to about 2 km in the rear of packet. The SAR images suggest a tidal generation mechanism. The tide at the sill region is predominantly semi-diurnal and estimates of the waves' phase speed can be derived by measuring the distance between the first solitons in the successive packets. Detailed examination shows that the phase velocity increases from 2.4–2.6 m/s (propagation direction WNW-NNW) to 2.9–3.1 m/s (N-NNE) and then decreases to 2.7 m/s (NE). These variations can be caused by the differences in water properties along propagation directions.

Similar results were obtained by the analysis of medium-resolution MERIS, and MODIS data and high-resolution SPOT and Landsat data. Figure 7.12a shows IWs in the Banda Sea as seen in a MODIS image acquired on 30 September 2009. The same area was sensed by Landsat-7 ETM+ 25 min before MODIS. The velocity of the IWs, estimated by an analysis of the co-located MODIS and ETM+ images, was about 3–3.2 m/s for leading solitons in the packets. Very likely, the detailed study of the brightness field spatial structure in panchromatic Landsat images (resolution 15 m × 15 m) will provide information about both the slope distribution of the surface waves and intensity of wave breaking. In turn, data on wave breaking and whitecap/foaming induced by wind action, variable currents and other causes are required for advancing radar backscatter theory.

7.5 Conclusions

With the recent launch of the new generation SARs (ALOS PALSAR L-band, TerraSAR-X, and COSMO-SkyMed X-band and RADARSAT-2 C-band) and the availability of high- and low-resolution visible, IR and passive/active microwave sensors (Landsat-ETM+, MODIS, MERIS, AMSR-E, QuikSCAT, etc.), multi-sensor technology has become a reality in a growing series of oceanic and atmospheric applications. SAR data in synergy with optical, IR and microwave data can provide a better insight into oceanic and atmospheric processes at different scales with improved temporal resolution. Optical data are relatively easily available, the measured parameters are more familiar and an extensive established knowledge base for processing exists. Spaceborne SAR provides finely detailed imagery of the ocean's surface, which is the most complex and least understood data provided by remote sensors. The sea surface can contain the signatures of surface currents, eddies, sea ice, IWs, examples of which were considered in the paper, as well as such diverse phenomena as upwelling, shallow water bathymetry, wind, mesoscale convective rolls and cells, atmospheric gravity waves, rains, storms, etc. Variations of radar contrasts and superposition of signatures of various nature, depending on several environmental factors, hinder their interpretation.

The fusion of multi-sensor and multi-resolution satellite imagery has become an effective mean of exploiting the complementary nature of various images types, improving our understanding of the geophysical content of SAR images and the identification of physical and biological processes controlling the main SAR signatures. The all-weather and high-resolution SAR imagery can clearly serve as a complement both to lower resolution satellite-derived SST, chl-*a* and sea surface wind fields and to in-situ observations. The increased availability and use of SAR data merged with optical and microwave sensors, with distinct swaths, spatial and temporal resolutions can provide a powerful observational tool to link large, mesoscale and small-scale ocean dynamics. This synergy also offers opportunities to advance the application of remote sensing techniques in areas such as physical/biological oceanography, ocean–atmosphere interaction, fisheries management, and pollution monitoring.

Acknowledgements The Authors wish to thank the European Space Agency for providing Envisat ASAR images (project AO-BEAR-2775) and the Japan Aerospace Exploration Agency (projects 356 and 364). This work is partly supported by the Russian Fund for Basic Research, project 09-05-00487-a.

References

Apel JR (2004) Oceanic internal waves and solitons. In: Jackson CR, Apel JR (eds.) Synthetic Aperture Radar Marine Users Manual, National Oceanic and Atmospheric Administration, Washington, DC, pp. 189–206

Atlas of Internal Solitary-like Waves and their Properties (2004) 2nd edition, Global Ocean Associates Prepared for Office of Naval Research – Code 322 PO, available at http://www.internalwaveatlas.com /

Cox C, Munk W (1954) Statistics of the sea surface derived from sunglitter. J Mar Res 13:198–227

Danchenkov MA, Aubrey D, Riser SC (2003) Oceanographic features of LaPerouse Strait. PICES Sci Rep 12:159–171

Dubina VA, Mitnik LM (2005) Evidence of the West Sakhalin Current in ERS-2 and Envisat-1 collocated SAR images. Proceedings of the 2004 Envisat & ERS Symposium, Salzburg, Austria 6–10 September 2004. ESA SP-572

Ebuchi N, Fukamachi Y, Ohshima K, Shirasawa K, Ishikawa M, Takatsuka T, Daibo T, Wakatsuchi M (2006). Observation of the soya warm current using HF ocean radar. J Oceanogr 62(1):47–61

Ebuchi N, Fukamachi Y, Ohshima KI, Wakatsuchi M (2009) Subinertial and seasonal variations in the Soya Warm Current revealed by HF ocean radars, coastal tide gauges and bottom-mounted ADCP. J Oceanogr 65(1):31–43

Gade M, Fiedler G, Dreschler-Fischer L (2003) Two-dimensional sea surface current fields derived from multi-sensor satellite data. Proceedings of the IGARSS'03 6:3540–3542

Gagliardini DA, Clemente Colón P (2004) Ocean feature detection using microwave backscatter and sun glint observations. Gayana 68(2):180–185

Ichie T (1984) Some problems of circulation and hydrography of the Japan Sea and the Tsushima Currents. In: Ichie T (ed.) Ocean Hydrodynamics of the Japan and East China Seas, Elsevier, Amsterdam, pp. 15–54

Jackson C (2007) Internal wave detection using the Moderate Resolution Imaging Spectroradiometer (MODIS). J Geophys Res 112:C11012, doi:10.1029/2007JC004220

Johannessen JA, Pettersson LH, Eldevik T, Durand D, Evensen G, Winther N, Breivik O (2006) Coastal physical and biochemical processes. In: Gower JFR (ed.) Remote Sensing of the Marine Environment, American Society for Photogrammetry and Remote Sensing, Bethesda, Maryland, pp. 179–196

Johannessen JA, Shuchman RA, Digranes G, Lyzenga DR, Wackerman C, Johannessen OM, Vachon PW (1996) Coastal ocean fronts and eddies imaged with ERS-1 SAR. J Geophys Res 101(C3):6651–6667

Kudryavtsev VN, Akimov DB, Johannessen OM, Shapron B (2005) On radar imaging of current features: 1. Model and comparison with observations. J Geophys Res 110:C07016, doi:10.1029/2004JC002505

Liu AK, Hsu MK (2009) Deriving ocean surface drift using multiple SAR sensors. Remote Sens 1:266–277

Mitnik L (2006) Dynamic phenomena in the Lombok Strait and in surrounding waters: imprints on ERS and Envisat SAR images. Earth Observation and Remote Sensing 6:83–91 (in Russian)

Mitnik L, Alpers W (2000) Sea surface circulations through the Lombok Strait studied by ERS SAR. Proceedings of the 5th Pacific Ocean Remote Sensing Conference (PORSEC 2000), Goa, India, 5–8 December, I:313–317

Mitnik L, Alpers W, Chen KS, Chen AJ (2000) Manifestation of internal solitary waves on ERS SAR and SPOT images: similarities and differences. Proc. IGARSS'00 5:1857–1859

Mitnik LM, Dubina VA (2005) Synoptic-scale, mesoscale and fine-scale oceanic features in the Japan/East Sea: study with ERS-1/2 SAR and Envisat ASAR. Geoscience and Remote Sening Symposium. IGARSS '05. Proceedings of the IEEE International, vol 7, Seoul, Korea, pp. 4788–4791

Mitnik LM, Dubina VA (2009) Non-linear internal waves in the Banda Sea on satellite synthetic aperture radar and visible images. Proceedings of the IGARSS'09, vol 3, Cape Town, South Africa, pp. 192–195

Ohshima KI, Wakatsuchi M (1990) A numerical study of barotropic instability associated with the soya warm current in the sea of Okhotsk. J Phys Oceanogr 20:570–584

Ufermann S, Robinson IS, da Silva JCB, Johannessen JA (2002) The role of synergy in developing a marine SAR analysis and interpretation system. Proc. IGARSS'02 3:1582–1584

Chapter 8
Perspectives on Oil Spill Detection Using Synthetic Aperture Radar

Michele Vespe, Monica Posada, Guido Ferraro, and Harm Greidanus

8.1 Introduction

Deliberate oil discharging is a well known source of marine pollution, calling for automatic and reliable detection to allow timely intervention. On account of this large coverage, all-weather and day and night capabilities, SAR technology has long been recognised as one of the most suitable instruments for oil spill monitoring especially over wide areas.

Following the European Directive 2005/35/EC, which entered into force in September 2005, the European Maritime Safety Agency (EMSA) is required to provide assistance to the European Union in tracing discharges by satellite monitoring and surveillance. The CleanSeaNet service has been developed from research activities on oil spill detection, becoming an operational asset within the European Union Coastal Member States, Norway, and probably soon also Iceland. This service is currently implemented delivering Near Real Time (NRT) oil spill alerts and positions. As a consequence, oil spill detection from SAR images is required to deliver reliable results that can be efficiently used by the Coastal States as decision support tools to efficiently coordinate the operations for combating and monitoring oil pollution.[1]

In this paper we present a set of advances in SAR satellite based oil spill detection that relate to the integration of additional information into the classification process. The work reflects the JRC experience in the field (see Muellenhoff et al., 2007; Ferraro et al., 2007; Vespe et al., 2009; Ferraro et al., 2009). The primary End User's need in this field is the assurance of truthfulness of the delivered information, which translates into the ability to associate a degree of confidence of the detection to each oil spill candidate as presented in this work. This is crucial in order to evaluate the status of alert to be considered by the competent authorities for

M. Vespe (✉)
Institute for the Protection and Security of the Citizen, Joint Research Centre, European Commission, Ispra 21027, Italy
e-mail: michele.vespe@jrc.ec.europa.eu

[1] See CLEANSEANET website http://cleanseanet.emsa.europa.eu (accessed 01.12.2009).

V. Barale et al. (eds.), *Oceanography from Space*,
DOI 10.1007/978-90-481-8681-5_8, © Springer Science+Business Media B.V. 2010

legal infringements detection and potential environmental damage estimation. After a brief description of common methodologies applied to SAR based oil spill detection, the oil spill detection problem is introduced through the description of data quality analysis. This process aims at defining the suitability of the image data for the service application. The portions of the data considered "usable", i.e. exploitable for the derivation of meaningful results, can be further analysed. As described in the following, the detected oil spill candidates can be cross checked with other ancillary data (metocean, contextual and maritime traffic data). This shall subsequently increase the reliability of the service by combining information on oil spill candidate location with the relevant degree of "risk" and "detection capability" properties of the area of interest. For instance a traffic lane shall increase the likelihood ("risk") of having and oil spill, as opposed to low wind areas that reduce the performance of the SAR based algorithm ("detection capability"). Ultimately, this paper introduces a possible way of automatically fusing the mentioned heterogeneous information.

8.2 On the Use of SAR Imagery to Detect Oil Spills

Some oil on the sea surface dampens the short gravity-capillary waves generated by the wind (Alpers and Hühnerfuss, 1988), leading to reduced Bragg scattering in radar images. This results in a reduced radar backscatter to the sensor, thus creating a darker signature in the image over the area of interest. This can be observed if the local wind is greater than a threshold (typically 2–3 m/s) which, amongst other factors, is dependent on the water salinity and temperature. When the wind speed is too high, on the other hand, the short waves receive enough energy to counterbalance the damping effect of the oil film, and if the sea-state is fully developed, the turbulence of the upper sea layer may break and/or sink the spill or a part of it. As a result the oil spill is not detectable from the image.

The reliability of spill detection based on SAR data only is not fully robust to guarantee consistent and automatic satellite-based detection of oil spills. This is a consequence to the distinctive variability of radar based oil signatures, their spatial features, and the interaction with the local environment. Moreover, other than oil spills, a number of phenomena originate Bragg scattering reduction, making it difficult to discriminate between the so called "look-alikes". Such phenomena can be grouped as follows:

- Man-related (e.g. ship wakes, floating production facilities drain emissions);
- Atmospheric (e.g. wind sheltering, rain cells, atmospheric instability areas);
- Oceanographic (e.g. internal waves, coastal upwelling, eddies, current shears, bathymetry/currents interaction, grease ice);
- Natural/Biological (e.g. natural seepage, fish oil in cold waters, algae blooms, pollen from plants and trees, coral spawn, natural surfactants).

Many of these false alarm sources become even more pronounced in low wind conditions, as a consequence of the reduced Bragg waves and therefore backscatter of the sea surface. Several research investigations on techniques and algorithms to detect oil spills from SAR data have been conducted in the last years (e.g. Solberg et al., 1999; Del Frate et al., 2000; Fiscella et al., 2000; Pavlakis et al., 2001; Karathanassi et al., 2006), as also recently reviewed by Topouzelis (2008). The typical functions implemented by SAR based oil spill detection are based on magnitude detected products and can be summarised as follows.

1. *Image pre-processing*: the image is pre-processed through well known algorithms especially tailored on the particular application. For oil spill detection, the radiometric resolution of the image is a key parameter and often is advantaged against spatial resolution using 2-D edge preserving filters (e.g. Lee, Local Region, Frost etc). Such filters increase the Equivalent Number of Looks (ENL) with the effect of reducing the speckle noise and therefore enhancing the contrast of dark patches over the image. The image is commonly land-masked in order to reduce the number of processed pixels, thus decreasing the computational burden of the algorithm.
2. *Image segmentation*: the pre-processed image is then segmented in order to extract the dark patches of interest. The segmentation process can be thought of as a Constant False Alarm Rate (CFAR) algorithm that, on the basis of the local sea backscattering statistics, applies a NRCS (Normalised Radar Cross Section) threshold below which the pixels are indicated. The pixels are then organised in adjacent sets commonly referred to as image objects.
3. *Feature extraction*: the image objects are analysed in detail, leading to the estimation of a set of features previously selected. Such features are related to shape, internal texture and neighbourhood contrast. The *feature selection* process is a key process for any classification algorithm. As regards oil spill classification, the problem is often reduced to the discrimination between "oil spill" and "look-alike". Separating these two classes in feature space requires an accurate optimisation.
4. *Classification*: for each object, the values related to the selected features are compared with a database of templates in order to evaluate the class to which the object belongs. A preliminary screen on size of the objects is commonly performed in order to exclude residual speckle objects that are of no interest to the analysis. The classification function has been approached using a variety of algorithms spanning from statistical, to associative mapping techniques. An overview, as well as a review of the oil spill classification literature, can be found in Brekke and Solberg (2008), and Topouzelis (2008).

The use of satellite based SAR oil spill detection has been considered taking into account operational sensors, e.g. Radarsat-1 and 2, ERS-2, Envisat, Alos, Terrasar-X and CosmoSkyMed. The primary products used are ScanSAR magnitude detected images. However, the advances made by polarimetric oil spill detection foresee the developing of relevant algorithms for operational use. Among them, fully polarimetric data have been demonstrated to be useful in discriminating look-alikes from oil spills using circular polarization coherence (CPC) and polarimetric anisotropy

(Fortuny-Guash, 2003) based on Cloude and Pottier developed polarimetric decomposition (Cloude and Pottier, 1997). Nunziata et al (2008) propose the use of the Co-polarised Phase Difference (CPD) between HH and VV returns to discriminate biogenic from mineral oil because of the relevant different damping properties.

ScanSAR modes have been predominantly selected for oil spill detection because of the large covered swath and less stringent spatial resolution requirements. Nevertheless, there are issues related to residual scalloping and elevation beam boundary that often reduce the data "usability" for this application. Novel products shall be operationally available for this purpose. In particular, the TopSAR radar mode (Zan et al., 2006) has been recently implemented by TerraSAR-X by azimuth beam steering within the different sub-swaths. Differently from ScanSAR, TopSAR allows a wide swath coverage with nearly uniform signal to noise ratio, therefore reducing scalloping and range dependent radiometric errors (Meta et al., 2007).

8.3 From Research to Operational Services

In this section, a number of aspects related to operational oil spill detection are presented. The main parameters to be taken into account, when designing operational SAR based oil spill detection, are the following:

- Minimum detectable dark feature contrast under variable wind speed conditions (which primarily translates to radiometric resolution but is influenced by noise level, incidence angle, polarisation, ENL);
- Instantaneous swath (large swath is usually preferred for global routine monitoring);
- Ability to describe geometric properties of the detected dark patch (shape structure and area features);
- Ability to estimate the age of the spill (as derived from the shape);
- Revisit frequency (no repeat-pass interferometry over the sea, so not necessarily with the same beam);
- Tasking lead time (i.e., how fast can an acquisition be planned);
- Delay in delivering the results (detected spills positions at the end user after overpass);
- Ability to acquire imagery at any time of the day (and not just e.g. at dawn or dusk passes, because some activities at sea do not occur at those hours);
- Availability (related to scheduling conflicts) and accessibility (how easy is for the End User to task satellites);
- Reliability (primarily in the sense of reliability of detection, i.e. detection probability and false alarm rate).

Most of the above mentioned aspects are strictly related to the specific SAR instrument performance and cannot be varied from an application point of view. Nevertheless, the "reliability" of the oil spill detection service can be improved through (*i*) the accurate analysis of the delivered image, as described in the section on quality analysis, and (*ii*) the investigation of ancillary data describing the area of interest, as discussed in the section on the use of ancillary data.

8.3.1 Legal Framework

Scientific studies focusing on remote sensors capacities to detect oil spills generally do not take into consideration the legal framework. Marine oil pollution by vessels, termed "operational oil pollution", includes various types of discharges of oil and oily mixtures, as a result of ships' daily routine operations. Some of these, such as oily ballast water and tank washing residues, relate to tankers only. However, all types of ships may cause pollution discharging into the sea oil coming from engine room wastes, bilge waters and, only in rare cases, used oil.

The International Convention for the Prevention of Pollution from Ships, 1973, as amended by the Protocol of 1978 thereto (MARPOL 73/78) is aimed at minimizing and eliminating pollution from ships. The Convention covers two main subjects:

(i) the special construction and equipment rules for the prevention of accidental pollution;
(ii) the circumstances in which discharges into the sea are authorized.

The general provision in art.6 of MARPOL 73/78 contains the obligation to co-operate in the detection of violations and the enforcement of the provisions of the Convention, using all appropriate and practicable measures of detection and environmental monitoring, adequate procedures for reporting and accumulation of evidence. Each Contracting Party to MARPOL 73/78 is obliged to incorporate the regulations in its national legislation, including provisions for prosecution of any discharge above legal limits.

The oil discharge regulations in the Convention apply differently depending on whether or not the sea area has been designated a "special area". Almost all the seas around Europe have been designated Special Areas for oil pollution. Only the Norwegian Sea, the Bay of Biscay and the Atlantic Iberian Coast are not covered by the Special Area status.

The limit of 15 parts per million (ppm) is the key parameter to identify legal discharges of ballast and from machinery spaces, in special areas, and legal discharges from machinery spaces, outside special areas. Requirements to legally discharge from cargo tanks outside special areas are less strict. In this last case, oil discharges are allowed above the 15 ppm limit.

Following a detailed study, the International Maritime Organization (IMO) recognized that it is not possible to see oily mixtures at sea with oil content below 15 ppm (Resolution MEPC.61(34) of 9 July 1993 on Visibility Limits of Oil Discharges). Therefore, not all visible (by eye or by Remote Sensing) oil spills are necessarily illegal. But, visible and/or detectable oil discharges from ships, observed in a MARPOL special area, are illegal without any doubt.

Following the detection of a spill, the major issue is the prosecution of the polluter, but two main principles co-exist: the Nationality of the ship and the Geographical position of the ship. The principle of the Nationality of the ship is also defined as principle of jurisdiction of the Flag State. On the other hand, the possibility to apply the jurisdiction relevant for the position of the ship could concern: the jurisdiction of the State where the ship is sailing (principle of Coastal State) or the jurisdiction of the port where the ship is in (principle of Port State).

There are two fundamental instruments available to the international community for taking action against the perpetrators of illegal acts of marine pollution:

- MARPOL 73/78 Convention;
- 1982 United Nations Convention on the Law of the Sea (UNCLOS).

MARPOL 73/78 does not deviate from the exclusive jurisdiction of the Flag State. On the other side, UNCLOS enlarges the jurisdiction of the Coastal States in the Exclusive Economic Zone (EEZ), in addition to the sovereign rights of the territorial sea. UNCLOS provisions break through the traditional supremacy of the jurisdiction of the Flag State in respect of discharge violations in areas outside the jurisdiction of a Coastal State: when a ship is voluntarily within a port, the Port State is granted extensive judicial powers in respect of a discharge violation committed by that ship outside its internal waters, territorial sea or EEZ. The provisions of UNCLOS concerning this universal Port State jurisdiction represent a key factor to the successful prosecution of the polluters.

Also, the European Union has taken specific actions to strongly combat ship-source pollution through a system of sanctions in case of intentional acts or serious negligence. Directive 2005/35/EC of 7 September 2005 on ship-source pollution and the introduction of penalties for infringements, recently amended by Directive 2009/123/EC of 21 October 2009, establishes that sanctions will be applicable to any person – including the master, the owner, the operator, the charterer of a ship or the classification society – who has been found to have caused or contributed to illegal pollution intentionally or by means of serious negligence. This Directive aims also to enhance cooperation among Member States to detect illegal discharges and to develop methods to identify a particular discharge as originating from a particular ship.

8.3.2 Image Quality

In Europe, oil spill detection from satellite SAR has become fully operational with the CleanSeaNet service. The image analysis for slicks is carried out mostly by visual inspection. For ship detection, on the other hand, automatic detection software is used, still with a final visual verification. In recent years, Europe has seen a widespread build-up of close-to-operational ship detection capabilities. This increased (near-) operational use of SAR images for ship and oil spill detection is confronting service providers with several SAR image quality issues that impact on the final product quality. A number of these issues yield an increase of the false positives, others introduce "non-exploitable" areas within the image leading to potential false negatives in the detection process.

Some SAR image particularities, well known to SAR engineers but perhaps less so to the downstream service providers, can cause serious problems with the service product, such as azimuth and range ambiguities giving rise to false ship detections. In such cases, the SAR image may be in accordance with specifications, but the impact of some image artifacts on the final product is still problematic, e.g. when interference, or nadir ambiguities (Fig. 8.1) are present. In other cases, operationally delivered images are not in accordance with specifications, e.g. when processing errors occur due to wrong Doppler parameters estimation or when the radiometric instrument performance is degraded (e.g. scalloping effect, Fig. 8.2). This will obviously also lead to wrong outcomes, since the degree of reliable exploitation of SAR data is commensurate to the instrument performance compliance with the expected requirements. The distinction between *specification compliance* and *application suitability* can be defined when comparing the expected image quality with the specific sensor performance or the application requirements respectively. However, here below a brief list of parameters relevant to maritime applications is presented.

The detection of oil spills often prefers coverage extension over spatial resolution performance. This is mainly because of the large size of operational or accidental oil spills of interest. As a consequence, the fulfillment of radiometric indicators has a higher priority than geometric indicators. Nonetheless, geolocation accuracy has to

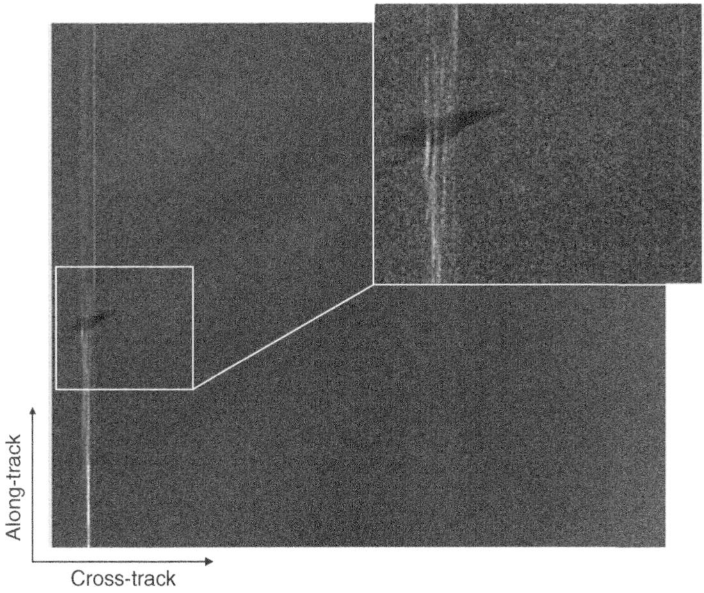

Fig. 8.1 Nadir return partially covering an oil spill in a Radarsat-1 Wide Mode image © MDA/CSA 2008

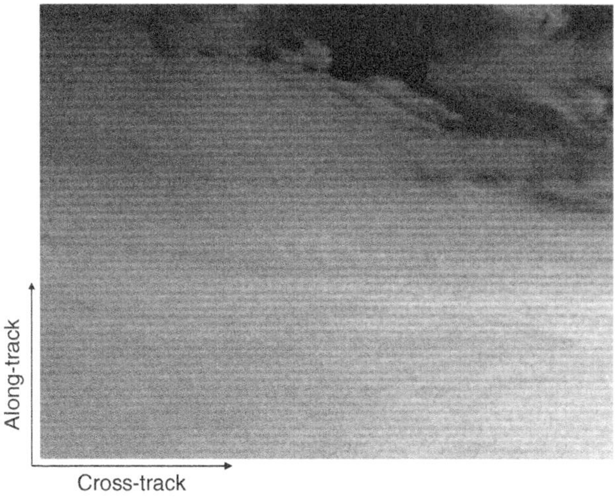

Fig. 8.2 Pronounced scalloping from Envisat ASAR Wide Swath Mode © ESA 2008

be considered as crucial for operational oil spill detection services. Also the spatial resolution performance reduction might lower the capability to describe the shape of detected dark patches and small features useful to correctly classify oil spills from look-alikes.

Radiometric resolution and radiometric error deviations from the specifications may lead to missed oil spill detections. In particular, radiometric resolution degradation leads to reduced contrast between Bragg scattering and potential dark features. Moreover, the hydrophobic nature of mineral oil leads to specific backscattering textures undetectable with reduced radiometric resolution. Finally, reduced radiometric resolution performance decreases the potential age indications of the oil spill.

Image radiometric error degradation might bias the backscattering statistics of the area of interest, yielding either missed detections or false positives. The sensitivity degradation is also a general key issue that affects this application: a reduced sensitivity performance progressively decreases the backscattering transitions between Bragg scattering and potential oil spills with the result of reducing the detection capabilities.

Data integrity issues, other than affecting the backscatter statistics and therefore the detection process, may mask actual spills. For this reason they have to be taken into account in the image quality estimation process by identifying non exploitable areas.

All these parameters have to be properly checked before operating oil spill detection, as a component of the application quality assessment.

8.3.3 From Confidence Level to an Alert System

In order to make the SAR image analysis objective and to provide decision support to the user, the detected oil spills are usually classified according to confidence levels. Such levels are supposed to describe the probability that an observed dark feature in the satellite image is related to the presence of an oil spill. The SAR derived oil spill detection probability estimation has been mathematically explored as an intrinsic aspect of oil spill classification, which fundamentally computes the likelihood that the detected dark area and its extracted features are related to oil spill. In particular, Fiscella et al. (2000) introduced a nearest neighbours classifier based on the Mahalanobis metric applied to the feature vector, and compared its results to a statistical classifier. Nirchio et al. (2005) investigated a Fischer discriminant analysis approach applied to the selected features. Nevertheless, the SAR based probability estimation should be integrated with additional criteria in order to become a more effective tool for the End User. It is worth mentioning also that some of these criteria and their correspondent weight within the decision making process, are very National and Regional dependent.

New trends in confidence level estimation involve different sources of data, linking the SAR derived information to ship traffic routes, metocean and other context-specific information. For this reason we are observing a progressive transition from a three-level SAR based Confidence output to a more articulated Alert System that includes additional criteria. This is summarised by Ferraro et al. (2009). The application output shall consist of a product that delivers additional information not only in terms of reliability but also concerning the related alarm extent of the detection (i.e. potential impact and polluter detection capability).

8.3.4 Ancillary Data

In order to increase the reliability of oil spill classification, environmental data have been used as additional help to the human operator (Tahvonen and Pyhalahti, 2006; Muellenhoff et al., 2007).

The ancillary data can be grouped into two main conceptual datasets, *Risk of Pollution* and *Detection Capability*. The former contains information about areas at risk due to its proximity to potential pollution sources. These data contain information about distance from main traffic lanes, world ports distribution, pipeline runs, wreckage risk, oil rig locations. *Detection Capability Dataset* holds information about specific ocean and atmospheric phenomena that decrease the detection capability of oil spills based on SAR imagery. This dataset includes among others: low wind conditions, Sea Surface Temperature (SST) derived cold water fronts, air-sea interactions, current shears, biogenic oily films, etc.

The generated maps of *Risk of Pollution* and *Detection Capability* degree could be used for cross checking with the automated detected outcomes, in order to effectively assign the correspondent level of reliability to oil spill candidates. The data described here are not related to the specific time of SAR image acquisition. On the contrary, such maps aim at describing monthly and seasonal trends in the areas of interest.

These maps could be considered a potential decision making support and validation tool that could be easily adapted to stakeholders requirements.

8.3.4.1 Contextual Data – Risk of Pollution

The contextual data are exploited to identify and properly map areas of influence of both ports and main ship corridors. Specifically, data sources used for the traffic density estimations are (i) ICOADS data for long term observational records, and (ii) Automated information System (AIS) for almost NRT and more accurate assessments. Whereas for port influence areas, the World Port Index data are injected, weighted on the basis of the availability of oil terminals. Once processed and identified the main areas at risk, both data sources are integrated and mapped into a GIS layer to present the risk information integrated output.

8.3.4.2 Metocean Data – Detection Capability

The information considered here is related to metocean data that directly or indirectly could be exploited to identify phenomena affecting SAR imagery, in particular the oil spill detection process. Specifically, the data sources used by the authors are described below:

- *Wind Data*: can be thought of as the main driving force generating Bragg waves, therefore directly affecting the detection capabilities. The data are retrieved from SeaWinds on QuikSCAT monthly averaged products, containing information about wind speed and direction 10 m above the sea surface. The data are processed and parameterised then in two variables (1) wind speed range for mapping areas according to their suitability for oil spill detection, and (2) wind fronts for mapping wind masses interaction associated with potential decrease of wind velocity and thus yielding false positives.
- *Sea Surface Temperature (SST)*: variations on the SST may affect the stability of the atmosphere above. This effect may have an impact on the wind stress and indirectly modify the sea surface modulation. False alarms are likely to happen in the oil spill detection process under this specific circumstance. Water temperature strong gradients are therefore mapped over SST layer. The Data for the mapping of water fronts was retrieved from (1) AVHRR PathFinder V5 product providing temporal monthly averages as a result of the re-analysis of older AVHRR data, and (2) MODIS/AQUA: derived from the MODIS IR channels using two channels in either the thermal IR (11–12 μm) or channels in the mid-IR region (3.8–4.1 μm).
- *Chlorophyll-A*: such information is injected as an indicator of biogenic activity that, under certain favourable circumstances of quantity of light energy and inorganic nutrients, may produce the release of biogenic oil films, i.e. look-alikes. The relevant data are retrieved from the JRC Mersea (Mersea website) monthly averaged regional products. The detection capability degree is derived using Chlorophyll-A concentration threshold. This threshold is set on the basis of the extensive analysis of ENVISAT SAR imagery over the Mediterranean Basins (oligothrophic environment) with synchronous Chlorophyll-A measurements from MERIS sensor.

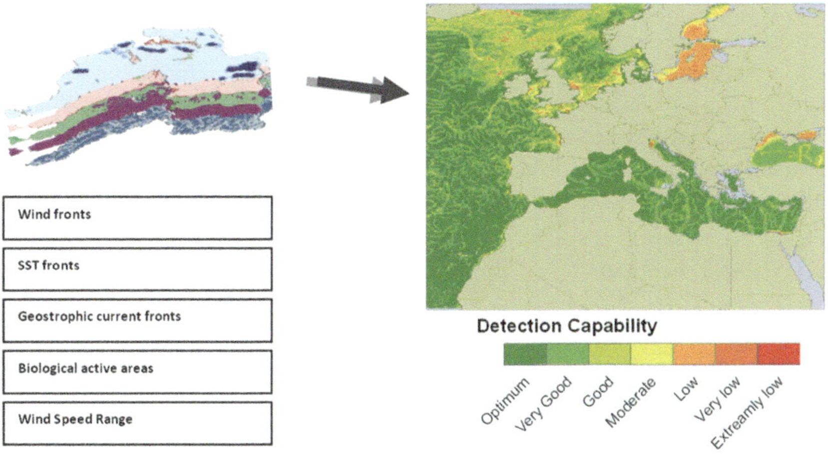

Fig. 8.3 Integrated *Detection Capability* map for the month of July

- *Geostrophic Currents:* sea currents alter short waves patterns mainly in areas where currents interact (e.g. gyres, current shears, eddies, etc.) by accumulating surfactants on certain areas of the converging/diverging waters. The approach intends to identify sea currents fronts where their interaction may yield the generation of dark/bright patches. Geostrophic velocity anomalies derived from AVISO altimeter information are used, specifically Absolute Dynamic Topography products both for Mediterranean region and Global extent.

Once all relevant ancillary sources are processed and potential affecting phenomena isolated, the outcome is integrated into a single map. Given the ocean-atmosphere interface interactions complexity and the interrelation between measurable variables, a superposition effect was approached for the detection capability derivation. The more affecting undesired phenomena occur over an area at the same time, the lower the detection capability. Figure 8.3 depicts the results of the methodology used when mapping the detection capability degrees over EU Areas.

8.3.4.3 Joint Ship-Oil Spill Detection

Ship detection is considered a key asset for oil spill detection. This is because it can allow the derivation of the potential polluter position within the illuminated area if the acquisition is close in time with the actual discharge. Satellites SAR images are in fact unable to identify the pollution culprit (i.e. the name of the ship that polluted). At best, satellite can detect the position of the probable pollution culprit. Nonetheless, by using back-propagation techniques, the SAR detected ship can be also identified by correlating other maritime surveillance sensors, e.g. Automatic Identification System (AIS), Long-Range Identification and Tracking (LRIT), Vessel Monitoring System (VMS), Satellite-AIS and coastal radar tracks. Such techniques are based on oil spill dispersion and drift models (Ferraro et al.,

2007). The degree of confidence of such identification is proportional to the time lag between the relevant information and the sensors intrinsic uncertainties. Ship detection is sometimes used (Solberg et al., 1999) as additional information to increase the likelihood of the detection being an oil spill, although this does not increase the probability of true positive. Conversely, connected ship and oil spill events shall raise the alert level of the detection since it can be thought of as a relevant factor for follow-up activities concerning the prosecution of potential polluters.

8.3.5 Data Fusion

The oil spill features extracted from SAR images can be combined with the ancillary data described in the previous section using data fusion algorithms. This process is expected to augment the "reliability" of the detection simply by adding into the automatic application the information related to the likelihood of the specific context to present oil spills and look-alikes. In this section, a possible way to approach data fusion of SAR and ancillary data for oil spill detection is presented focussing on the detection of operational oil spills, i.e. oily bilge water, deliberate tank washing residues, and unsegregated ballast water discharge.

The ancillary dataset can be organised into homogeneous tiles presenting a certain degree of spatial uniformity in terms of information content. Artificial Neural Networks can then be used to ultimately link the SAR data to the ancillary data, ultimately leading to the estimation of a reliability index of the oil spill classification. This is based on the vector X whose elements X_i are in the range [0,1] and represent the different levels of information related to the area of interest. Specifically, X_1 is the oil spill classification level derived from SAR image processing only. The particular classification algorithm implemented follows a fuzzy logic approach, where the set of fuzzy rules correspond to each detected feature. The next element of the information vector, X_2, is based on the homogeneity of the backscattering surrounding the dark area, measured by the standard deviation of the NRCS data. Reduced backscatter homogeneity would increase the likelihood of observing look-alikes. However, high data homogeneity, in conjunction with low wind conditions, still decreases the detection capabilities. This shows the non-linear relationship between the elements of the vector X in the evaluation of the final output, motivating the selection of associative mapping methodology for the data fusion algorithm.

The ancillary data information content (X_3, X_4 and X_5) is then mapped using empirical functions. For instance, the wind intensity information is retrieved from QuikSCAT data and processed according to an empirical function taking into account that:

- A threshold at 2 m/s is set since it is the minimum wind speed to generate gravity-capillary waves,
- 15 m/s can be thought of as the wind speed upper bound for oil spill detectability,
- At moderate to high wind speed conditions, the probability of observing look-alike decreases (Pavlakis et al., 2001).

In similar ways, it is possible to derive the oil spill high risk probability areas (X_4) and reduced oil spill detectability areas (X_5). Specifically, X_4 is generated from contextual (distance from traffic lanes and world ports distribution) and from the fusion of different levels of metocean information (SST gradient, wind turbulences, Chlorophyll-a concentration). Low values of X_5 represent a high probability of the detection being a look-alike, whereas high X_4 values identify areas with high probability of having an oil spill.

The SAR detected dark patches labelled by X_1 and the relevant surrounding homogeneity X_2 are then fused with the set of described ancillary data X_3, X_4 and X_5, forming the heterogeneous information vector X that represents the input for the data fusion engine as illustrated in Fig. 8.4. The probability $P(H|X)$ estimation of the events H_1 "oil-spill" or H_2 "look-alike" given the input vector X is performed using supervised Feed-forward Artificial Neural Networks (FANNs) approach, whose output can be interpreted as the posterior probability $P(Event|input)$. The training phase of the FANN is achieved using a set of positive association examples obtained from validated spills.

The performance of the algorithm is illustrated for the Radarsat-1 image in Fig. 8.5. The image has been previously pre-processed (i.e. de-speckled, radiometrically normalised in order to equalise the incidence angle sea backscatter influence, and land masked).

The oil spill detection based on SAR data only exhibits a number of look-alikes related to low wind and atmospheric instability. Such phenomena can be in this case automatically resolved with the aid of ancillary data and the data fusion approach described. The classification reliability $P(H_1|X) = 0.96$ for oil spills is increased with respect to the SAR based performance only $X_1 = 0.94$, while the false positive classification output level $[1 - P(H_2|X)] = P(H_1|X) = 0.62$ for look-alikes is decreased in relation to $X_1 = 0.62$.

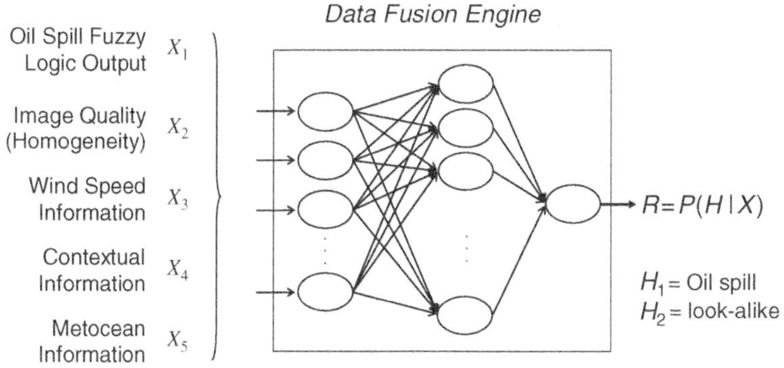

Fig. 8.4 Associative mapping, the network learns to associate a new input layer vector X to a reliability R through positive and negative examples

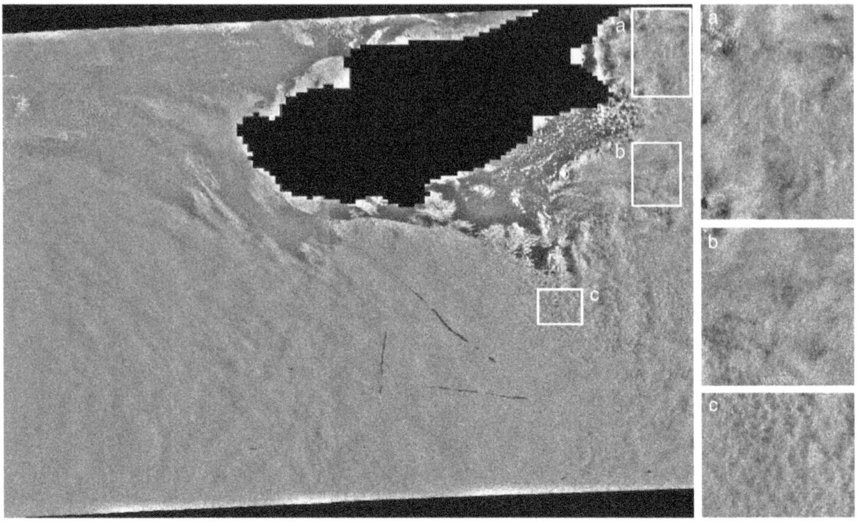

Fig. 8.5 Oil spill detections (*red*) having $P(H_1|X) > 0.85$, and false alarms (*pink*) discriminated as output of the Feed Forward Artificial Neural Network (Radarsat-1 image © CSA/MDA/EMSA 2007)

8.4 Conclusion

The evolution of SAR based oil spill detection towards integrated information systems has been presented. Starting from the data quality issue, and based on a consolidated legal framework, it is possible to increase the "reliability" of currently operational services with the aid of ancillary data. The composite output, which takes into account *risk of pollution* and *detection capability* maps, can eventually lead to the generation of higher level alarms. An example of SAR image analysis using ancillary information has also been illustrated showing how the system separation capabilities can increase in the feature space, improving the performance over the classic SAR based detection. This shall provide in the future more effective decision making tools for End Users supporting the formulation of follow-up strategies after detections.

One of the major limitations of SAR based oil spill detection is the poor quantity and typology estimation of the detection. This is a consequence of the impossibility of determining the film thickness and class from the SAR image in an accurate way. The amount and type of spilled oil is essential information in order to accurately quantify the nature of the pollution. Further investigation on additional technologies should be conducted in order to improve operational services in the future.

Acknowledgements This work has been partly funded within the scope of the European Maritime Safety Agency – EC Joint Research Centre collaboration for the development and support of satellite monitoring techniques for oil spill detection.

References

Alpers W, Hühnerfuss H (1988) Radar signatures of oil films floating on the sea and the Marangoni effect. J Geophys Res 93:3642–3648

Brekke C, Solberg A (2008) Classifiers and confidence estimation for oil spill detection in ENVISAT ASAR images. Geosci Rem Sens Lett IEEE 5(1):65–69

Cloude SR, Pottier E (1997) An entropy based classification scheme for land applications of polarimetric SAR. IEEE Trans Geosci Rem Sens 35(1):68–78

Del Frate F, Petrocchi A, Lichtenegger J, Calabresi G (2000) Neural networks for oil spill detection using ERS-SAR data. IEEE Trans Geosci Rem Sens 38:2282–2287

Ferraro G, Baschek B, de Montpellier G, Njoten O, Perkovic M, Vespe M (2009) On the SAR derived alert in the detection of oil spills according to the analysis of the EGEMP. Mar Pollut Bull 60(1):91–102

Ferraro G, Bernardini A, David M, Meyer-Roux S, Muellenhoff O, Perkovic M, Tarchi D, Topouzelis K (2007) Towards an operational use of space imagery for oil pollution monitoring in the Mediterranean Basin: a demonstration in the Adriatic Sea. Mar Pollut Bull 54:403–422

Fiscella B, Giancaspro A, Nirchio F, Pavese P, Trivero P (2000) Oil spill monitoring in the Mediterranean sea using ERS SAR data ERS-Envisat Symposium, Göteborg, 16–20 October

Fortuny-Guasch J (2003) Improved oil slick detection and classification with polarimetric SAR, Proceedings of the POLinSAR 2003, ESA-ESRIN, Frascati, Italy, 14–16 January

Karathanassi V, Topouzelis K, Pavlakis P, Rokos D (2006) An object-oriented methodology to detect oil spills. Int J Rem Sens 27(23):5235–5251

Mersea Ocean Colour Portal, http://mersea.jrc.ec.europa.eu/ (accessed 01.12.2009)

Meta A, Mittermayer J, Steinbrecher U, Prats P (2007) Investigations on the TOPSAR acquisition mode with TerraSAR-X. IEEE IGARSS'07, Barcelona (E)

Muellenhoff O, Bulgarelli B, Ferraro G, Topouzelis K (2007) The use of ancillary metocean data for the oil spill probability assessment in SAR images. 14th Symposium on Environmental Pollution and Its Impact on Life in the Mediterranean Region, Seville (E)

Nirchio F, Sorgente M, Giancaspro A, Biamino W, Parisato E, Ravera R, Trivero P (2005) Automatic detection of oil spills from SAR images. Int J Rem Sens 26(6):1157–1174

Nunziata F, Gambardella A, Migliaccio M (2008) On the use of dual-polarized sar data for oil spill observation, IEEE IGARSS'08, Boston, MA

Pavlakis P, Tarchi D, Sieber A, Ferraro G, Vincent G (2001) On the monitoring of illicit discharges – a reconnaissance study in the Mediterranean sea. European Commission, EUR 19906 EN. Available on: http://serac.jrc.it/midiv/pub/jrc_illicit_study.pdf

Solberg A, Storvik G, Solberg R, Volden E (1999) Automatic detection of oil spills in ERS SAR images. IEEE Trans Geosci Rem Sens 37:1916–1924

Tahvonen K, Pyhalahti T (2006) The use of environmental data in reliability assessment of oil spill detection by SAR imagery. IEEE IGARSS'06 3688–3691

Topouzelis KN (2008) Oil spill detection by SAR images: dark formation detection, feature extraction and classification algorithms. Sensors 8(10):6642–6659

Vespe M, Posada M, Ferraro G, Bulgarelli B, Greidanus H, Djavidnia S (2009) Oil spill detection based on SAR and Metocean/Contextual Data Fusion. 33rd International Symposium on Remote Sensing of Environment, Stresa (I), 4–8 May

Zan FD, Monti Guarnieri A (2006) Topsar: terrain observation by progressive scan. IEEE Trans Geosci Rem Sens 44(9):2352–2360

Chapter 9
Determining Ocean Circulation and Sea Level from Satellite Altimetry: Progress and Challenges

Lee-Lueng Fu

9.1 Introduction

The idea of flying a radar altimeter in space to measure the height of sea surface for a variety of geophysical studies was quickly developed after the first launch of artificial satellite. Seasat, launched in 1978, carried the first radar altimeter with a precision capable of revealing the variability of ocean currents. After the premature demise of the mission after only 3 months' data collection, the oceanographic community realized the potential of satellite altimetry for making global observation of the ocean. The military community also realized the utility of the measurement for naval operations and hence launched Geosat in 1985. The instrument itself was a copy of Seasat, but without the correction for the effects of tropospheric water vapor due to the lack of an onboard radiometer. The oceanographic community was more ambitious in developing a mission specifically designed for studying the ocean circulation and its variability, leading to the Franco-American collaboration in the TOPEX/Poseidon Mission (T/P), launched in 1992.

While T/P was in development, the data from Geosat, although not as accurate as oceanographers had desired, provided nonetheless the first multi-year altimetry data for the study of ocean dynamics as well as a test bed for processing altimetry data. As part of the payload of the European Remote Sensing satellite (ERS-1), a radar altimeter similar to Seasat was launched by the European Space Agency (ESA) in 1991. The launch of T/P then started an era in which at least two satellite altimeters were flying simultaneously. The era has continued as of the writing of the article today. Results from these missions cover a wide range of earth sciences: oceanography, geophysics, geodesy, and hydrology. Fu and Cazenave (2001) provided a comprehensive review. There has been tremendous progress made since that review. The scope of this paper, however, is focused on recent progress in the study of ocean circulation and sea level change, including some recap of earlier results to provide a historic perspective. At the end, the limitation of the present approach and the challenges for future development are addressed.

L.-L. Fu (✉)
Jet Propulsion Laboratory, California Institute of Technology, Pasadena, CA 91109-8099, USA
e-mail: lee-lueng.fu@jpl.nasa.gov

V. Barale et al. (eds.), *Oceanography from Space*,
DOI 10.1007/978-90-481-8681-5_9,

9.2 Ocean General Circulation

A major driving goal for developing ocean altimetry mission was the vision of determining the ocean general circulation from space (Wunsch and Gaposchkin, 1980). In parallel to the development of T/P, a satellite gravity mission was also conceived in the early 1980s for mapping the geoid, which in combination with altimetry measurement would yield the ocean circulation. The gravity mission did not materialize until the launch of the GRACE Mission in 2002. The accuracy of the GRACE geoid is estimated to be 2–3 mm at scales as small as 400 km (Tapley et al., 2004). This represents more than an order of magnitude improvement over the results based on previous geoid models, which have errors on the order of 10 cm at scales larger than 1,000 km (Fu and Chelton, 2001).

Most of the swift boundary currents in the ocean have cross-stream scales less than 100 km and thus are not fully resolved by using the GRACE geoid. As part of the World Ocean Circulation Experiment conducted in the 1990s, thousands of surface drifters were deployed in the ocean to measure surface current velocity. Such drifters suffer from sampling errors for computing the time mean velocity. This problem has been mitigated by combining the densely populated altimetry data with the drifter data to correct for the sampling bias in the latter. Using this technique, the surface mean circulation was estimated based on decade-long observations (Niiler et al., 2003; Rio and Hernandez, 2004). Many more details of the mean circulation have been revealed down to scales of 50–100 km. The spatial resolution is highly heterogeneous due to the drifters' uneven distribution. The most interesting finding is the prominent zonal striations in the surface mean circulation (Maximenko et al., 2005, 2008; see Fig. 9.1). The appearance of the striations was enhanced by applying a two-dimensional high-pass filter to the mean ocean dynamic topography.

Simulations by high-resolution ocean general circulation models (Richards et al., 2006) have shown similar features but the details (exact locations, orientations,

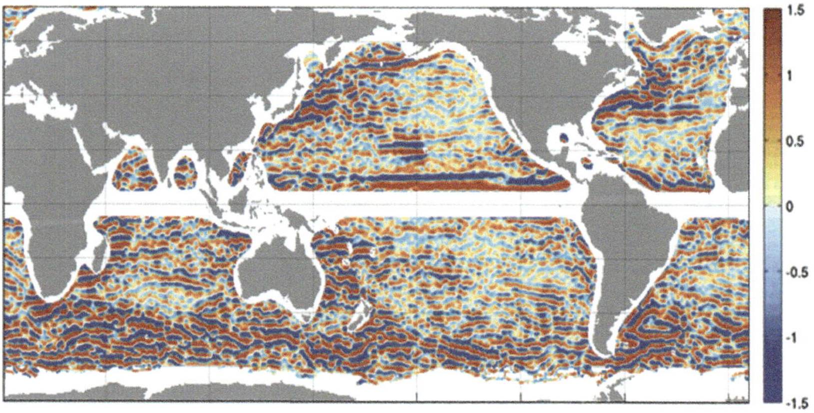

Fig. 9.1 1993–2002 mean zonal surface geostrophic velocity [cm/s] from drifter/altimetry based mean dynamic topography high-pass filtered with a two-dimensional Hanning filter of 4° half-width (from Maximenko et al., 2008)

strengths, etc.) exhibit discrepancies. The dynamic mechanisms for the striations are not yet well understood. Using a kinematic model of randomly distributed field of eddies, Schlax and Chelton (2008) showed that such striations could be caused by the migration of eddies that were not completely averaged out over a finite time period. The residual speed of the eddy currents after 10-year averaging is on the order of 1 cm/s, comparable to the results of Maximenko et al. (2008). As the residual currents go down with 1/T, where T is the averaging time, a data set of multiple decades is needed to average out the eddy effects.

A testimony of the quality of the surface mean dynamic topography of Niiler et al. (2003) was the study of the vorticity balance of the Antarctic Circumpolar Current (ACC) by Hughes (2005). Estimating the vorticity of the flow by differentiating the dynamic topography, Hughes (2005) discovered two modes of flow behaviors: Meanders in which a balance was achieved between the advection of relative and planetary vorticity as in a stationary equivalent-barotropic Rossby wave, and a flow in which the advection of total vorticity was related to bottom topographic steering.

9.3 Large Scale Low Frequency Variability

A series of El Niño Southern Oscillation events in the 1990s including the phenomenal event of 1997–1998 provided a focus for demonstrating the power of satellite altimetry to study large-scale climate variability. Fu and Smith (1996) demonstrated an early comparison of altimetry observation with a model simulation of a Pacific warming event. The success of satellite altimetry in providing global ocean observations was a major motivation for the advancement in global ocean modeling and data assimilation in the 1990s (Stammer et al., 1996, 2002). This development has establish a new framework for performing ocean reanalysis using modern state estimation approach by integrating data from a observing network into ocean general circulation models (Wunsch et al., 2009).

As the altimetry data record extended into its second decade, oceanographers for the first time had a continuous global data set for studying ocean variability beyond the seasonal-to-interannual scales. Hakkinen and Rhines (2004) reported a slow-down of the subpolar gyre circulation of the North Atlantic Ocean from analysis of T/P data in combination with earlier altimeter data. They attributed this change to weakened thermohaline forcing. Based on satellite altimetry data in combination with a variety of in-situ observations, Roemmich et al. (2007) discovered a decadal intensification of the subtropical gyre of the South Pacific Ocean from 1993 to 2004. The gyre circulation increased by 20%, resulting from a decadal strengthening of wind forcing east of New Zealand as part of a circumpolar change of climatic state. On the other hand, Lee (2004) found that the upper ocean overturning circulation of the Indian Ocean decreased by 70% from 1992 to 2000, caused by the weakening of the trade winds. Subsequently Lee and McPhaden (2008) found a larger-scale linkage of the decadal variability of the Indian and Pacific Oceans.

In the Pacific Ocean, sea surface temperature, wind stress, and ocean circulation are involved in a decadal variability called the Pacific Decadal Oscillation (PDO). Qiu and Chen (2009) analyzed 16 years' worth of altimetry data and noted that the phase transition of PDO triggered westward baroclinic Rossby waves, which affected the stability of the Kuroshio Extension upon arrival in that region. Shifting between stable and unstable regimes, the eddy energy and its interaction with the Kuroshio Extension is linked to the larger-scale PDO.

Westward propagation of large scale variability is a ubiquitous phenomenon well documented by numerous papers since the seminal paper by Chelton and Schlax (1996). Fu and Chelton (2001) provided a review of the subject. The conclusion then was that the predominant westward propagation was associated with baroclinic Rossby waves (see the Chapter 12 by Cipollini et al., this volume). Fu (2004) discussed the latitudinal variation of the frequency content of the propagation and identified cases in which the wave frequency was higher than allowed by the conventional Rossby wave theory. Some of the cases were attributable to barotropic Rossby waves.

9.4 Large Scale High Frequency Variability

A big surprise when the T/P data were first analyzed was the presence of large-scale variability of the ocean at periods on the order of 10 days. These turned out to be barotropic response of the ocean to rapid changes in wind forcing. Chao and Fu (1995) showed that the observed variability could be simulated by ocean general circulation models. The results were further confirmed by Fu and Smith (1996). A modeling study by Fukumori et al. (1998) suggested that up to 50% of the variance of such large-scale variability could have periods shorter than 20 days, the Nyquist period of T/P. This raised concerns for aliasing these high-frequency signals to low frequencies in altimetry data. Model simulations forced by good quality wind were then used to de-alias the high-frequency signals in altimetry data (e.g., Stammer et al., 2000; Carrere and Lyard, 2003).

To a large extent the high-frequency variability is influenced by bottom topography. Fu et al. (2001) found a 25-day oscillation of the Argentine Basin over the Zapiola Rise and explained it as a free barotropic mode of the basin (also see Weijer et al., 2007). Using a simple wind-driven linear vorticity model, Fu (2003) illustrated the intraseasonal variability of the Southern Ocean and the North Pacific Ocean could be explained as a balance between wind stress curl and relative vorticity with bottom friction. In the Indian Ocean, the highly periodic monsoon wind generates intraseasonal variability at periods of 180, 120, 90, 75 days (Fu, 2007). Some of these could be explained as resonant basin modes.

9.5 Mesoscale Eddies

Mesoscale variability was observed even by the GEOS-3 altimeter (Huang et al., 1978) with a noise level of 25 cm. The strength of the signals and the relatively small scales made them survive the substantial instrument noise and orbit errors of the GEOS-3 altimeter. However, the trade-off between temporal and spatial resolution of a single altimeter has made it difficult to map the two-dimensional evolution of mesoscale eddies. Despite this difficulty, many statistical properties of mesoscale variability were ascertained from altimetry: geographic distribution of energy, spatial and temporal scales, wavenumber-frequency spectrum, eddy transports, viscosity and diffusivity (see Le Traon and Morrow, 2000 for a review).

The merging of multiple altimeter data into a gridded data set (Ducet et al., 2000) created a first opportunity to study the two-dimensional movement of mesoscale eddies. The intrinsic resolution of the data set is about 150–200 km, allowing to map eddies of sizes larger than these scales. The data set has created a surge of effort in tracking eddies, e.g. off the Central America coast (Palacios and Bograd, 2005) and in the Oyashio (Isoguchi and Kawamura, 2006) and studying their behavior. Advanced methods exist for tracking certain vorticity properties of eddies and allow automatic tracking (Isern-Fontanet et al., 2006; Morrow et al., 2004; Chelton et al., 2007). Figure 9.2 summarizes the properties of large eddies with lifetime grater 12 weeks surveyed by Chelton et al. (2007). A notable feature is that the propagation of cyclonic eddies has a slight tendency for a poleward deflection from purely westward. Anti-cyclonic eddies have a slight tendency for an equatorward deflection. At low latitudes, the westward speed of eddy propagation is somewhat less than that of non-dispersive baroclinic Rossby waves represented by the large-scale variability. At mid and high latitudes, the eddy speed is indistinguishable from the Rossby wave speed.

Using the maximum correlation method, Fu (2006, 2009) mapped the propagation velocity vector of ocean eddy variability. The method used the spatial and temporal lags of the maximum correlation of time series of sea surface height anomalies to compute the velocity of propagation of the dominant variability in the time series. The results correspond to the energy containing variability mostly associated with mesoscale motions, but cannot distinguish isolated eddies from other forms of variability like meandering currents and fronts. Displayed in Fig. 9.3 is an example in the Southern Ocean (from Fu, 2009), showing the vectors of propagation superimposed on the bottom topography (color shade) along with the tracks of the Subtropical Front, the Sub-Antarctic Front, and the Polar Front.

Within the Antarctic Circumpolar Current (ACC) which basically flows eastward parallel to these fronts, eddy propagation is steered from westward to eastward by the mean currents. A prominent feature is the "U" shaped turn of eddy propagation over a fracture zone centered at 55°S and 240°E (the Menard Fracture Zone). Another notable influence of bottom topography and ACC is the deflection of eddy paths over the Mid-Atlantic Ridge between 340° and 360°E. In the Argentine Basin, a counter-clockwise gyre-like pattern of eddy propagation is centered over a topographic feature called the Zapiola Rise. The center of the "gyre" seems to be a region where eddies tend to dissipate. This may be evidence for eddies being a source of energy driving the large-scale counter-clockwise (anti-cyclonic) barotropic circulation over the Zapiola Rise (de Miranda et al., 1999).

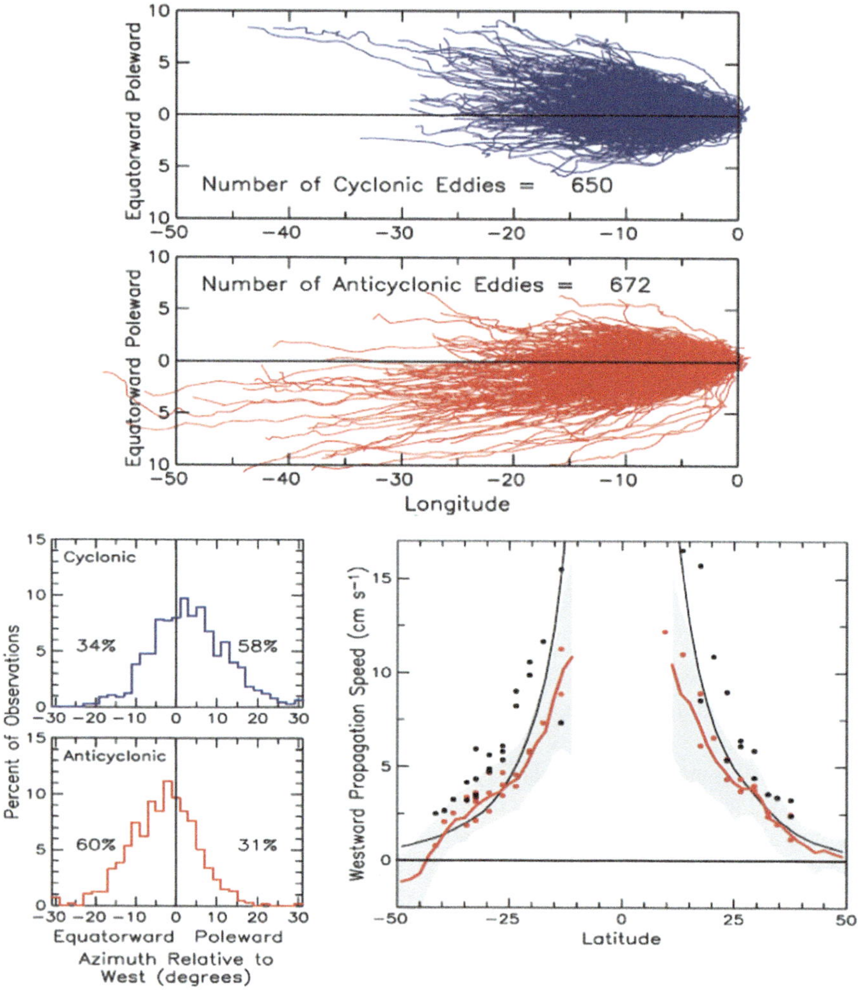

Fig. 9.2 The global propagation characteristics of long-lived cyclonic and anti-cyclonic eddies with lifetimes longer than 12 weeks. *Upper panels*: relative changes in longitude (negative westward) and latitude (poleward vs. equatorward). *Lower left panels*: histograms of the mean propagation angle relative to due west. *Lower right panel*: latitudinal variation of the westward zonal propagation speeds of large-scale sea surface height (*black dots*) and small-scale eddies (*red dots*) along the selected zonal sections considered previously by Chelton and Schlax (1996). The global zonal average of the propagation speeds is shown in the *right panel* by the *red line*, with *gray shading* to indicate the central 68% of the distribution in each latitude band, and the propagation speed of nondispersive baroclinic Rossby waves is shown by the *black line* (from Chelton et al., 2007)

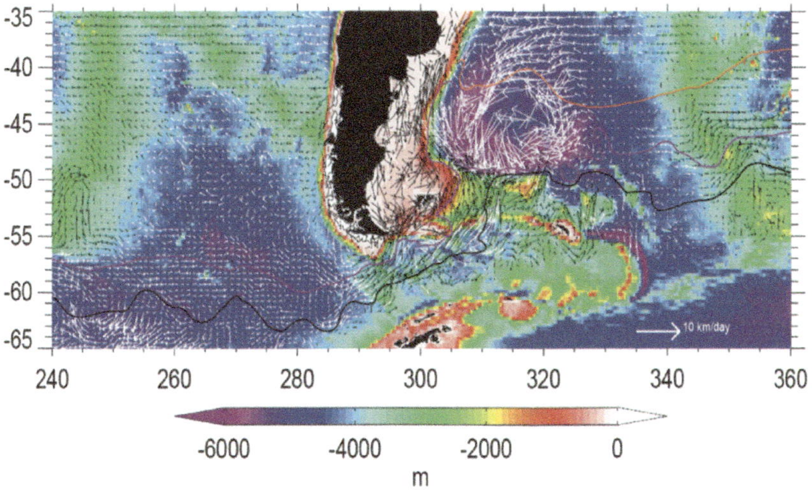

Fig. 9.3 The velocity of eddy propagation in the Southern Ocean between 240° and 360°E, superimposed on the ocean bathymetry. The different colors of the *arrows* are for the ease of viewing. The *three colored curves* are (from north to south) the Subtropical Front, the Sub-Antarctic Front, and the Polar Front, respectively (from Fu, 2009)

9.6 Energy Cascade and Eddy-Mean Flow Interaction

A fundamental issue in ocean dynamics is the flux of kinetic energy in wavenumber spectral space. Namely, the rate of energy transfer across different scales. Theoretical and modeling work suggests that baroclinic energy would cascade from high-order modes to the first mode, where the energy is transferred into barotropic mode, which then goes through an inverse cascade to larger scales (Rhines, 1977; Fu and Flierl, 1980). There was no direct observational evidence for the validity of these ideas until Scott and Wang (2005) computed the spectral energy flux directly using the gridded data set from merged multiple altimeter observations. As shown in Fig. 9.4, there is an inverse cascade of energy (negative flux) from the first baroclinic deformation scale to larger scales. Because altimetry observations are primarily related to the energy in the first baroclinic mode, this result indicates that the baroclinic energy in the ocean has an inverse cascade, in contrast to previous theoretical predictions.

Scott and Arbic (2007) then used a quasi-geostrophic model to illustrate that the inverse cascade is dominated by baroclinic modes in the model simulations. Altimetry observations have been used to revise a long-held theoretical concept about ocean dynamics. Using the same technique, Qiu et al. (2008) studied the seasonally modulated energy exchange process between the mean flow and eddy field of the Subtropical Counter Current in the South Pacific Ocean. They found a transfer of energy from meridionally-oriented modes to zonally-oriented modes through baroclinic instability. The energy is then transferred to larger zonal scales in an anisotropic inverse cascade process reflecting the effect of the meridionally changing Coriolis force (the beta effect). This is the first demonstration of the

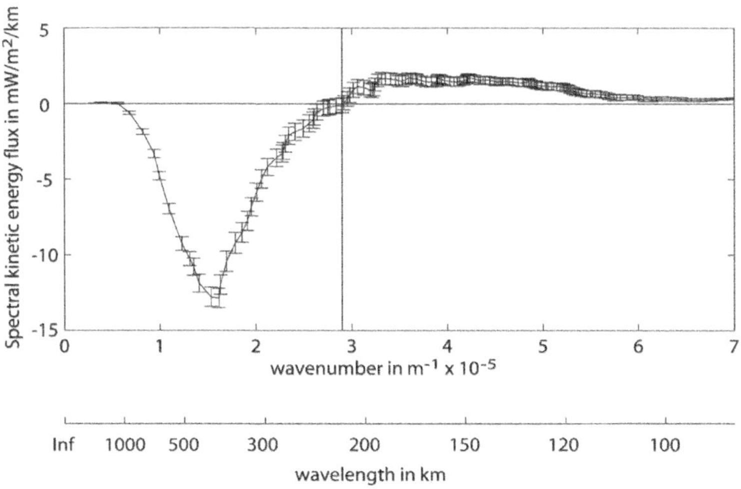

Fig. 9.4 The spectral kinetic energy flux associated with the near-surface geostrophic flow in the Kuroshio Extension region from altimeter data (~ 24°–46°N, 156°–174°E). The *straight vertical line* indicates the wavenumber of the first baroclinic deformation scale averaged over the region (Scott and Arbic, 2007)

detailed mechanism of energy exchange between mean flow and eddy variability and the transformation of scales.

9.7 Tides

T/P was the only altimeter mission that was designed to fly in a orbit optimized for resolving tidal signals for separation from those of ocean circulation. This effort has led to the most accurate information of the barotropic tides in the open ocean (Le Provost, 2001). Using the tide models derived from T/P, Egbert and Ray (2000) computed the flux of tidal energy and concluded that up to 30% of the total tidal dissipation took place in the deep ocean, in contrast to the traditional notion that more than 90% of tidal dissipation occurred over the shelves and shallow seas. This finding has confirmed the conjecture of Munk and Wunsch (1998) that half of the energy (about one terawatt) required to mix the ocean waters to maintain the thermohaline circulation comes from the tidal dissipation in the deep ocean. A major mechanism for converting tidal energy to mixing energy is through scattering of barotropic tides into internal tides over rough topography. Ray and Mitchum (1997) demonstrated that surface manifestations of internal tides could be detected in altimetry data. This work has led to a surge of studies of ocean internal tides, their sources, pathways and energetics (e.g. Merrifield et al., 2001).

9.8 Global Sea Level Change

The ability of measuring the change of the global mean sea level with uncertainty on the order of 1 mm/year represents the culmination of the development of precision altimetry. Although T/P and its follow-ons were not designed for meeting performance requirement at this level, through a dedicated effort of a large team of scientists and engineers, it has been demonstrated that the measurement accuracy has reached sub mm/year level (Ablain et al., 2009). The record of altimetry data from T/P and its follow-ons, supplemented by data from other missions, has provided the foundation for the determination of present-day sea level change (Cazenave and Nerem, 2004). The global coverage of altimetry observation has provided not only a mean value of the sea level rise, estimated at a rate of 3.4 ± 0.6 mm/year, from 1993 to 2008 (Ablain et al., 2009), but also a map of the geographical pattern of sea level change.

Shown in Fig. 9.5 is such a map (from Merrifield et al., 2009), which illustrates the complexity of the decadal variability of sea level as discussed in Section 9.3. Also shown are the locations of well-surveyed tide gauge locations. It is clear that tide gauges alone cannot capture the complicated geographic variability of sea level. As the planet is warming up, the potential threat of the collapse of polar ice sheets creates a geographically uneven risk of inundation (Bamber et al., 2009). It is therefore important to closely monitor not only the global mean sea level rise, but also its geographic pattern for early warnings and adaptations.

The launch of the GRACE satellite and the deployment of the Argo floats in the world's oceans, together with satellite altimetry, have provided an observing system for separating the contributions to sea level change into its components- the change of steric sea level and ocean mass. The three independent measurements allow a consistency check of the individual measurements. The total sea level can be determined by altimetry directly or by the combination of Argo with GRACE. The steric sea level can be determined by Argo directly or by the combination of

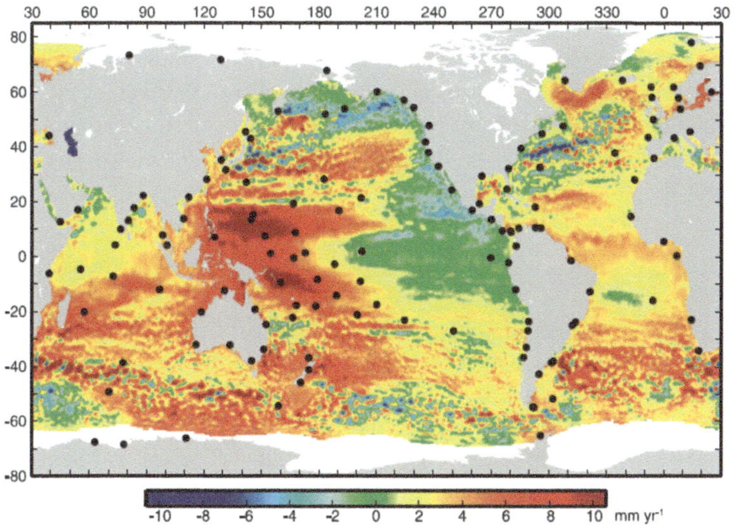

Fig. 9.5 Sea level trends (1993–2007) from multi-mission gridded sea level anomalies. The *black dots* are locations of well-surveyed tide gauges (from Merrifield et al., 2009)

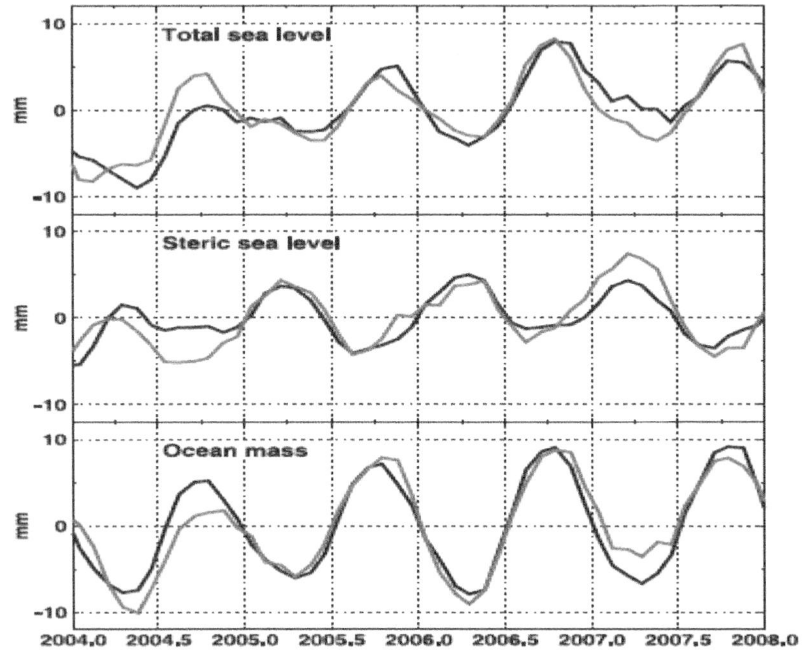

Fig. 9.6 Variability in total global mean sea level and its steric and mass components. The *black lines* are the observed (*top*) total sea level from Jason-1, (*middle*) steric sea level from Argo, and (*bottom*) ocean mass from GRACE. The *gray lines* show the inferred variability from the complementary observations. A 3-month boxcar smoothing is applied to each time series (from Leuliette and Miller, 2009)

altimetry and GRACE. The ocean mass can be determined directly by GRACE or by the combination of altimetry and Argo.

Figure 9.6 (from Leuliette and Miller, 2009) shows that the two estimates of each of the three components are consistent with each other to the extent of the errors of each. Maintaining the three measurement systems for monitoring and understanding future sea level change is crucial for preparing for the impact of a warming climate to our society.

9.9 Future Challenges

The sampling of a single altimeter presents a trade-off between spatial and temporal resolutions. Since the launch of T/P and ERS-1 in the early 1990s, it has been fortunate to have at least 2 altimeters fly simultaneously.

The spatial resolution of the merged data set from two altimeters is estimated to be from 150 km (Ducet et al., 2000) to 3° in longitude (Chelton and Schlax, 2003). This has prevented the observation of ocean variability smaller than these scales that contains a substantial amount of kinetic energy and plays significant roles in mixing and dissipation in the energy cycle of ocean circulation.

For example, Lapeyre and Klein (2006) illustrated that up to 50% of the oceanic tracer content is found at scales shorter than 100 km. Analyzing simulations by a high-resolution ocean general circulation model, Klein et al. (2009) showed that the vertical velocity of upper ocean currents can be estimated from sea surface height at the sub-mesoscales (wavelengths shorter than 100 km). Therefore, a significant part of ocean circulation and variability that has a fundamental role in the vertical exchange process for transporting nutrients, CO_2, and heat has been missed in the current altimetry observations.

To extend altimetry observation to higher resolution over a wide swath, radar interferometry has been developed since the early 1990s (Rodriguez and Martin, 1992). An instrument called Wide-Swath Ocean Altimeter (WSOA) was developed for flight on the Ocean Surface Topography Mission/Jason-2 in the early 2000s (Fu and Rodriguez, 2004).

Because of funding problems, WSOA was cancelled after substantial development had been conducted. A new mission concept called Surface Water and Ocean Topography (SWOT) was recommended by the US National Research Council Decadal Survey for addressing the need of high-resolution observation of water elevation in both the oceans and land surface water (Alsdorf et al., 2007). SWOT is currently being developed by NASA and CNES for flight in the late 2010s.

A challenge SWOT is facing is illustrated in Fig. 9.7 (from Fu and Ferrari, 2008). The wavenumber spectrum of sea surface height anomaly observed by the Jason-1 altimeter shows the domination of instrument noise at wavelengths shorter than

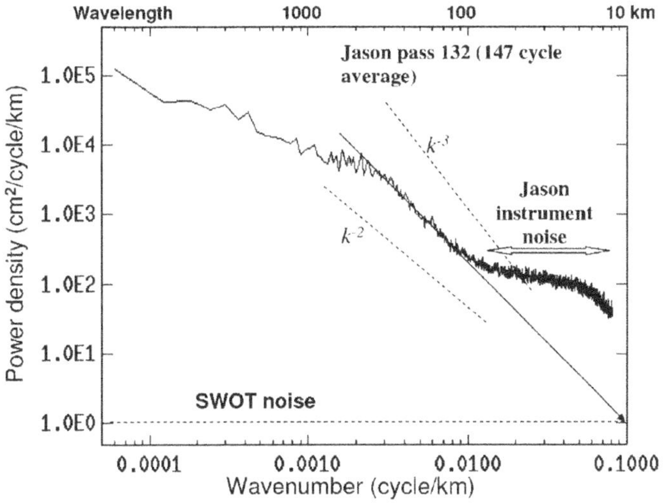

Fig. 9.7 Spectrum of sea surface height anomaly from Jason altimeter data (*solid line*). The *two slanted dashed lines* represent two spectral power laws with k as wavenumber. The *horizontal dashed line* represents the SWOT measurement noise at 1/km sampling rate. The *slanting solid straight line* represents a linear fit of the spectrum between 0.002 and 0.01 cycles/km. It intersects with the SWOT noise level at 10 km wavelength (from Fu and Ferrari, 2008)

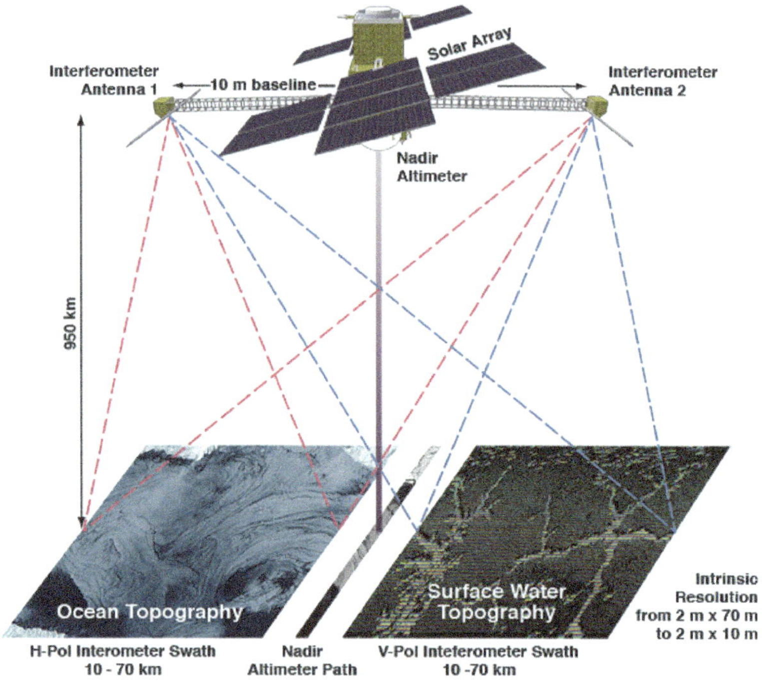

Fig. 9.8 The configuration of the SWOT[1] measurement system

100 km. Assuming that the spectrum of ocean signals continues to follow a power law extrapolation from the observation, the noise level of SWOT must be lower than that of Jason-1 by two orders of magnitude in order to resolve ocean signals down to a wavelength of 10 km.

Shown in Fig. 9.8 is the configuration of the SWOT measurement approach. The basic instrument payload is composed of two Ka-band synthetic aperture radars (SAR) with their antennae separated by a 10-m long mast. The intrinsic resolution of the SAR is on the order of a few meters to tens of meters. The backscatter of the ocean surface is received by the two antennae. Through the technique of interferometry and precision orbit determination, the height of the surface elevation of the backscatters can be determined. Using different polarizations, the two beams of radar transmission can be analyzed without interference. Over the ocean, the raw data are processed and smoothed onboard over cells of 1 km × 1 km to achieve the required low noise level. Over land, the raw data are transmitted to ground stations for processing according to the requirements for hydrological applications. The nadir track where interferometry does not work is to be covered by a conventional

[1] For more information on SWOT, see: http://bprc.osu.edu/water

nadir-looking Jason-class altimeter. The combined two swaths plus the nadir observation will provide a swath of width of 130 km for studying the mesoscale and submesoscale processes over the ocean and the storage and discharge of freshwater on land.

Because of the finite swath coverage, it takes at least 22 days to cover the Earth between the inclination latitudes without gaps. The repeat period of SWOT is set to 22 days. To avoid sun-synchronous orbits for sampling tides, the inclination of the orbit is set to 78°. At the oceanic mesoscales and submesoscales, shallow water tides and internal tides become key concerns for studying ocean circulation. These tides have not been well resolved by conventional altimetry and pose new challenges for SWOT. As noted in Section 9.7, tides in their own right are important to the understanding of ocean circulation. The choice of Ka-band for SWOT is primarily for meeting the interferometry measurement requirement because the height errors are proportional to the ratio of radar wavelength to the length of the mast. Furthermore, at Ka-band, the range delay caused by the ionospheric free electrons becomes negligible. To correct for the errors cause by the tropospheric water vapor, a multi-frequency microwave radiometer will be included in the payload.

9.10 Conclusions

Satellite altimetry has revolutionized oceanography since the 1990s. Precision missions like T/P and its follow-ons have provided the first view of large-scale ocean circulation, its variability, and the global mean sea level. The new observations have motivated the advancement in ocean modeling and data assimilation leading to the development of ocean state estimation for a variety of applications. The discovery of the high-frequency large-scale variability led to a new view and appreciation of the barotropic processes in the ocean. The decade-long data record provides the first global view of the decadal change in ocean circulation and its geographic variability. The capability of detecting the rate of global mean sea level change at a level of uncertainty less than 1 mm/year represents the state-of-the-art of precision altimetry. One must realize that T/P and its follow-ons were not designed for reaching this level of performance. The achievement was made by a dedicated effort of a large team of scientists and engineers to push the limit of the measurement system. While altimetry system is being transitioned from research to operation, we must recognize the critical importance of maintaining such a team effort to ensure the precision and stability of the measurement into the future.

Combined data from multiple altimeters have enabled a wide range of advances in ocean dynamics. Further, the combination with surface drifter data has led to the most detailed knowledge of the global ocean general circulation, revealing the unexpected ubiquitous presence of small-scale striations in ocean currents. It is now possible to conduct detailed analysis of the balance of vorticity, a high-order computation, of large-scale ocean currents. For the first time, one can track the movement of ocean eddies around the world's oceans and determine their pathways and interaction with mean circulation and ocean topography. The analysis of the balance of energy in spectral space has led to a surprising finding of inverse cascade of energy associated with the first baroclinic mode of the ocean.

The specific orbit choice of T/P and its follow-ons has led to the most accurate knowledge of the surface (or barotropic) tides in the open ocean. The calculation of the energy flux of the tides proves that half of the energy required to mix ocean waters to maintain the large-scale thermohaline circulation comes from tidal dissipation in the deep ocean. The finding of internal tides in altimetry data is another surprise that creates a new surge of studies of the subject of fundamental importance in ocean mixing and tidal energy cycle.

A limitation of the conventional nadir-looking altimeter is its sampling in space and time. With data merged from two altimeters, the decade-long data set that has made great strides in advancing our knowledge of ocean circulation has a spatial resolution that prevents observation of the important sub-mesoscale processes at scales shorter than 100 km. A way to advance the capability of future altimetry is the use of radar interferometry for making high-resolution wide-swath altimetry measurement. The SWOT Mission recommended by the US National Research Council's Decadal Survey is taking on this challenge by developing a Ka-band radar interferometry system for flight in the late 2010s. SWOT measurement will significantly advance both oceanography and land hydrology and address two key aspects of climate change: improving the prediction of the rate of warming through improved understanding of the oceanic submesoscale processes, and improving the capability of monitoring and managing the shifting water resources caused by a warming climate.

By providing wide-swath coverage, a single mission like SWOT is equivalent to more than 10 conventional nadir-looking altimeters. After the demonstration of its workings, radar interferometry is potentially a candidate for replacing nadir-looking altimetry as a standard tool for oceanographic and hydrological applications.

Acknowledgements The research presented in the paper was carried out at the Jet Propulsion Laboratory, California Institute of Technology, under contract with the National Aeronautic and Space Administration. Support from the Jason-1 and OSTM/Jason-2 Projects is acknowledged.

References

Ablain M, Cazenave A, Valladeau G, Guinehut S (2009) A new assessment of the error budget of global mean sea level rate estimated by satellite altimetry over 1993–2008. Ocean Sci 5: 193–201

Alsdorf D, Fu LL, Mognard N, Cazenave A, Rodriguez E, Chelton D, Lettenmaier D (2007) Measuring the global oceans and terrestrial fresh water fromspace. EOS Trans AGU 88(24):253

Bamber JL, Riva REM, Vermeersen BLA, LeBrocq AM (2009) Reassessment of the potential sea-level rise from a collapse of the West Antarctic ice sheet. Science 324:901–903, doi:10.1126/science.1169335

Carrère L, Lyard F (2003) Modeling the barotropic response of the global ocean to atmospheric wind and pressure forcing – comparisons with observations. Geophys Res Lett 30:1275, doi:10.1029/2002GL016473

Cazenave A, Nerem RS (2004) Present-day sea level change: observations and causes. Rev Geophys 42:RG3001, doi:10.1029/2003RG000139

Chao Y, Fu LL (1995) A comparison between the TOPEX/POSEIDON data and a global ocean general circulation model during 1992–1993. J Geophys Res 100:24965–24976
Chelton DB, Schlax MG (1996) Global observations of oceanic Rossby waves. Science 272: 234–238
Chelton DB, Schlax MG (2003) The accuracies of smoothed sea surface height fields constructed from tandem satellite altimeter datasets. J Atmos Ocean Tech 20:1276–1302
Chelton DB, Schlax MG, Samelson RM, de Szoeke RA (2007) Global observations of large oceanic eddies. Geophys Res Lett 34:L15606, doi:10.1029/2007GL030812
de Miranda AP, Barnier B, Dewar WK (1999) On the dynamics of the Zapiola Anticyclone. J Geophys Res 104:21137–21150
Ducet N, Le Traon PY, Reverdin G (2000) Global high resolution mapping of ocean circulation from the combination of TOPEX/POSEIDON and ERS-1/2. J Geophys Res 105: 19477–19498
Egbert GD, Ray RD (2000) Significant dissipation of tidal energy in the deep ocean inferred from satellite altimeter data. Nature 405:775–778
Fu LL (2003) Wind-forced intraseasonal sea level variability of the extratropical oceans. J Phys Oceanogr 33:436–449
Fu LL (2004) Latitudinal and frequency characteristics of the westward propagation of large-scale oceanic variability. J Phys Oceanogr 34:1907–1921
Fu LL (2006) Pathways of eddies in the South Atlantic revealed from satellite altimeter observations. Geophys Res Letts 33:L14610, doi:10.1029/2006GL026245
Fu LL (2007) Intraseasonal basin modes of the equatorial Indian Ocean observed from sea surface height, wind, and temperature data. J Phys Oceanogr 37:188–202
Fu LL (2009) Pattern and velocity of propagation of the global ocean eddy variability. J Geophys Res 114:C11017, doi:10.1029/2009JC005349
Fu LL, Cazenave A (eds.) (2001) Satellite Altimetry and Earth Sciences: A Handbook of Techniques and Applications, Academic Press, San Diego, 463 pp
Fu LL, Chelton DB (2001) Large-scale ocean circulation. In: Fu LL, Cazenave A (eds.) Satellite Altimetry and Earth Sciences: A Handbook for Techniques and Applications, Academic Press, San Diego, pp. 133–169, 423pp
Fu LL, Cheng B, Qiu B (2001) 25-Day period large-scale oscillations in the Argentine Basin revealed by the TOPEX/POSEIDON altimeter. J Phys Oceogr 31:506–517
Fu LL, Ferrari R (2008) Observing oceanic submesoscale processes from space. Eos Trans AGU 89(48):488
Fu LL, Flierl GR (1980) Nonlinear energy and enstrophy transfers in a realistically stratified ocean. Dyn Atmos Ocean 4:219–246
Fu LL, Rodriguez R (2004) High-resolution measurement of ocean surface topography by radar interferometry for oceanographic and geophysical applications. In: Sparks RSJ, Hawkesworth CJ (eds.) State of the Planet: Frontiers and Challenges, AGU Geophysical Monograph 150, IUGG 19:209–224
Fu LL, Smith RD (1996) Global ocean circulation from satellite altimetry and high-resolution computer simulation. Bull Amer Meteorol Soc 77:2625–2636
Fukumori I, Raghunath R, Fu LL (1998) The nature of global large-scale sea level variability in relation to atmospheric forcing: a modeling study. J Geophys Res 103: 5493–5512
Hakkinen S, Rhines P (2004) Decline of subpolar North Atlantic circulation during the 1990s. Science 304:555–559, doi:10.1126/science.1094917
Huang NE, Leitao CD, Para CG (1978) Large-scale Gulf Stream frontal study using GEOS3 radar altimeter data. J Geophys Res 83:4673–4682
Hughes CW (2005) Nonlinear vorticity balance of the Antarctic Circumpolar Current. J Geophys Res 110:C1108, doi:10.1029/2004JC002753
Isern-Fontanet J, Garcia-Ladona E, Font J (2006) Vortices of the Mediterranean Sea: an altimetric perspective. J Phys Oceanogr 36:87–103, doi:10.1175/JPO2826.1
Isoguchi O, Kawamura H (2006) Seasonal to interannual variations of the western boundary current of the subarctic North Pacific by a combination of the altimeter and tide gauge sea levels. J Geophys Res 111:C04013, doi:10.1029/2005JC003080

Klein P, Isern-Fontanet J, Lapeyre G, Roullet G, Danioux E, Chapron B, Le Gentil S, Sasaki H (2009) Diagnosis of vertical velocities in the upper ocean from high resolution sea surface height. Geophys Res Lett 36:L12603, doi:10.1029/2009GL038359

Lapeyre G, Klein P (2006) Impact of the small-scale elongated filaments on the oceanic vertical pump. J Mar Res 64:835–851

Lee T (2004) Decadal weakening of the shallow overturning circulation in the South Indian Ocean. Geophys Res Lett 31:L18305, doi:10.1029/2004GL020884

Lee T, McPhaden MJ (2008) Decadal phase change in large-scale sea level and winds in the Indo-Pacific region at the end of the 20th century. Geophys Res Lett 35:L01605, doi:10.1029/2007GL032419

Le Provost C (2001) Ocean tides. In: Fu LL, Cazenave A (eds.) Satellite Altimetry and Earth Sciences: A Handbook for Techniques and Applications, Academic Press, San Diego, pp. 267–301, 423 pp

Le Traon PY, Morrow R (2001) Ocean currents and eddies. In: Fu LL, Cazenave A (eds.) Satellite Altimetry and Earth Sciences: A Handbook for Techniques and Applications, Academic Press, San Diego, pp. 171–210, 423 pp

Leuliette EW, Miller L (2009) Closing the sea level rise budget with altimetry, Argo, and GRACE. Geophys Res Lett 36:L04608, doi:10.1029/2008GL036010

Maximenko NA, Bang B, Sasaki H (2005) Observational evidence of alternating zonal jets in the world ocean. Geophys Res Lett 32:L12607, doi:10.1029/2005GL022728

Maximenko NA, Melnichenko OV, Niiler PP, Sasaki H (2008) Stationary mesoscale jet-like features in the ocean. Geophys Res Lett 35:L08603, doi:10.1029/2008GL033267

Merrifield MA, Holloway P, Johnston T (2001) The generation of internal tides at the Hawaiian Ridge. Geophys Res Let 28:559–562

Merrifield MA, Merrifield ST, Mitchum GT (2009) An anomalous recent acceleration of global sea level rise. J Climate 22:5772–5781

Morrow R, Florence B, Griffin D, Sudre J (2004) Divergent pathways of cyclonic and anti-cyclonic ocean eddies. Geophys Res Lett 31:L24311, doi:10.1029/2004GL020974

Munk WH, Wunsch C (1998) Abyssal recipes II: energetics of tidal and wind mixing. Deep Sea Res 45:1977–2010

Niiler PP, Maximenko NA, McWilliams JC (2003) Dynamically balanced absolute sea level of the global ocean derived from near-surface current. J Geophys Res 110:C11008, doi:10.1029/2004JC002753

Palacios DM, Bograd SJ (2005) A census of Tehuantepec and Papagayo eddies in the northeastern tropical Pacific. Geophys Res Lett 32:L23606, doi:10.1029/2005GL024324

Qiu B, Chen S (2010) Eddy-mean flow interaction in the decadally-modulating Kuroshio extension system. Deep Sea Res 57, doi:10.1016/j.dsr2.2008.11.036

Qiu B, Scott RB, Chen S (2008) Length scales of eddy generation and nonlinear evolution of the seasonally-modulated South Pacific Subtropical Countercurrent. J Phys Oceanogr 38: 1515–1528

Ray RD, Mitchum GT (1997) Surface manifestation of internal tides in the deep ocean: observations from altimetry and island gauges. Prog Oceanogr 40:135–162

Rhines PB (1977) The dynamics of unsteady currents. In: Goldberg E et al (eds.) The Sea, John Wiley and Sons, New York, pp. 189–318

Richards KJ, Maximenko NA, Bryan FO, Sasaki H (2006) Zonal jets in the Pacific Ocean. Geophys Res Lett 33:L03605, doi:10.1029/2005GL024645

Rio MH, Hernandez F (2004) A mean dynamic topography computed over the world ocean from altimetry, in situ measurements, and a geoid model. J Geophys Res 109:C12032, doi:10.1029/2003JC002226

Rodriguez E, Martin J (1992) Theory and design of interferometric synthetic aperture radars. IEEE Proc 139(2):147–159

Roemmich J, Gilson D, Davis R, Sutton P, Wijffels S, Riser S (2007) Decadal spinup of the South Pacific Subtropical Gyre. J Phys Oceanogr 37:162–173, doi:10.1175/JPO3004.1

Schlax MG, Chelton DB (2008) The influence of mesoscale eddies on the detection of quasi-zonal jets in the ocean. Geophys Res Lett 35:L24602, doi:10.1029/2008GL035998

Scott RB, Arbic BK (2007) Spectral energy fuxes in geostrophic turbulence: implications for ocean energetics. J Phys Oceanogr 37:673–688.

Scott RB, Wang F (2005) Direct evidence of an oceanic inverse kinetic energy cascade from satellite altimetry. J Phys Oceanogr 35:1650–1666

Stammer D, Tokmakian R, Semtner A, Wunsch C (1996) How well does a 1/4 degree global circulation model simulate large-scale oceanic observations? J Geophys Res 101:25779–25811

Stammer D, Wunsch C, Giering R, Eckert C, Heimbach P, Marotzke J, Adcroft A, Hill CN, Marshall J (2002) The global ocean circulation during 1992–1997, estimated from ocean observations and a general circulation model. J Geophys Res 107:3118, doi:10.1029/2001JC000888

Stammer D, Wunsch C, Ponte R (2000) De-aliasing of global high frequency barotropic motions in altimeter observations. Geophys Res Lett 27:1175–1178

Tapley BD, Bettadpur S, Ries JC, Thompson PF, Watkins MM (2004) GRACE measurements of massvariability in the earth system. Science 305:503–505, doi:10.1126/science.1099192

Weijer W, Viver F, Gille ST, Dijkstra H (2007) Multiple oscillatory modes of the Argentine Basin. Part I: Statistical analysis. J Phys Oceanogr 37:2855–2868

Wunsch C, Gaposchkin EM (1980) On using satellite altimetry to determine the general circulation of the oceans with application to geoid improvement. Rev Geophys Space Phys 18:725–745

Wunsch C, Heimbach P, Ponte RM, Fukumori I, ECCO-GODAE Consortium Members (2009) The global general circulation of the ocean estimated by the ECCO-Consortium. Oceanography 22:88–103

Chapter 10
Absolute Dynamic Topography from Altimetry: Status and Prospects in the Upcoming GOCE Era

Marie-Helene Rio

10.1 Introduction

In the oceans, all major current systems, which transport heat and mass all around the globe, regulating our climate, can be considered, in a first approximation, to be in geostrophic equilibrium, meaning that they can be simply derived by the knowledge of the ocean absolute dynamic topography which is the sea level above an hypothetical ocean at rest (the geoid). Up to a certain point, the ocean dynamic topography can be measured from space by satellite altimetry. The limitation stands in the fact that the actual quantity η measured by an altimeter is the sea level above the ellipsoid, which differs from the absolute dynamic topography h by the value of the geoid height N:

$$h = \eta - N \tag{10.1}$$

While the computation and exploitation of the full ocean dynamical signal h has been for long hampered by the lack of an accurate geoid, the successful launch, on March 2009, of the GOCE (Gravity Field and Steady-State Ocean Circulation) satellite has opened a new prospect for the use of altimetric data. The exploitation of GOCE data will allow to resolve for the first time the spatial scales of the geoid down to around 100 km with centimetric accuracy. Together with GRACE data, this will lead to the estimation of an accurate geoid at all spatial scales greater than 100 km.

The objective of this paper is, shortly before actual GOCE data become available, to draw an overview of what has been achieved in the last 20 years for absolute dynamic topography estimation and exploitation, as well as to discuss the benefits and limits of future GOCE data for further improving our knowledge of the ocean absolute sea level.

Since the very beginning of altimetry, the so-called "repeat-track" method (Cheney et al., 1983) has allowed to cope with the uncertainty on the geoid: Sea Level Anomalies are derived along the altimetric satellite tracks subtracting from

M.-H. Rio (✉)
CLS-DOS, Parc Technologique du Canal, Ramonville Saint Agne 31526, France
e-mail: mrio@cls.fr

V. Barale et al. (eds.), *Oceanography from Space*,
DOI 10.1007/978-90-481-8681-5_10,

the altimetric height η a mean profile computed over a given period of time P. Under the assumption of a stationary geoid, the residual quantity is the sea level anomaly $h' = \eta'$ relative to the time period P. The routine computation and exploitation of the Sea level anomalies has led to numerous advances in our knowledge of the ocean mesoscale (Fu and Cazenave, 2001).

In order to reconstruct the full dynamical signal from the SLA, the missing quantity is the Mean Dynamic Topography, i.e., the average over the period P of the sea level above the geoid. Up to a given spatial scale, governed by the geoid model accuracy, the Mean Dynamic Topography can be derived by subtracting the geoid from an altimetric Mean Sea Surface. This is the so-called "direct method", whose actual applicability has been greatly improved since the very first altimetric missions, as will be described in Section 10.2 of the present chapter. However, in order to retrieve the shortest spatial scales of the Mean Dynamic Topography, and therefore compute the ocean absolute dynamic topography with sufficient accuracy, a number of methods have been developed which are further described in Section 10.3. In Section 10.4, we will make a (non exhaustive) review of scientific advances that have been made possible by the use of these Mean Dynamic Topography estimates and in Section 10.5 we will finally discuss the potential and limits of the use of future GOCE data.

10.2 Twenty Years of Geoid Improvements and Its Impact for MDT Determination

Since the very first global geoid model computation, based on the analysis of satellite's orbit perturbation, important technological and conceptual steps have been made for computing the Earth gravity field from space leading to the launch of gravity dedicated space missions as CHAMP (2000), GRACE (2002) and more recently GOCE (2009).

Figure 10.1 show the RMS difference, over the oceans only, and for a common resolution of 300 km, between the latest static GRACE geoid models EIGEN-GRGS.RL02 computed at GRGS from 4½ years of GRACE data (http://bgi.cnes.fr:8110/geoid-variations/static.html) and previous models computed from 1995 to

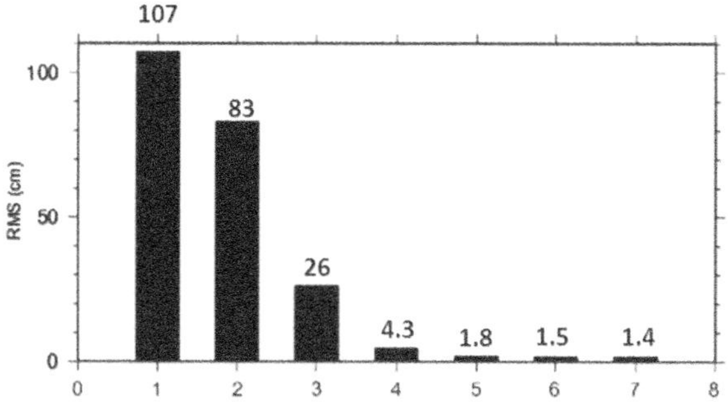

Fig. 10.1 RMS differences (in cm) between the EIGEN-GRGS.RL02 geoid model and different geoid models available from 1995 to 2008 and listed in Table 10.1

Table 10.1 Different global geoid models computed from 1995 to 2009

Model	Year	HS	Data
GRIM4S4	1995	70	Geodetic sat
GRIM5S1	1999	99	Geodetic sat
CHAMP3S	2003	140	33 months of CHAMP
GGM02S/EIGEN3S	2005	150	2 years of GRACE
EIGEN4S	2006	150	3 years of GRACE
ITG-GRACE03S	2007	180	$4\frac{1}{2}$ years of GRACE
GGM03S/EIGEN5S	2008	150–180	4 years of GRACE
EIGEN-GRGS.RL02	2009	160	$4\frac{1}{2}$ years of GRACE

2008 and listed in Table 10.1. From more than 1 m in 1995, the difference has dropped to 26 cm in 2003 with the use of CHAMP data, and further down to 4.3 cm in 2005 with the use of the first geoid models based on GRACE data.

This of course has had significant impact for the Mean Dynamic Topography determination. As stated in introduction, the most direct method consists in subtracting a geoid model from an altimetric Mean Sea Surface to get an estimate of the MDT following Equation (10.1) averaged over a given period. The application of the method however is not as straightforward as it seems: A number of key points have to be taken into account to make sure that the MSS and the geoid model are consistent (same reference ellipsoid, same tide system . . .).

An exhaustive review of all geodetic concepts relevant to oceanographers for a correct use of satellite gravity data has been done by (Hughes and Bingham, 2006). Other useful information can be found in the GUT (GOCE User Toolbox) tutorial (http://earth.esa.int/gut/). Altimetric Mean Sea Surfaces, like the CLS01 (Hernandez and Schaeffer, 2001) or the DNSC08 MSS (Andersen, 2008) resolve much shorter spatial scales (down to 10–20 km) than recent satellite-only geoid models (300–400 km). Once the geoid model has been subtracted from the altimetric MSS, further filtering is thus needed in order to match the spectral content of both surfaces. This can be done either in the spatial or in the spectral domain. A comparison of the two approaches has been done by (Bingham et al., 2008). In the spatial domain, simple to more complex filter can be used: Jayne (2006) applied a hamming window smoother while Vianna et al. (2007) developed an adaptative filter, based on principal components analysis techniques, in order to extract as much noise as possible minimizing signal attenuation. Figure 10.2 shows the MDT computed from the direct method using the CLS01 MSS and different satellite-only geoid models and applying a 300 km Gaussian low pass filter. Whereas in 1995, one could hardly distinguish between noise and the main ocean circulation structure, the use of CHAMP allowed to resolve the main features of the ocean circulation, with higher (resp. lower) values of the mean sea level above the geoid in the center of the subtropical (resp. subpolar) gyres. In 2005, with the first GRACE-based geoids, the pictures of the ocean circulation became even clearer, all major currents starting to be resolved (the Antarctic circumpolar current (ACC), the Gulfstream, the Kuroshio, the Aghulas current, the Falkland current. . .) although still polluted by some noise which was finally almost

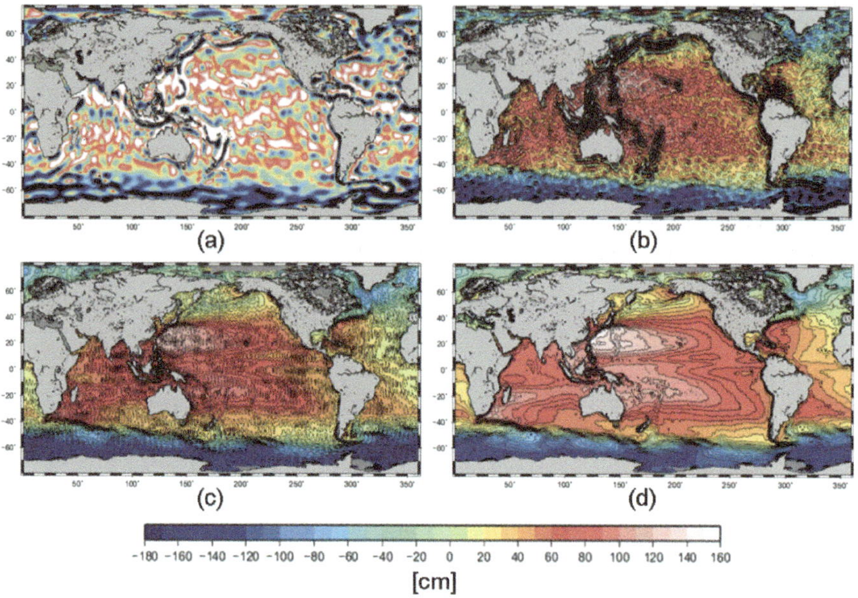

Fig. 10.2 Mean dynamic topography computed at 300 km resolution through the direct method using the CLS01 MSS and (**a**) GRIM5S1 (**b**) CHAMP3S (**c**) GGM02S (**d**) EIGEN-GRGS.RL02.MEAN-FIELD

entirely cancelled out in 2009 at scales greater than 300 km with the use of more than 4 years of GRACE data.

In order to better quantify the quality of these geoid models for altimetric use, e.g. for computing absolute dynamic topography values and associated geostrophic currents, Rio et al. (2006) advocate to compare the MDT solutions, computed at

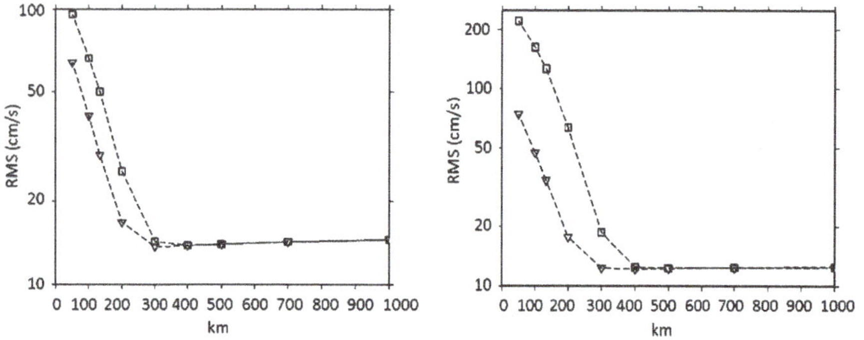

Fig. 10.3 RMS differences (*left*: zonal component; *right*: meridian component) between synthetic estimates of the mean geostrophic circulation and mean geostrophic velocities computed from the direct MDT filtered at spatial scales ranging from 133 to 1,000 km. *Squares*: GGM02S *Inverted triangles*: EIGEN-GRGS.RL02.MEAN-FIELD

various wavelengths, using various geoid models, to independent synthetic estimates of the ocean Mean Dynamic Topography. These synthetic estimates of the ocean Mean Dynamic Topography are simply obtained by subtracting to instantaneous in-situ measurements of the ocean dynamic topography h or the ocean geostrophic surface current u_g,v_g, the time variable $\left(h'_a,u'_a,v'_a\right)$ component as measured from altimetry (Equation 10.2).

$$
\begin{aligned}
\bar{h}(x,y) &= h(t,x,y) - h'_{\mathrm{a}}(t,x,y) \\
\bar{u}_{\mathrm{g}}(x,y) &= u_{\mathrm{g}}(t,x,y) - u'_{\mathrm{a}}(t,x,y) \\
\bar{v}_{\mathrm{g}}(x,y) &= v_{\mathrm{g}}(t,x,y) - v'_{\mathrm{a}}(t,x,y)
\end{aligned}
\tag{10.2}
$$

Figure 10.3 shows the Root Mean Square (RMS) differences between the synthetic velocities computed from the combination of in-situ 15 m-drogued drifting buoy velocities deployed from 1993 to 2008 in the framework of international programs (SVP, WOCE-TOGA...) and the mean geostrophic currents deduced from two direct MDT (one based on GGM02S, computed from 2 years of GRACE data, the other based on EIGEN-GRGS.RL02, computed from 4½ years of GRACE data), filtered at different spatial scales. An optimal filtering length of 300 km is found where the RMS value is minimum. At scales greater than 300 km the RMS values are dominated by omission error: MDT scales contained in the synthetic estimates

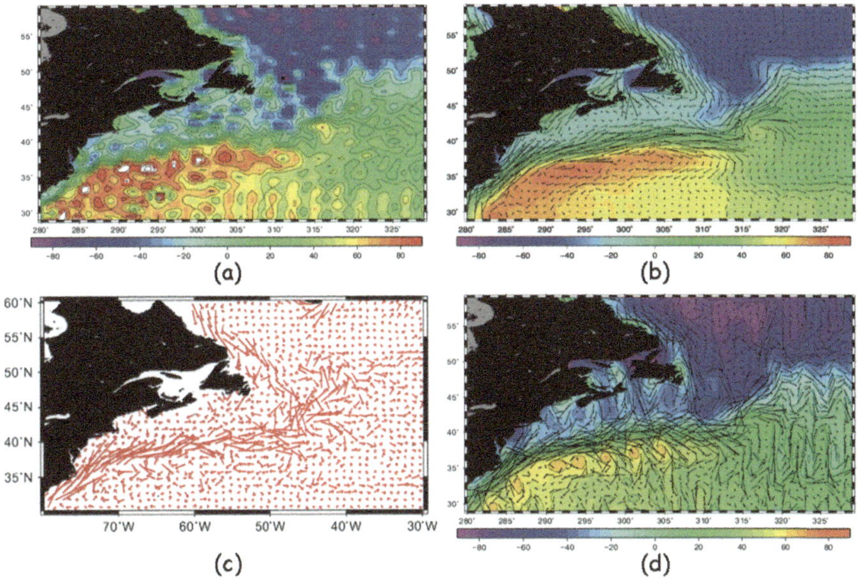

Fig. 10.4 Mean Dynamic Topography in the Gulfstream area: (**a**) MSS CLS01 minus EIGEN-GRGS.RL02 filtered at 133 km; (**b**) MSS CLS01 minus EIGEN-GRGS.RL02.MEAN-FIELD filtered at 300 km; (**c**) synthetic mean velocity estimates; (**d**) MSS CLS01 minus GGM02S filtered at 300 km

but not in the direct MDT from which they have been removed by increasing filtering (resulting in slightly increasing RMS values). At scales shorter than 300 km, the RMS values are dominated by both the geoid omission and commission errors.

The use of almost 3 supplementary years of GRACE data for the computation of the EIGEN-GRGS.RL02.MEAN-FIELD compared to the GGM02S model results in a reduction of the RMS difference to synthetic estimates at scales shorter than 300 km for the zonal component and 400 km for the meridian component.

To further illustrate this point, a focus is chosen in the Gulfstream area (see Fig. 10.4). Figure 10.4a shows the direct MDT obtained at 133 km resolution with EIGEN-GRGS.RL02. The signal is quite noisy. When a 300 km filter is applied (Fig. 10.4b), the noise is reduced and the velocities computed by geostrophy from the direct MDT are quite consistent with the synthetic velocity estimates (Fig. 10.4c). This is a strong improvement compared to the circulation obtained, at 300 km resolution, with the GGM02S geoid (Fig. 10.4d), and for which a lot of noise, mainly on the meridional component of the velocity, is observed.

10.3 Toward Higher Resolution of the Geoid and the MDT

As we have seen in the previous section, huge improvements have been made during the last 20 years for the estimation of the geoid and consequently the ocean Mean Dynamic Topography. However, the quality of the latest "satellite-only" geoid models still limits the spatial resolution of the ocean MDT to scales larger than around 300 km. On the other hand, altimetric Sea Level Anomalies are available along-track every 7 km and, when gridded, with an approximate resolution of 50–100 km.

The spatial resolution of altimetric Mean Sea Surfaces is much higher, down to 20–30 km. Various methods have therefore been developed in order to increase the resolution of the ocean Mean Dynamic Topography, for a better exploitation of altimetric data. These methods can be divided into two main categories. In the first method, the geoid resolution is improved, and then a higher resolution MDT is computed using the direct method. In the second method, a large scale MDT is first computed and further improved using external oceanographic data to resolve the shorter scales.

In the first case, the quantity to improve is the geoid. This can be done using in-situ gravimetric data (Hunegnaw et al., 2009; Thompson et al., 2009), most often limited in spatial extension, resulting in local to regional improvement of the geoid. Global improvement can be achieved using the shortest scales information of the altimetric Mean Sea Surface (the spatial resolution of the ocean MDT being coarser than the spatial resolution of the MSS, the shortest spatial scales of the altimetric Mean Sea Surface are only due to the shortest spatial scales of the geoid, and can therefore be used to enhance this latter). This method is commonly used to enhance the resolution of the satellite-only solution, resulting in the so-called combined geoid models (GGM02C, EIGEN3C, EIGEN5C, ...), which are developed to a higher degree and order than their satellite-only counterpart. In the case of the

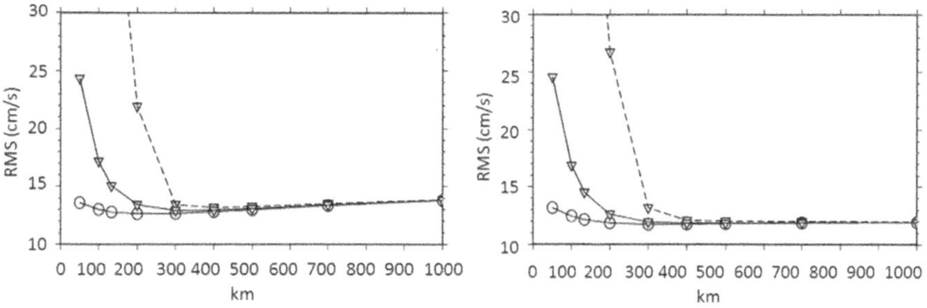

Fig. 10.5 RMS differences (*top*: zonal component; *bottom*: meridian component) between synthetic estimates of the mean geostrophic circulation and mean geostrophic velocities computed from the direct MDT filtered at spatial scales ranging from 133 to 1,000 km. *Squares*: GGM02S *Inverted triangles*: EIGEN-GRGS.RL02.MEAN-FIELD. *Circles*: DNSC08 MDT

recent EGM08 geoid model (Pavlis et al., 2008), both in-situ gravimetric data and altimetry-derived gravity anomalies have been used to compute the spherical harmonics coefficients of the geoid up to degree and order 2,400 ($\sim$ 8 km resolution). Andersen (2008) used this geoid model, together with the DNSC08 altimetric MSS to compute the spatial scales of the ocean Mean Dynamic Topography greater than around 75 km (Fig. 10.6b). Hence, the RMS difference to independent synthetic estimates of the obtained ocean MDT and the associated geostrophic velocities is much reduced at scales shorter than 300 km (circles on Fig. 10.5).

The second way of getting higher resolution estimates of the ocean Mean Dynamic Topography is to enhance the large-scale fields based on the satellite-only geoid models thanks to the use of in-situ oceanographic measurements. In (Rio and Hernandez, 2004; Rio et al., 2005, 2007, 2010), the synthetic estimates of the mean heights and mean velocities computed in Section 10.2 are used to improve the resolution of the large-scale GRACE based solution. The more recent MDT CNES-CLS09 is shown on Fig. 10.6a. A very similar approach was developed by (Niiler et al., 2003; Maximenko and Niiler, 2005; Maximenko et al., 2009), based on drifting buoy velocities only. The resulting MDT fields however were shown to be very close one to each other (Vossepoel, 2007; Maximenko et al., 2009). The RMS differences between the mean geostrophic velocities from the CNES-CLS09 MDT and independent synthetic velocity estimates (computed using drifting buoy velocities available in 2009 and not used in the CNES-CLS09 MDT computation) is reduced to 14.7 cm/s (resp. 12.8 cm/s) for the zonal component (resp. meridian component) compared to the use of the DNSC08 MDT (16.9 cm/s for the zonal component and 14.8 cm/s for the meridian component).

Alternatively, the synthesis of all available information (in-situ oceanographic data, altimetry) can be done in a dynamically consistent way through inverse modeling (Legrand et al., 2003), or through data assimilation into ocean general circulation models, whose outputs are then averaged to obtain an estimate of the ocean Mean Dynamic Topography. Figure 10.6c, d show the MDT obtained respectively by Legrand et al. (2003) and by the GLORYS1V1 $\frac{1}{4}°$ reanalysis from the Mercator

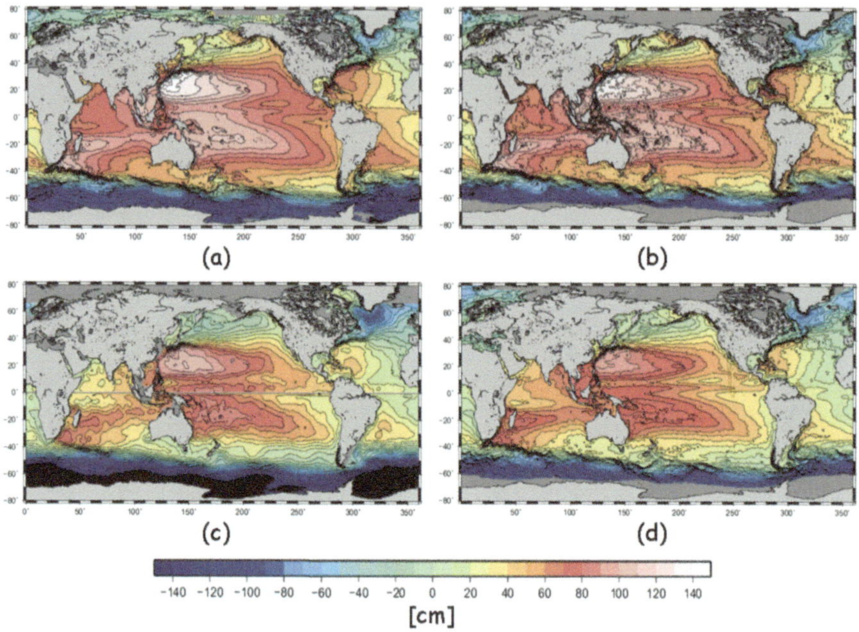

Fig. 10.6 Mean Dynamic Topographies computed from various methods: (**a**) CNES-CLS09 MDT; (**b**) DNSC08 MDT; (**c**) GLORYS1v1 MDT; (**d**) Legrand MDT

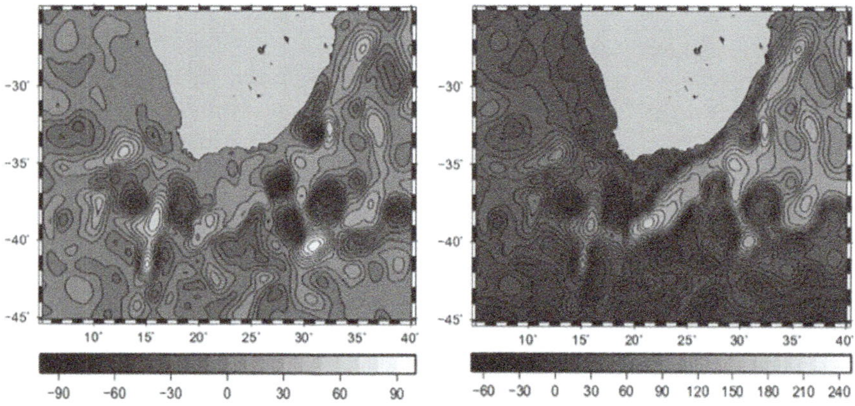

Fig. 10.7 *Left*: Sea level anomalies in the Aghulas current area on 18 February 2009. *Right*: matching absolute dynamic topography from CNES-CLS09 MDT

system. The use of these high resolution MDT solutions has drastically changed our vision of the ocean, as illustrated in the Aghulas current area on Fig. 10.7. Furthermore, as we will now be discussed in Section 10.4, it has led to a number of scientific advances that had not been permitted by the only use of altimetric Sea Level Anomalies.

10.4 Scientific Advances Allowed by Recent Improvements in MDT Determination

Significant improvements have been made for absolute dynamic topography computation since the launch of the first altimetric satellites more than 20 years ago. Although most studies based on altimetric data have looked at the variable part of dynamic topography, the recent improvements have generated an increasing number of papers looking not anymore at ocean mesoscale variability but at the interpretation of the full dynamical signal.

By taking a deeper look into their high resolution Mean Dynamic Topography, Maximenko et al. (2008) reveal the presence of new stationary jet-like striations, which they validated against historical XBT profiles, highlighting a coherent vertical structure until at least 700 m depth. Although the dynamics of these structures is not well understood yet, the emergence of high resolution MDT has made possible the discovery of such new features.

The good quality of the (Niiler et al., 2003) MDT has allowed (Hughes, 2005) to compute the near-surface vorticity balance of the Antarctic Circumpolar Current. They found a compensation of relative and planetary advection at wavelengths greater than 300–500 km. The resulting total vorticity advection was clearly related to features in the bottom topography. The strong mean flow in the ACC allowed the non-linear terms of the vorticity balance to be well resolved. In weaker flow however, the author insisted on the need of higher resolution MDT as well as error covariance estimates in order to better quantify the errors.

The improved knowledge of the ocean MDT has also allowed to improve our understanding of the interaction between the ocean mean circulation and the mesoscale eddy field. Several studies have recently investigated this issue. Using AVISO Sea Level Anomalies together with the MDT from Niiler et al. (2003), Fu, 2006 showed that in areas where the mean flow is mainly eastward (as the Antarctic Circumpolar Current), the intrinsic westward eddy propagation can be compensated leading locally to eastward propagating eddies.

On the other hand, in areas where the mean flow is mainly westward, the intrinsic westward eddy propagation may be reinforced by the mean flow. Inversely, the role of mesoscale eddies for reinforcing the time-mean circulation was investigating in the Kuroshio extension by Qiu and Chen (2010) using AVISO SLA and a mean filed by Teague et al. (1990). They showed that eddy forcing was sustaining the time-mean meanders of the Kuroshio Extension against dissipation.

The better estimation of the ocean's MDT and hence the ocean's absolute dynamic topography has also led to the emergence of a number of ocean surface current's products as OSCAR (Bonjean and Lagerloef, 2002), SURCOUF (Larnicol et al., 2006) or the currents computed by Sudre and Morrow (2008). These surface currents are computed as a sum of the geostrophic component from altimetry and an estimate of the Ekman component. These currents estimates are commonly used for a number of applications as ocean model validation, support to offshore activity, search and rescue, oil spill monitoring...

Another application of getting absolute estimates of the ocean dynamic topography is the estimation and monitoring of the ocean mass transports all around the globe.

In areas where the ocean circulation is known to be mostly barotropic, the absolute dynamic topography values obtained adding altimetric Sea Level Anomalies to the Mean Dynamic Topography, can be used to compute the surface geostrophic current transport, that, multiplicated by the ocean depth, results in an estimate of the volume transport. This approximation was made by Saraceno et al. (2009) to compute the time series of the Zapiola anticyclonic flow transport over the 1993–2007 period. They found a mean transport value for the full time period of 50 Sv when using the Rio et al. (2005) MDT, compared to 40 Sv with the Maximenko and Niiler (2005) MDT. Using model outputs, Volkov and Fu (2008) found a mean value of 77.5 Sv for the 1993–2006 period.

In areas where the ocean circulation is both barotropic and baroclinic, this approximation is not anymore valid. However, provided the vertical structure of the ocean density is known, the geostrophic surface velocity field inferred from altimetric absolute dynamic topography can be used as reference level velocity so as to reconstruct, through the thermal wind equation, the 3D structure of the geostrophic flow (Hunegnaw et al., 2009).

Last but not least, a key application of higher resolution MDT is the assimilation of altimetric anomalies into operational ocean forecasting systems, whose development has been greatly accelerated in the last decade thanks to international projects as GODAE, MERSEA, or MyOcean.

Studies have been carried out to quantify the impact of using an observed Mean Dynamic Topography (as opposed to the model mean) to assimilate altimetric anomalies. For example, a twin experiment was done by the MERCATOR team, covering 8 months (starting in September 2001) to compare the model outputs (analyses and forecasts) using the model MDT (Reference run) or an observed, Combined MDT (CMDT run). The Reference MDT is derived from the mean sea surface height from a forced model run (i.e. no assimilation) covering January 1992–December 1995. The CMDT was computed by (Rio and Hernandez, 2004) through a combination of in-situ data, altimetry and a geoid model. The study showed the strong impact of a more realistic MDT for getting improved analysis and forecasts. Figure 10.8 shows the Mean Eddy Kinetic Energy (EKE) for the Reference run, the CMDT run, and from observed drifters in the Gulf stream region. The CMDT enhances eddy activity suggesting that constraining the modeled mean current to be in the correct place, allows the model to generate eddies that are more consistent with assimilated sea level anomalies, thereby enhancing EKE.

A similar experiment was done in the framework of the European GOCINA (Geoid an Ocean CIrculation in the North Atlantic) project: a specific Mean Dynamic Topography was computed for the North-East Atlantic region, combining information from ocean general circulation models, altimetry and gravimetry, and studies were carried out in three European operational forecasting systems (FOAM, TOPAZ, MERCATOR) to quantify the impact of using this new MDT when assimilating altimetric anomalies. It was shown that the use of the improved MDT led to

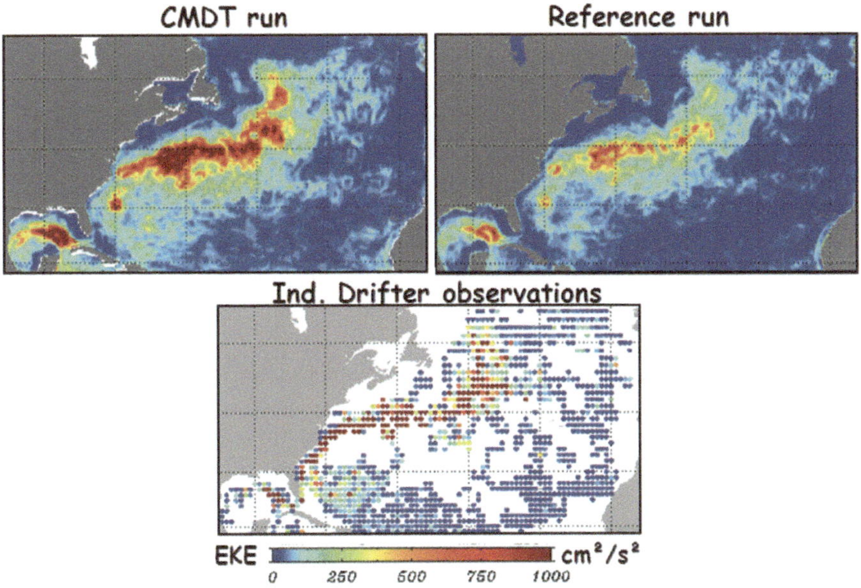

Fig. 10.8 Eight months EKE in the Gulfstream area computed from the reference run (*top right*), the run using an observed MDT (*top left*) and from independent drifter observations (*bottom*)

changes in mass transport of up to 10–20% and of changes in heat transport of about 30%. Overall, increased agreement versus observations was found.

Finally, in the framework of the ECCO consortium (Stammer et al., 2002), Stammer et al. (2007) showed that an improved geoid model would lead to a modification of 10% of the Atlantic Meridional Overturning Circulation. They concluded however that better assessment of the ocean MDT error was mandatory for its optimal use into operational forecasting systems.

The support to operational forecasting centers for developing strategies to make best use of the future GOCE data for altimetry assimilation was the objectives of the European FP7 GOCINO project whose main outcomes are described in (Haines et al., 2010).

10.5 Benefits and Limits of GOCE for Oceanographic Applications

The GOCE (Gravity Field and Steady-State Ocean Circulation) satellite, launched on March, 17th 2009, is the first Earth Explorer Core Missions identified by ESA in the framework of the Living Planet Programme. The principle of the measurement is based on satellite gradiometry: the gradiometer embarked on GOCE satellite consists of three pairs of three axis accelerometers, each of which will measure the Earth's varying gravity field caused by the passage of the spacecraft over varying masses of mountains, ocean ridges, subduction zones... The basic gradiometric measurement is the difference of acceleration measured between two accelerometers in the direction joining them, which correspond to the second derivatives of the gravitational potential.

The gradiometer will provide information for the short and medium wavelengths of the Earth Gravity field while longer wavelengths will be obtained by high-accuracy Satellite-to Satellite measurements and star tracker information. The primary objectives of the GOCE mission are to determine, at a spatial resolution of 100 km, the Earth's gravity field and the geoid height with unprecedented accuracy of respectively 1 mGal and 1 cm. A high number of scientific studies based on altimetry uses gridded maps of Sea Level Anomalies computed from the merging of data from different missions (Ducet et al., 2000).

The actual resolution of these gridded maps depends both on the number of satellites and of their across-track distances. In a 2-satellite configuration (Topex + ERS for instance), the maximum resolution achieved, for a 10 days temporal resolution, is 250–300 km. In a 3-satellites configuration, the spatial resolution is increased to 200 km. In the future, a configuration made of a wide-swath altimeter like SWOT (Surface Water and Ocean Topography) and 2 nadir altimeters would further increase the spatial resolution to 100 km. At that resolution, the accuracy of future GOCE geoids, and therefore of future GOCE-based Mean Dynamic Topography will be sufficient to compute 100 km resolution maps of absolute dynamic topography, and associated geostrophic surface currents.

However, nadir altimeters measure the sea level above the ellipsoid with a maximum frequency of 20 Hz (1 measurement every 350 m). In order to reduce the measurement noise, single pulses are often averaged to a reduced frequency of 1 Hz (1 measurement every 7 km). In the future, even higher spatial resolution will be achieved thanks to the use of wide-swath altimeters ($\sim$1 km). At this spatial resolution, the accuracy of future geoid models based on GOCE data will be insufficient to directly apply Equation (10.1). Applying the repeat track method will therefore be mandatory to compute along-track Sea Level Anomalies. In order to take major benefit from the raw altimetric measurements, and extract the sea level anomalies with maximum accuracy and resolution, high resolution mean profiles are needed or, in the case of non repetitive orbit missions, high resolution Mean Sea Surfaces. Then, to obtain absolute dynamic topography values from the sea level anomalies, the computation of high resolution Mean Dynamic Topography estimates will be mandatory. The required MDT resolution will be highly dependent on the ocean region of interest (higher in western boundary currents, in semi-enclosed seas, in short straits, gyres, where GOCE resolution will not be sufficient. . .lower in the center of suptropical gyres, where GOCE resolution may be sufficient. . .). To achieve the higher resolution required, methods similar to those described in Section 10.4 will be applied, either using in-situ gravimetric data to enhance the geoid resolution, or using in-situ oceanographic data to enhance the resolution of GOCE-based Mean Dynamic Topography.

10.6 Conclusion

While altimetric measurements have been exploited for a long time to study the ocean mesoscale dynamics, the launch of dedicated space gravity missions as GRACE, as well as the development of merging techniques allowing to estimate with increasing accuracy and resolution the ocean Mean Dynamic Topography, has led in the recent years to the better exploitation of the absolute dynamic topography signal. This is of crucial importance for the better understanding and monitoring of the ocean circulation. The availability in the near future of the new GOCE geoid, whose accuracy at 100 km resolution is expected to be close to 1–2 cm, will definitively make the altimetric sciences enter a new era. However, new technology as wide swath altimetry are emerging, pushing back the limits. Similarly, we assist to an increasing need for coastal products (requiring higher spatial and temporal resolution). As a consequence, the optimal use of GOCE data for altimetry will require a sustained effort of combination with other kind of data (gravimetric/oceanographic), in order to keep on improving the estimation of the ocean absolute dynamic topography, for the better understanding, modelling, and forecasting of the ocean currents.

References

Andersen O (2008) The DNSC08 Global Mean Sea Surface and Bathymetry. Presented EGU-2008, Vienna, Austria, April, 2008

Bingham RJ, Haines K, Hughes CW (2008) Calculating the ocean's mean dynamic topography from a mean sea surface and a geoid. J Atmos Ocean Tech 25(10):1808–1822, doi:10.1175/2008JTECHO568.1

Bonjean F, Lagerloef GSE (2002) Diagnostic model and analysis of the surface currents in the Tropical Pacific Ocean. J Phys Oceanogr 32:2938–2954

Cheney RE, Marsh JG, Beckley BD (1983) Global mesoscale variability from collinear tracks of Seasat altimeter data. J Geophys Res 88(C7):4343

Ducet N, Le Traon PY, Reverdin G (2000) Global high resolution mapping of ocean circulation from Topex/Poseidon and ERS-1 and -2. J Geophys Res 105(C8):19477–19498

Fu LL (2006) Pathways of eddies in the South Atlantic Ocean revealed from satellite altimeter observations. Geophys Res Lett 33:L14610, doi:10.1029/2006GL026245

Fu LL, Cazenave A (2001) Satellite Altimetry and Earth Sciences: A Handbook of Techniques and Applications. International Geophysics Series, Vol. 69, Academic Press, San Diego

Haines K, Johannessen J, Knudsen P, Rio MH (2010) An ocean modelling and assimilation guide to using GOCE geoid products (in preparation)

Hernandez F, Schaeffer P (2001) Surface Moyenne Oceanique: support Scientifique à la mission altimetrique Jason-1, et à une mission micro-satellite altimétrique. Contrat SSALTO 2945 – Lot2 – A.1. Rapport final n° CLS/DOS/NT/00.341, CLS, Ramonville St Agne

Hughes CW (2005) Nonlinear vorticity balance of the Antarctic Circumpolar Current. J Geophys Res 110:C11008, doi:10.1029/2004JC002753

Hughes CW, Bingham RJ (2006) An oceanographer's guide to GOCE and the geoid. Ocean Sci Discuss 3:1543–1568

Hunegnaw A, Siegismund F, Hipkin R, Mork KA (2009) Absolute flow field estimation for the Nordic seas from combined gravimetric, altimetric, and in situ data. J Geophys Res 114:C02022, doi:10.1029/2008JC004797

Jayne SR (2006) Circulation of the North Atlantic Ocean from altimetry and the gravity recovery and climate experiment geoid. J Geophys Res 111:C03005, doi:10.1029/2005JC003128

Larnicol G, Guinehut S, Rio MH, Drevillon M, Faugere Y, Nicolas G (2006) The global observed ocean products of the French Mercator project. Proceedings of the "15 Years of Progress in Radar Altimetry" ESA symposium, Venice, March 2006

LeGrand P, Schrama EJO, Tournadre J (2003) An inverse estimate of the dynamic topography of the ocean. Geophys Res Lett 30:1062, doi:10.1029/2002GL014917

Maximenko NA, Melnichenko OV, Niiler PP, Sasaki H (2008) Stationary mesoscale jet-like features in the ocean. Geophys Res Lett 35:L08603, doi:10.1029/2008GL033267

Maximenko NA, Niiler PP (2005) Hybrid decade-mean global sea level with mesoscale resolution. In: Saxena N (ed.) Recent Advances in Marine Science and Technology (2004), PACON International 2005, Honolulu, pp. 55–59

Maximenko NA, Niiler P, Rio MH, Melnichenko OV, Centurioni L, Chambers D, Zlotnicki V, Galepin B (2009) Mean dynamic topography of the ocean derived from satellite and drifting buoy data using three different techniques. J Atmos Ocean Tech 26(9):1910, doi:10.1175/2009JTECHO672.1

Niiler PP, Maximenko NA, McWilliams JC (2003) Dynamically balanced absolute sea level of the global ocean derived from near-surface velocity observations. Geophys Res Lett 30(22):2164, doi:10.1029/2003GL018628

Pavlis N, Holmes SA, Kenyon SC, Factor JK (2008) An Earh Gravitational Model to Degree 2160: EGM2008. Presented EGU-2008, Vienna, Austria, April 2008

Qiu B, Chen S (2010) Eddy-mean flow interaction in the decadally-modulating kuroshio extension system. Deep Sea Res 57, doi:10.1016/j.dsr2.2008.11.036

Rio MH, Hernandez F (2004) A mean dynamic topography computed over the world ocean from altimetry, in situ measurements and a geoid model. J Geophys Res 109:C12032, doi:10.1029/2003JC002226

Rio MH, Poulain PM et al. (2007) A mean dynamic topography of the Mediterranean sea computed from altimetric data, in situ measurements and a general circulation model. J Mar Syst 65: 484–508

Rio MH, Schaeffer P, Hernandez F, Lemoine JM (2006) From the altimetric sea level measurement to the ocean absolute dynamic topography: mean sea surface, geoid, mean dynamic topography, a three-component challenge. Proceedings of the "15 years of progress in Radar altimetry" ESA symposium, Venice, March

Rio MH, Schaeffer P, Moreaux G, Lemoine JM (2010). A new mean dynamic topography computed over the global ocean from GRACE data, altimetry and in situ measurements (in preparation)

Rio MH, Schaeffer P et al. (2005) The estimation of the ocean mean dynamic topography through the combination of altimetric data, in situ measurements and GRACE geoid: from global to regional studies. Proceedings of the GOCINA International Workshop, Luxembourg

Saraceno M, Provost C, Zajaczkovski U (2009) Long-term variation in the anticyclonic ocean circulation over the Zapiola Rise as observed by satellite altimetry: evidence of possible collapses. Deep Sea Res Part I Oceanogr Res Pap 56(7):1077–1092

Stammer D, Köhl A, Wunsch C (2007) Impact of accurate geoid fields on estimates of the ocean circulation. J Atmos Ocean Tech 24:1464–1478

Stammer D, Wunsch C, Fukumori I, Marshall J (2002) State estimation improves prospects for ocean research. EOS Trans Amer Geophys Union 83(289):294–295

Sudre J, Morrow RA (2008) Global surface currents: a high-resolution product for investigating ocean dynamics. Ocean Dyn 58:101–118

Teague WJ, Carron MJ, Hogan PJ (1990) A comparison between the generalized digital environmental model and levitus climatologies. J Geophys Res 95:7167–7183
Thompson KR, Huang J, Véronneau M, Wright DG, Lu Y (2009) Mean surface topography of the northwest Atlantic: comparison of estimates based on satellite, terrestrial gravity, and oceanographic observations. J Geophys Res 114:C07015, doi:10.1029/2008JC004859
Vianna ML, Menezes VV, Chambers DP (2007) A high resolution satellite-only GRACE-based mean dynamic topography of the South Atlantic Ocean. Geophys Res Lett 34:L24604, doi:10.1029/2007GL031912
Volkov DL, Fu LL (2008) The role of vorticity fluxes in the dynamics of the Zapiola Anticyclone. J Geophys Res 113:C11015, doi:10.1029/2008JC004841
Vossepoel FC (2007) Uncertainties in the mean ocean dynamic topography before the launch of the Gravity Field and Steady-State Ocean Circulation Explorer (GOCE). J Geophys Res 112:C05010, doi:10.1029/2006JC003891

Chapter 11
The Marine Geoid and Satellite Altimetry

Walter H.F. Smith

11.1 Introduction

If the tides and currents in the ocean and atmosphere ceased their motion, so that the fluid parts of the Earth came to rest on the solid parts, with all parts rotating together uniformly as a rigid body, then hydrostatic equilibrium would require that the ocean-atmosphere interface (that is, sea level) must lie on a surface of constant potential energy of the Earth's gravity field. (Gravity in this sense is what is experienced by an observer rotating with the Earth, so that it includes the effects of a uniform rigid body rotation added to the Newtonian gravitational attraction.) This equipotential surface, the hydrostatic equilibrium shape for sea level in the absence of tides, currents and winds, is called the marine geoid.

Satellite altimeters measure the instantaneous sea surface height above a reference Earth ellipsoid. This height is the sum of the geoid height plus the dynamic topography associated with the ocean's flows and responses to tidal and atmospheric forcings. If the geoid height can be removed from the altimetric observations, then the residual height can be directly interpreted in terms of ocean dynamics. Until recently, however, geoid heights were not known with sufficient accuracy to be used directly in this way.

In some altimetric applications to ocean dynamics, it is the horizontal gradient of the dynamic height, that is, the dynamic slope, that is most relevant. In the geostrophic approximation, e.g., the dynamic slope is related to the surface current velocity through the Coriolis parameter; at mid-latitudes, a 1 m/s current produces a dynamic slope of $\sim$10 μrad (1 μrad is 1 mm change in height per 1 km of horizontal distance). If a geoid model is to be used to obtain the dynamic ocean signal, one must consider not only errors in geoid height, but also in geoid slope. The geoid slope is coupled to gravity anomalies via Laplace's equation. Therefore one can use gravity anomaly data to verify the accuracy of a geoid slope model.

W.H.F. Smith (✉)
Laboratory for Satellite Altimetry, National Oceanic and Atmospheric Administration, Silver Spring, MD 20910, USA
e-mail: Walter.HF.Smith@noaa.gov

V. Barale et al. (eds.), *Oceanography from Space*,
DOI 10.1007/978-90-481-8681-5_11,

This chapter reviews those aspects of potential theory, gravity and geodesy that are relevant for the present accuracy and error in the marine geoid. These include the relation between gravity anomaly potential and geoid height, upward continuation, and the use of ship gravimetry to verify marine gravity field models. The power spectrum of the gravity field is shown in comparison with Earth topography, to establish that the marine geoid has considerable signal due to sea floor topography. This signal cannot be sensed at orbital altitude by space gravity missions, due to upward continuation. The geoid slope at short scales can be determined from altimetry, and I present an assessment of this using marine gravity surveys, as well as a summary of other tests and results presented elsewhere.

11.2 The Geoid in Classical Potential Theory

In classical geodesy, the geoid height is related to the gravity potential as follows. Here, "potential" is used to mean potential energy per unit mass. Let V be the Newtonian gravitational attraction potential of the Earth. Let $W = V + \Phi$ be the gravity potential, with Φ the potential of the Earth's rotation, assumed uniform.

Let a reference ellipsoidal gravity field model $U = E + \Phi$ be defined, with E the attraction potential and Φ again accounting for the rotation; E is chosen so that when $U = U_0$, a constant, the shape of the $U = U_0$ surface is an oblate ellipsoid of revolution corresponding closely to the Earth's actual geoid. The reference gravity acceleration is $\vec{\gamma} = \nabla U$. Thus on the ellipsoid the normal to the ellipsoid would be the direction of the vertical in the absence of gravity anomalies.

Let ϕ, λ, z indicate the latitude, longitude and height above the reference ellipsoid, and make a first-order Taylor series expansion of the gravity potential at a height z in terms of the value on the ellipsoid:

$$W(\phi,\lambda,z) = W(\phi,\lambda,0) + z\hat{n} \cdot \nabla W(\phi,\lambda,0) + \ldots \qquad (11.1)$$

Setting $W(\phi, \lambda, z) = U_0$ in this equation gives an implicit equation for the geoid height anomaly, $z = N(\phi, \lambda)$.

The potential of the actual field departs from the ellipsoidal model field by an amount, $T = W - U = V - E$, known as the disturbing potential in the literature. T is the potential of the anomalies in the gravity field. Since T is smaller than 10^{-3} times W, in solving (11.1) for the geoid height we may make the approximations $\vec{\gamma} \approx \nabla W$ and $\hat{n} \cdot \nabla W \approx -\gamma_0$, with γ_0 the magnitude of reference (standard) gravity on the ellipsoid. Then the expression for the geoid height reduces to

$$N(\phi, \lambda) \approx \frac{T(\phi, \lambda, 0)}{\gamma_0(\phi)} \qquad (11.2)$$

which is known as Bruns' formula (Bruns, 1828).

Two approximations are made in Bruns' formula. First, the Taylor series (11.1) is truncated after the first-order term. Second, the actual gravity is replaced by the standard gravity in the denominator on the right hand side of (11.2). Since gravity anomalies have peak values less than 10^{-3} of standard gravity, while the geoid height has peak values around 100 m, use of (11.2) results in a geoid height estimate that is accurate to better than a decimetre in the worst case. For use with altimetry, higher accuracy is required, and modern geoid calculations employ higher order iterative refinement. Geoid height errors in the EGM2008 gravity field model, for example, are around 5 cm (Pavlis et al., 2008), and these should be due to errors in the model, not in the calculations.

When classical geodesy developed in the nineteenth century it was assumed that the geoid would correspond to mean sea level, and topographic elevations on land are given as heights above sea level by classical levelling surveys. However, today one recognizes that a long-term average of sea level would include the dynamic topography of steady flow and the permanent deformation of the solid and fluid Earth in response to the tides. Because classical geodesy expects the disturbing potential, T, to arise from mass anomalies within the Earth, care must be taken to define the reference ellipsoidal model for the field appropriately.

Modern geoid models now come in various flavours, depending on how the zero-frequency tidal effects are reckoned. Since tide models are also used in altimetry, both for correcting sea surface heights and for estimating satellite orbit ephemeredes, one must choose a geoid definition consistent with the sea surface height corrections being used (Ekman, 1989; Rapp, 1994; Pavlis et al., 2008). Classical geodesy also assumes that the reference ellipsoid can be defined so that the geometric center of the ellipsoid coincides with the center of mass of the Earth. Modern observations reveal that the center of mass is in motion by small amounts, primarily on an annual cycle. Since the satellite altimeters are also orbiting the center of mass, these geocenter variations are handled as coordinate reference frame issues, and will not be treated here.

11.3 Upward Continuation

The potential of the anomaly field, T, obeys Laplace's equation outside any volume enclosing its sources. If its sources are confined within a sphere of radius a, then T may be expanded in spherical harmonics:

$$T(r,\theta,\lambda) = \frac{GM}{r} \sum_{n} \left(\frac{a}{r}\right)^{n} \sum_{m=-n}^{n} \alpha_{nm} Y_{nm}(\theta,\lambda). \tag{11.3}$$

In (11.3), r, θ, λ are the spherical coordinates radius, geocentric colatitude, and longitude, G is the Newtonian gravitational constant, M the mass of the Earth, Y is a surface spherical harmonic of degree n and order m, and α is a dimensionless coefficient on Y.

In dealing with anomalies whose spatial extent is small compared with the mean radius of the Earth, a flat Earth approximation may be useful. Letting x, y, z be coordinates in the east, north and upward directions, and assuming the sources of the field are confined to the half-space $z < 0$, the potential may be written:

$$T(x,y,z) = \iint \tilde{T}_0(u,v) \exp\left[i2\pi(ux+vy) - 2\pi z\sqrt{u^2+v^2}\right] du\,dv. \qquad (11.4)$$

In (11.4), u, v are spatial wavenumbers in the x and y directions with units of cycles per length, and $\tilde{T}_0$ is the Fourier transform of T in the $z = 0$ plane.

Equations (11.3) and (11.4) are examples of solutions to boundary value problems for Laplace's equation. Each of these equations includes a term that diminishes the amplitude of T as one moves farther away from the source of the field. This diminishing of the field as one moves away from the source is scale, or wavelength, dependent in each case. This phenomenon is called "upward continuation".

In (11.3), a component with spherical harmonic degree n is diminished by an amount $(a/r)^n$ when $r > a$. In (11.4), a component with wavenumber $q = \sqrt{u^2+v^2}$ is diminished by an amount $\exp(-2\pi zq)$. The wavelength associated with q is $L = 1/q$. The wavelength associated with a spherical harmonic of degree n is $L = 2\pi R/\sqrt{n(n+1)}$, where R is a mean radius for the Earth (Backus et al., 1996). Expressing the upward continuation factors in terms of these wavelengths the spherical and flat-Earth formulae give essentially the same results.

Figure 11.1 plots the upward continuation attenuation factor for altitudes of 400 km, corresponding to the orbital altitude of the GRACE and GOCE satellite gravity missions, and 4 km, corresponding to the mean depth of the sea floor. The attenuation is $\exp(-\pi)$or more severe when the half-wavelength is shorter than the upward continuation altitude. Many geophysical signals are broad-band, and so one cannot assign a single wavelength, L, to them; however, roughly speaking, L is about twice the width of an anomaly. Thus one may expect that the marine geoid has little power at scales shorter than 4 km in the deep ocean. At the orbital altitude of GRACE and GOCE, an anomaly with a width less than 400 km will be less than 4% of the strength it would be on the surface of the Earth.

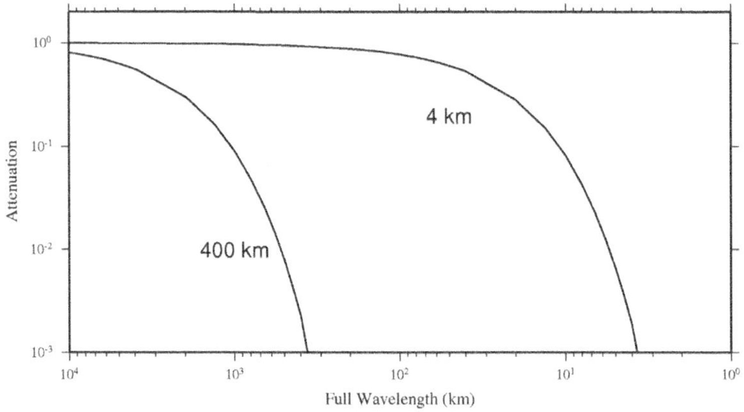

Fig. 11.1 The attenuation due to upward continuation to 4 km and to 400 km altitude, as a function of full-wavelength, L

Upward continuation prevents satellite gravity observing systems such as GRACE or GOCE from obtaining useful information on the marine geoid at scales shorter than a few hundred km. Although satellite altimeters orbit at 800 to more than 1,300 km above the Earth, their radars measure the sea surface height, which reflects gravity at sea level, not at orbital altitude. It turns out that altimetry is the best way to get information about marine gravity anomalies at scales shorter than a few hundred km.

11.4 Geoid Slopes and Gravity Anomalies

Anomalies in atmospheric mass contribute negligibly to marine geoid height anomalies, and so we can assume that T satisfies Laplace's equation on and above the sea surface. In Cartesian coordinates

$$\left[\frac{\partial^2}{\partial x^2} + \frac{\partial^2}{\partial y^2} + \frac{\partial^2}{\partial z^2}\right] T = 0. \tag{11.5}$$

Substituting the geoid slope components $\eta = -\partial N/\partial x$, $\xi = -\partial N/\partial y$, and the gravity anomaly $g = -\partial T/\partial z$ (the negative signs come from sign conventions in classical geodesy) one has

$$\gamma_0 \left[\frac{\partial \eta}{\partial x} + \frac{\partial \xi}{\partial y}\right] = \frac{\partial g}{\partial z} \tag{11.6}$$

which couples the geoid slopes to the gravity anomaly.

The Fourier transform of this expression

$$i2\pi \gamma_0 \left[u\tilde{\eta} + v\tilde{\xi}\right] = -2\pi q\tilde{g} \tag{11.7}$$

may be used to compute the gravity anomaly from geoid slopes, and vice versa (Haxby et al., 1983; Sandwell and Smith, 1997). The tilde over a quantity in (11.7) indicates the Fourier transform in the $z = 0$ plane, as in (11.4).

Geoid heights are measured from the ellipsoid, while the gravity anomaly is the difference between actual gravity on the geoid and reference gravity on the ellipsoid. If $z = 0$ is the local flat Earth approximation to the ellipsoid then the expressions (11.6) and (11.7) are correct apart from a small error in the gravity anomaly known as the "indirect effect" (Chapman and Bodine, 1979). This effect is negligible if the flat-Earth approximation is used only for local and short-wavelength anomalies, with geoid slopes and gravity anomalies of larger scale treated by spherical harmonics, as in the "remove-restore" procedure (Sandwell and Smith, 2009).

The geoid slopes η and ξ are known as "deflections of the vertical" in geodesy. Historically, on land, they were measured astronomically, and were given in arc-seconds (a.s.); in the context of altimetry of the marine geoid it is convenient to measure them in micro-radians (μrad, 1 a.s. is about 4.8 μrad). The μrad is dimensionless. Gravity anomalies are traditionally expressed in milliGals (mGal; 1 mGal = 10^{-5} m/s^2). Since γ_0 is about 9.8 m/s^2, or 0.98 times 10^6 mGal, the mGal is about one part per million of γ_0; if (11.6) or (11.7) are expressed in µrad and mGal, the scale factor γ_0 is effectively $0.98 \approx 1$ mGal/µrad. Thus geoid slopes in micro-radians are related to gravity anomalies in milliGals.

11.5 Root Mean Square Amplitude and Variance Spectrum

The expected root mean square (RMS) amplitude of the geoid slope may be related to the RMS amplitude of the gravity anomaly most easily by returning to the spherical harmonic expression. Here we will use a spherical approximation with an effective mean Earth radius, R. From the spherical harmonic expansion coefficients of T, α_{nm} in (11.3), the spherical approximations of the geoid height and gravity anomaly are

$$N(\theta, \lambda) = R \sum_{n} \sum_{m=-n}^{n} \alpha_{nm} Y_{nm}(\theta, \lambda) \tag{11.8}$$

$$g(\theta, \lambda) = \frac{GM}{R^2} \sum_{n} (n-1) \sum_{m=-n}^{n} \alpha_{nm} Y_{nm}(\theta, \lambda). \tag{11.9}$$

If the gravity anomaly were simply $-\partial T / \partial r$ then we would expect the factor $(n-1)$ in (11.9) should be $(n+1)$; the subtraction stems from the definition of gravity anomalies on the geoid (Heiskanen and Moritz, 1967).

Now define the degree variances:

$$\sigma_n^2 = \sum_{m=-n}^{n} |\alpha_{nm}|^2. \tag{11.10}$$

In terms of these,

$$\left\langle N^2 \right\rangle = R^2 \sum_{n} \sigma_n^2 \tag{11.11}$$

$$\left\langle g^2 \right\rangle = \left[\frac{GM}{R^2}\right]^2 \sum_{n} (n-1)^2 \sigma_n^2 \tag{11.12}$$

$$\left\langle |\nabla_1 N|^2 / R^2 \right\rangle = \sum_{n} n(n+1) \sigma_n^2. \tag{11.13}$$

The notation $\langle f^2 \rangle$ above indicates the squared quantity averaged over the surface of the sphere. In Equation (11.13), ∇_1 is the Beltrami operator that takes the gradient components in the surface of the unit sphere, so that $|\nabla_1 N| / R$ is the magnitude of the geoid slope; the factor $n(n+1)$ comes from the properties of the Beltrami operator (Backus et al., 1996).

For n large enough (wavelengths short enough), both (11.12) and (11.13) are dominated by their n^2 terms. If gravity and geoid slope are expressed in mGal and µrad, the degree variance spectrum of the geoid slope is essentially that of the gravity anomaly at large enough n. If a significant fraction of the total variance is at these large n, then the RMS amplitude of gravity anomalies can be used to infer the RMS amplitude of geoid slopes.

In the absence of isostatic compensation (Watts, 2001) and upward continuation, gravity is proportional to the product of topography and density contrast, to first order (Parker, 1973). By "density contrast" is meant the change in density across the topography, that is, from air to rock or sediment on land, and from water to rock or sediment on the sea floor.

Because the density contrasts across land and seafloor topography are very different, one should not expect a clear relationship in the spherical harmonic spectra of topography and gravity, as the spherical harmonic coefficients combine data on land and at sea. Even so, I have plotted these in Fig. 11.2, using spherical harmonic coefficients supplied in the EGM2008 model (Pavlis et al., 2008).

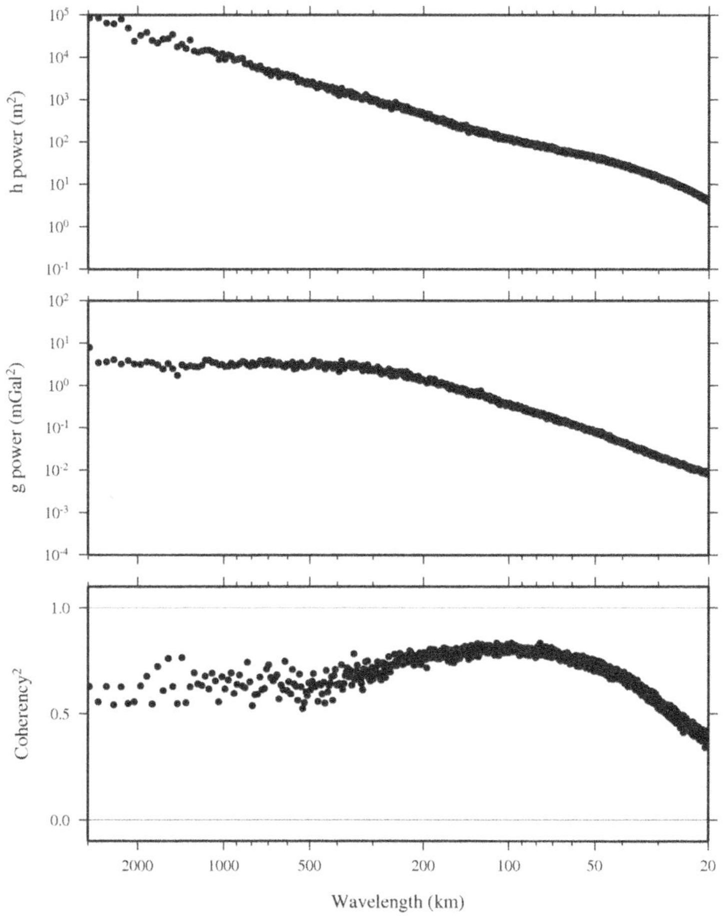

Fig. 11.2 Degree variance spectra of Earth topography (*top*), gravity anomaly (*middle*), and cross-spectral coherency between these (*bottom*), derived from the EGM2008 gravity field model

The degree variance spectrum for Earth's topography, h, decays as n^{-2} at all harmonic degrees, indicating that $|\nabla_1 h|/R$ would have a white spectrum. This is consistent with the notion that topography is a random walk or a fractal. The gravity spectrum shows the same trend at wavelengths shorter than a few hundred km, and gravity and topography are mostly coherent at these scales, indicating that the source of these gravity anomalies is primarily uncompensated topography. The root-mean-square amplitude of gravity anomalies averaged over the surface of the sphere is about 35 mGal. Since the gravity spectrum decays slowly with n, one may conjecture that the RMS geoid slope can be inferred from the RMS gravity anomaly, and that it is approximately 35 μrad.

11.6 Using Satellite Altimetry to Determine the Geoid

Figure 11.2 shows that there is significant power in the geoid at wavelengths too short to be determined by GRACE or GOCE, due to upward continuation. In order to determine the geoid at these scales one needs gravity anomalies measured on the geoid, not 400 km above it. If ships carrying gravity meters covered the oceans densely and uniformly enough, and if shipborne gravimetry were accurate enough, one could construct a geoid gravimetrically, independently of altimetry. Unfortunately, neither the accuracy nor the coverage of shipborne gravimetry is adequate to this task (Wessel and Watts, 1988).

The best hope of determining the short-wavelength marine geoid is therefore to use satellite altimetry itself. The question arises then, how can one be sure that the result will be the geoid and not some hybrid mean sea surface height? If a geoid determined by altimetry is used to obtain the dynamic topography, will one be accused of circular reasoning? How can this possibly work?

Geoid slopes come to the rescue and furnish a "trick". Although the satellite altimeters measure the instantaneous sea surface height, which is not the geoid, the slope along an altimeter height profile is very nearly (within 1 μrad, in most cases) equal to the slope of the geoid along the altimeter's ground track. Most errors in the altimeter measurement have long-enough correlation distances that they produce sub-microradian errors in sea surface slope. The slope of the tide and dynamic topography is likewise sub-microradian, except in areas where there are energetic mesoscale geostrophic currents or wide shallow shelves with large tide gradients.

Sandwell and Smith (2009) tabulate the various causes that may make the slope of an altimeter profile depart from the geoid slope, and show that nearly all are sub-microradian. They also describe an iterative procedure by which the significant departures can be filtered out. One first makes an initial model estimating north and east geoid slopes to best-fit many ascending and descending satellite tracks from many inclinations (Sandwell, 1984; Sandwell and Smith, 1997). These initial north and east slopes are used to build an initial gravity field model. Then each profile is compared against the initial model, and the residual is filtered to remove slopes due to ocean dynamics. The filtered residual slopes are then used to refine the initial model. After a few iterations, the results agree with adjusted altimeter slope profiles to 2–4 μrad and with the best available ship gravity to 2–3 mGal (Sandwell and Smith, 2009).

11.7 Verifying the Altimetric Geoid

An example of a marine gravity field constructed from altimetry by this method is shown in Fig. 11.3. Anomalies associated with sea floor topography and tectonic features are clearly evident: the rift valley and flanking mountains of the Mid-Atlantic Ridge; fracture zones; seamounts. Also evident are propagating rifts and changes in bottom roughness associated with changes in seafloor paleo-spreading rates (Smith, 1998; Goff et al., 2004) that have been observed to control changes in mixing rates in the water column above (Mauritzen et al., 2002). Some of these signals are very subtle, and their clear visibility attests to the high signal-to-noise ratio in the model at quite short wavelengths. There are no streaks evident that would suggest errors aligned with satellite ground tracks. If non-geoidal heights such as dynamic topography or systematic measurement errors had crept into the field, then there would be streaks appearing along satellite tracks (Olgiati et al., 1995).

The marine gravity field constructed from altimetry may be sampled along the tracks of ships carrying gravity meters, and the altimetric gravity compared with the shipborne gravimetry. Differences in the two will be due to errors in the ship gravity as well as in the altimetric gravity. In fact, ship gravity is prone to large errors (Wessel and Watts, 1988) and care must be taken to select modern cruises with GPS

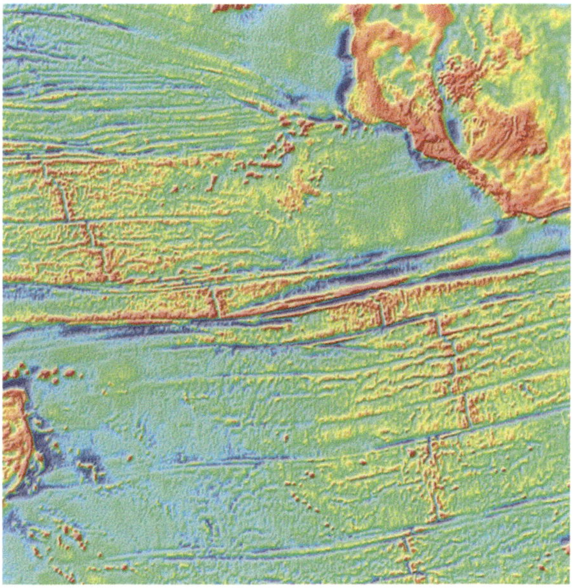

Fig. 11.3 Gravity anomaly field in the Equatorial Atlantic determined from altimetry by Sandwell and Smith (2009)

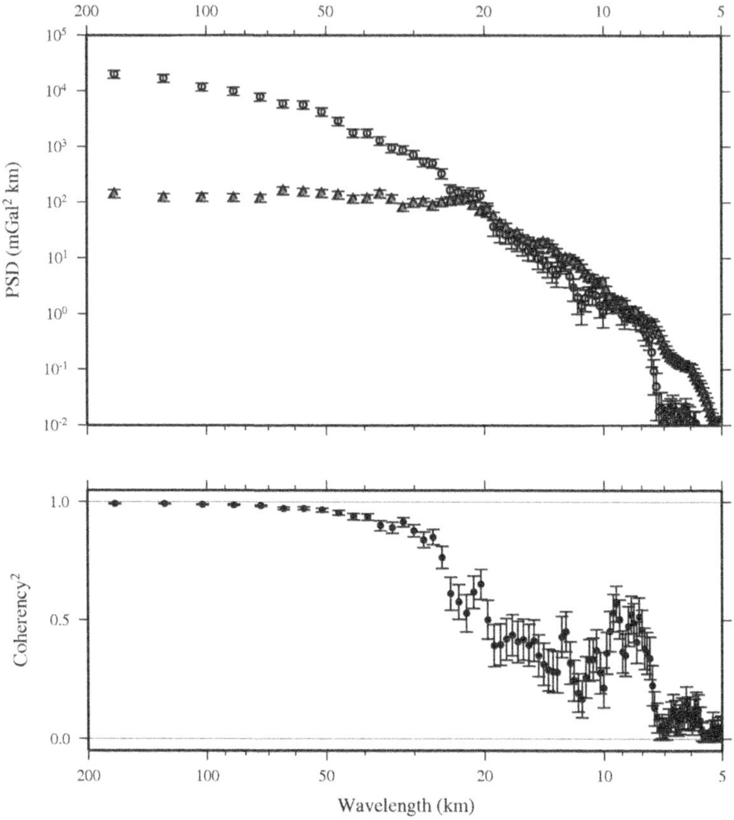

Fig. 11.4 *Top*: Coherent (*circles*) and incoherent (*triangles*) power spectral density from a cross-spectral analysis comparing ship and altimeter gravity. *Bottom*: Coherency between ship and altimeter gravity

navigation and state of the art gravimeters. I selected 9 ship gravity surveys globally distributed and totaling about 20,000 km of track line for the cross-spectral analysis (Fig. 11.4). The RMS differences between ship gravity and altimeter gravity were 2.3–3.6 mGal, depending on the ship used (RMS values determined from point measurements, without averaging the ship values). Without knowing the signal to noise ratio in the ship data as a function of wavelength, one cannot determine the noise in the altimetric field. However, Fig. 11.4 shows that the signals that are common to both exceed the differences at full wavelengths longer than 20 km.

One may conclude that the marine gravity anomalies determined from satellite altimetry are accurate at wavelengths of 20 km and longer, and overall error RMS is not more than 2–3 mGal.

Further verification comes from the EGM2008 model. Sandwell and Smith (2009) also iterated with the EGM2008 development team, for which the primary data source used over the ocean is the Sandwell and Smith model (Pavlis et al.,

2008). EGM2008 geoid heights were compared against land geodetic surveys, tide gauges, and long-term means of Topex/Poseidon/Jason altimeter data. RMS errors in geoid height are around 5 cm, and RMS errors in deflections of the vertical are around 0.3 a.s. (1.5 μrad). The geoid height at a point is sensitive to gravity anomalies over a wide area (Heiskanen and Moritz, 1967), and so the coastal tide gauge and land survey comparisons are sensitive to the accuracy of the deep ocean data.

It should be noted that resolution of the marine gravity field by altimetry requires spatially dense sampling. The presently available data come from the Geosat Geodetic Mission (GM) of 1985–1986 and the ERS-1 Geodetic Phases E and F of 1994–1995. All other precise altimeter data have been acquired along exact repeat mission ground tracks, the cross-track spacing of which is too large to resolve the gravity anomalies in Fig. 11.3.

11.8 Implications for Oceanographic Satellite Altimetry

Satellite altimetry developed as a tool for observing ocean dynamics at a time when the gravity field was not known well enough to furnish a geoid of sufficient accuracy to permit direct calculation of the dynamic topography. "Exact repeat mission" (ERM) orbits were designed so that the satellite's ground track would repeat within ±1 km after a fixed number of orbital revolutions and rotations of the Earth with respect to the satellite's orbital plane, called synodic "days". By making repeated observations on the same ground tracks, altimeters furnished temporal changes in sea surface height that could be interpreted as changes in tidal and dynamical signals, without requiring knowledge of the geoid height along the tracks.

Different satellite series have used different ERM orbits with different space and time sampling characteristics, with observations spanning different years and decades. The long-term average sea surface height from each series has different tidal aliasing and error characteristics, and different sampling of decadal and longer temporal variations, complicating the blending of observations from different ERM orbits. Even so, oceanographers are now in the habit of assuming that they must use ERM orbits to observe their signals, that geoids are not sufficient to allow direct observation of dynamic height signals, and that height anomalies must be referred to a long-term average of sea surface height, rather than a geoid model.

In fact, however, errors in current geoid models are small, around 5 cm in height and 1.5 μrad in slope (these are RMS values), and appear to be confined to full-wavelengths shorter than 20 km, that is, spatial scales much shorter than the correlation scales of some dynamical signals in the open ocean (Jacobs et al., 2001). These virtues permit the observation of dynamic ocean signals from orbits other than traditional oceanographic ERMs. For example, Scharroo and Smith (2009) find that mesoscale eddies can be observed equally well from ERM and non-ERM orbits. Scharroo and Smith's result has prompted others to examine whether non-ERM orbit data can be used to improve tide models (W. Bosch, personal communication, at the 2nd Coastal Altimetry Workshop, Pisa, Italy, 2008).

The ERM requirement of ground track repeat within ±1 km is a stringent requirement, necessitating maintenance of the orbit's semi-major axis to about 1 part per million, for example. Even so, cross-track geoid slopes can introduce apparent variability in sea surface height as the ERM track moves within the ±1 km band. The movement has the effect of shifting the track east or west as the satellite arrives early or late to its intended Equator crossing. If the global averages of the geoid slope components obey:

$$\left\langle |\nabla_1 N|^2 \right\rangle = \left\langle \eta^2 \right\rangle + \left\langle \xi^2 \right\rangle \tag{11.14}$$

and if the field is isotropic, on average, then each term on the right hand side of (11.14) is equal in magnitude, and it follows that if the RMS geoid slope is 35 μrad then the RMS east-west component of slope is $35/\sqrt{2}$, or 25 μrad. As the track moves ±1 km the observed sea surface height will appear to change by 2.5 cm typically, and much more in areas of strong gravity anomalies such as trenches, simply due to the east–west slope of the geoid. Thus use of an ERM does not completely eliminate the need for geoid knowledge, and geoid knowledge is perhaps good enough to make ERMs unnecessary in the future.

Altimetry cannot improve the geoid resolution further without new data collected along a spatially dense network of ground tracks. The Geosat GM and ERS-1 GM missions provide a network with a typical spacing of 4 km. To improve upon this will require a new altimeter with better signal-to-noise (more statistically independent looks), a long time between exact repeats, and a long mission duration. One may hope to achieve this with CryoSat-2.

Acknowledgements A. B. Watts and D. T. Sandwell introduced me to this topic. K. M. Marks, J. L. Lillibridge, and an anonymous reviewer kindly reviewed the manuscript. Any errors are my own. The manuscript contents are solely the opinions of the author and do not constitute a statement of policy, decision, or position on behalf of NOAA or the US Government.

References

Backus G, Parker R, Constable C (1996) Foundations of geomagnetism, Cambridge University Press, New York

Bruns EH (1828) Die figure der erde. Publikation Königl. Preussisch. Geodätisches Institut, P. Stankiewicz Buchdruckerei, Berlin

Chapman M, Bodine J (1979) Considerations of the indirect effect in marine gravity modelling. J Geophys Res 84(B8):3889–3892

Ekman M (1989) Impacts of geodynamic phenomena on systems for height and gravity. Bull Geodesique 63(3):281–296

Goff JA, Smith WHF, Marks KM (2004) The contributions of abyssal hill morphology and noise to altimetric gravity fabric. Oceanography 17(1):24–37

Haxby WF, Karner GD, LaBrecque JL, Weissel JK (1983) Digital images of combined oceanic and continental data sets and their use in tectonic studies. Eos Trans AGU 64(52):995–1004

Heiskanen WA, Moritz H (1967) Physical geodesy, W. H. Freeman, San Francisco

Jacobs GA, Barron CN, Rhodes RC (2001) Mesoscale characteristics. J Geophys Res 106(C9):19581–19595

Mauritzen C, Polzin KL, McCartney MS, Millard RC, West-Mack DE (2002) Evidence in hydrography and density finestructure for enhanced vertical mixing over the Mid-Atlantic ridge. J Geophys Res 107, doi:10.1019/2001JC000801

Olgiati A, Balmino G, Sarrailh M, Green CM (1995) Gravity anomalies from satellite altimetry: comparison between computation via geoid heights and via deflections of the vertical. Bull Geodesique 69:252–260

Parker RL (1973) The rapid calculation of potential anomalies. Geophys J R Astron Soc 31: 447–455

Pavlis NK, Holmes SA, Kenyon SC, Factor JK (2008) An Earth Gravipotential Model to Degree 2160: EGM2008. 2008 General Assembly of the European Geophysical Union, Vienna Austria, April 13–18, http://users.auth.gr/~kotsaki/IAG_JWG/EGM08/NPavlis&al_EGU2008.pdf

Rapp RH (1994) Global geoid determination. In: Vanicek X, Christou X (eds.) Geoid and its geophysical interpretations, CRC Press, Boca Raton, FL, pp. 57–73

Sandwell DT (1984) A detailed view of the South Pacific geoid from satellite altimetry. J Geophys Res 89(B2):1089–1104

Sandwell DT, Smith WHF (1997) Marine gravity anomaly from Geosat and ERS-1 satellite altimetry. J Geophys Res 102(B5):10039–10054

Sandwell DT, Smith WHF (2009) Global marine gravity from retracked Geosat and ERS-1 altimetry: ridge segmentation versus spreading rate. J Geophys Res 114:B01411, doi:10.1029/2008JB006008

Scharroo R, Smith HF (2009) Mesoscale ocean dynamics observed by satellite altimeters in non-repeat orbits. Geophys Res Lett 36:L06601, doi:10.1029/2008GL036530

Smith WHF (1998) Seafloor tectonic fabric from satellite altimetry. Annu Rev Earth Planet Sci 26:697–738

Watts AB (2001) Isostasy and flexure of the lithosphere, Cambridge University Press, Cambridge

Wessel P, Watts AB (1988) On the accuracy of marine gravity measurements. J Geophys Res 93(B1):393–413

Chapter 12
Oceanic Planetary Waves and Eddies: A Privileged View from Satellite Altimetry

Paolo Cipollini, Anna C. S. Sutcliffe, and Ian S. Robinson

12.1 Introduction

Satellite altimetry allows sustained observations of a wide range of ocean dynamics, from features spanning just a few tens of km to the mean global sea level spatially, and from a few days to decades temporally, with profound repercussions on our knowledge of the oceans and how they affect climate. Amongst the many success stories for altimetry, one particularly dear to ocean dynamicists is the systematic detection, characterization and tracking of large- and meso-scale propagating features, virtually ubiquitous in the world's oceans, and their classification as either *planetary waves* or *eddies*, which has involved significant efforts by several research groups. In this chapter we review (in Section 12.3) the main results of these efforts, and highlight the important implications that those findings have had on our understanding of how the ocean works. We also discuss (in Section 12.4) several intriguing questions that have been prompted by the satellite-based observations of propagating features – sometimes challenging the previous knowledge based on insufficient experimental data, other times opening completely new paths of investigation into scientifically uncharted waters. But first we start (in Section 12.2) with a brief explanation of the importance of propagating systems for oceans and climate.

12.2 The Importance of Oceanic Propagating Features

Oceans play a crucial role in the Earth's climate system and in mediating its response to the present radiative forcing imbalance, especially at the longer time scales. Ocean dynamics can be divided into a variety of components, each of which can be considered individually at its own scale, but it is important to keep in mind that it is the interactions between all the components at different scales that makes the

P. Cipollini (✉)
Ocean Observing and Climate Research Group, National Oceanography Centre,
Southampton SO14 3ZH, UK
e-mail: cipo@noc.soton.ac.uk

V. Barale et al. (eds.), *Oceanography from Space*,
DOI 10.1007/978-90-481-8681-5_12,

climate system so complex, intriguing and difficult to understand. Ocean mesoscale features are one of the most important modes of energy transfer in the oceans as well as one of the most important reflections of atmospheric forcing of the oceans.

Energy transfers can occur at a range of scales, from the smallest (such as the transfer of wind energy to the ripples on the sea surface) to the very large (such as the general overturning circulation that ensures that heat is transferred from the equator to the poles). Features of the mesoscale have spatial scales of the order of a few tens to 200 km, periods of the order of 10–100 days and speeds of the order of a few cm/s.

Features at larger scales than this are sensitive to the gradient of Coriolis parameter with latitude. They are characterized by meridional (north-south) oscillatory flow with a westward propagation of their phase. Their behaviour can be easily explained when considering the Earth's shape and rotation: when a parcel of water previously at rest, i.e. with no relative vorticity,[1] is displaced northwards (southward) then its planetary vorticity will increase (decrease). In order to conserve the absolute vorticity it must acquire a negative (positive) relative vorticity, which translates into a counter-clockwise (clockwise) rotation. If a line of particles is subjected to these motions then the changes in the relative vorticity will induce a net westward propagation of the disturbance (Killworth and John, 2001). That these features can propagate at all within the ocean is due to the fact that the ocean behaves as a waveguide: the ocean floor and surface effectively confine energy within these boundaries thus allowing for energy to propagate horizontally within them (Gill, 1982), although there is no westward translation of the water mass associated with the wave propagation. These are called planetary waves (also known as Rossby waves).

Features that propagate westwards can be linear (their propagation speed is largely independent of their amplitude) or non-linear (speed depends on their amplitude). Planetary waves are nearly linear and predominant at the larger scales, (300 km or longer). In contrast larger mesoscale eddies in the form of closed rings (normally of diameters around 100–200 km, see Chelton et al., 2007) propagate with non-linear characteristics, and transfer mass as they propagate. As we will illustrate in detail later, altimetry can be used very successfully to observe both classes of phenomena. Their energy dominates the ocean's energy spectrum at long timescales (Killworth and John, 2001); for example, the kinetic energy associated to mesoscale eddies alone is more than an order of magnitude greater than the ocean's mean (Chelton et al., 2007).

The importance of the westward propagating features within the climate system cannot be underestimated: they have been linked to major climate oscillations such as El Niño Southern Oscillation (ENSO) (Jacobson and Spiesberger, 1998; Fu and Qiu, 2002) and the North Atlantic Oscillation (NAO), they are known to interact with the Meridional Overturning Circulation (Hirschi et al., 2007), they interact

[1] *Relative vorticity* is the vertical component of the vorticity relative to the earth's rotating frame of reference; *planetary vorticity* is the vorticity due to the earth's rotation; *absolute vorticity* is the sum of the relative and planetary vorticity.

and maintain western boundary currents and they can also affect the phytoplankton distribution (hence the biology) of the oceans (Killworth et al., 2004), as we will discuss later.

12.3 Observational Evidence of Planetary Waves and Eddies

Of the two classes of ocean phenomena that we review here, eddies are the one that is more easily observed. Most eddies have a clear thermal signature, which makes them observable in SST images, as satellite-borne infrared radiometers have demonstrated since the late 1970s. Moreover, several other water properties in an eddy have significantly different values from the surrounding ocean (because an eddy tends to retain water in its core as it propagates), which along with their limited spatial scale (a few hundred km at maximum) makes them visible in hydrographic sections from ships and has allowed them to be studied extensively since the 1950s. Detecting planetary waves is more complicated. These waves are essentially internal waves, characterized by displacements of the isopycnals (levels of constant potential density) of a few tens of metres, and their signature in the surface elevation if of the order of just a few cm over length scales of hundreds of km. Until the early 1990s there had been only scarce observational evidence of these waves, despite a sound theoretical consensus on their existence for dynamical reasons (Anderson and Gill, 1975; Pedlosky, 1987; Fu and Chelton, 2001).

The advent of high-accuracy satellite altimetry in the early 1990s, with the TOPEX/Poseidon and ERS-1 missions, dramatically changed our viewing capabilities of planetary waves and also opened the way for great improvements in the observation and characterization of eddy dynamics. For the first time we were able to produce a synoptic, global view of the sea surface elevation, showing the ubiquity of westward propagating features at multiple scales, with the clearest signals being initially interpreted as planetary waves in a seminal paper by Chelton and Schlax (1996).

Due to the westward propagation of planetary waves and eddies, it is sufficient to plot east–west (zonal) sections of altimetric Sea Surface Height (SSH) at a chosen latitude (i.e. a section of the data cube in longitude and time, yielding a characteristic diagram called longitude/time plot or Hovmöller plot), to observe some noticeable signals that propagate to the west over time, with the range of speeds expected for eddies and planetary waves. More usually, the quantity plotted is the SSH anomaly with respect to the local mean, which removes the signature of residual geoid errors and of the mean ocean currents. Figure 12.1 shows one such longitude-time plot in the South Pacific at 14°S. The diagonal features apparent in the diagram (diagonal alignments of positive and negative anomalies moving to the west over time) are the surface signatures of planetary waves and eddies, with amplitudes in the range of $\sim$10 cm and propagation speeds of about 10–15 cm/s. The changing slope of the alignments in the longitude/time domain indicates some variability in the propagation speed. A closer look at the figure also reveals a multitude of horizontal scales for the westward propagating features – the most readily visible propagating features

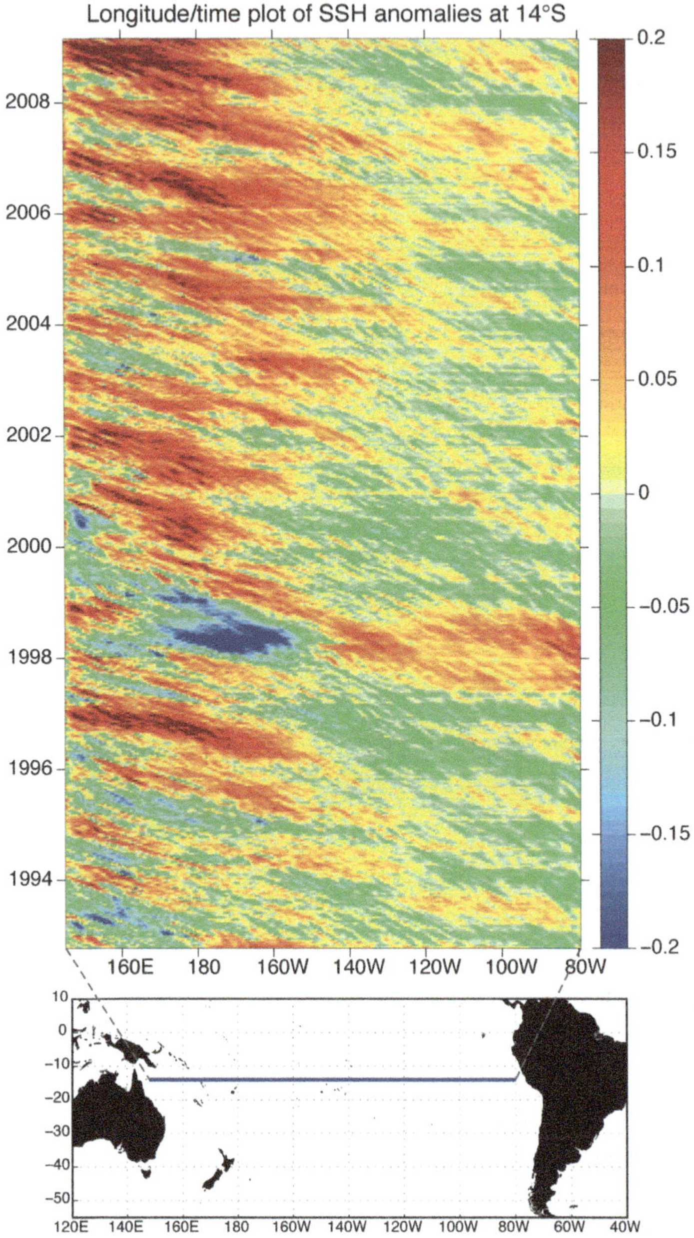

Fig. 12.1 Longitude/time diagram of SSH anomaly at 14°S in the Pacific Ocean (the section is indicated by the *blue line* on the map in the *bottom panel*), clearly showing westward propagating features. Data are SSALTO/DUACS merged multi-mission altimetric SSH distributed by AVISO

have scales of 100–300 km (the finer diagonal stripes in the diagram) but these are superimposed on "bands" of larger longitudinal scale, of the order of 1,000–2,000 km. Similar diagrams have been built for all latitudes in the world's ocean, and westward propagation at many scales is clearly seen almost everywhere. Such diagrams lend themselves to be analyzed with a number of statistical and signal processing techniques, in order to derive objective estimates of the main characteristics of the waves. The two most used analysis tools are the 2-D Fourier Transform, which decomposes the signal into its spectral components, providing an estimation of the zonal wavenumber and frequency for each of those components, and the Radon Transform, a particular image projection that directly offers an objective measure of the main propagation speed of the features. For a detailed description of these two analysis techniques see Cipollini et al. (2006b).

With the build-up of longer altimetric time series, made possible by the various missions that followed (ERS-2, Geosat Follow-On, Jason-1, Envisat and Jason-2, the last three still fully operative at the time of writing in November 2009), the amount of observational studies on planetary waves has increased considerably, giving rise to some important questions for the theoreticians. The most evident discrepancy was the mismatch between the observed propagation speed of the waves and the speeds predicted by the standard (or classic) linear theory of Rossby waves. The observed speeds were up to 2 times faster than the classic theory ones in several regions at mid-latitude, a disagreement already spotted by the early satellite-based studies. This resulted in a formidable amount of work on extending and improving the theoretical models by removing some of the assumptions of the classic one (Killworth et al., 1997; Killworth and Blundell, 1999, 2003a, b; Tailleux and McWilliams, 2000, 2001).

The discrepancy between theoretical and observed speeds is now much reduced with the latest theoretical models (Tailleux and McWilliams, 2001; Killworth and Blundell, 2004, 2005). Figure 12.2 illustrates this concept by showing the comparison of the observed speeds with the speeds predicted by the extended theory by Killworth and Blundell. The observed speeds, displayed in Fig. 12.2a, have been computed over more than 16 years (October 1992–February 2009) of multi-mission SSH anomaly data. The technique employed, which is completely automated, is based on the Radon Transform of longitude/time plots as explained by Cipollini et al. (2006b) using a moving longitude window of 30° and removing the mean value of each longitude-time plot prior to the analysis, as suggested by de la Rosa et al. (2007). Areas within 15° longitude from the coast, where land enters the longitude/time plots, have been blanked out. The theoretical speeds in Fig. 12.2b are from Killworth and Blundell's extended theory (Killworth and Blundell, 2003a, b, 2004, 2005), recomputed with the updated temperature and salinity climatologies of the 2005 World Ocean Atlas (Locarnini et al., 2006; Antonov et al., 2006). A 5° latitudinal band both sides of the equator has been left out of the calculations as the extended theory does not hold for equatorial dynamics; this band is blanked out in Fig. 12.2b.

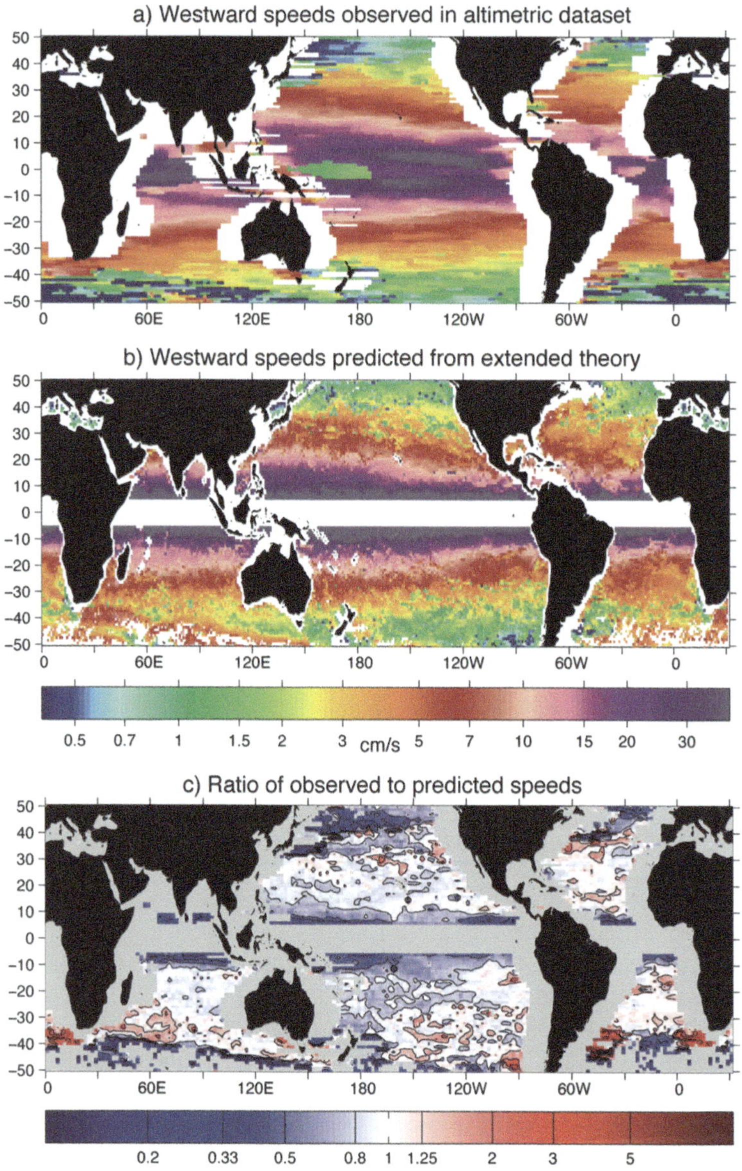

Fig. 12.2 Speeds of planetary waves: (**a**) observed in satellite altimetry; (**b**) predicted by the extended theory; (**c**) ratio of (**a**–**b**). See text for details of the derivation

The level of agreement between the observed and predicted speeds shown in Fig. 12.2a, b can be more easily assessed by looking at their ratio, shown in Fig. 12.2c. Observations and predictions are generally in good accord in the 10–35° latitude band, but slightly less satisfactorily in the South Pacific, for

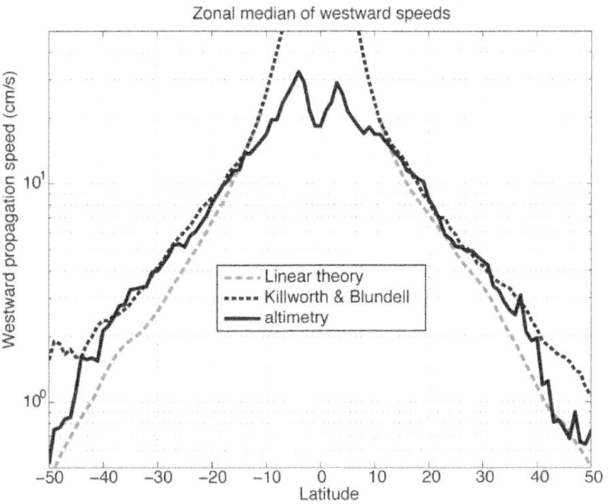

Fig. 12.3 Zonal median of the speeds of planetary waves from the old linear theory (*light gray dashed line*), Killworth and Blundell's extended theory (as in Fig. 12.2b, *dark gray dashed line*) and observed in satellite altimetry (as in Fig. 12.2a, *solid line*)

reasons not yet completely understood and that warrant further investigation. The agreement is also lower in the very energetic regions of the western boundary currents, where current-related instabilities affect the satellite estimates, as well as in the region within $\pm 10^\circ$ the equator, where the observed speeds are significantly lower than predicted. The discrepancy in this latter region is most likely due to a sampling problem: in fact a longitudinal window with a fixed width of 30° becomes too narrow to capture the dominant wavelengths of planetary waves, which may well exceed 2,000–3,000 km when approaching the equatorial band (Polito and Liu, 2003). Despite these problems over specific regions, the extension of the theory prompted by the altimetric observations represent a dramatic improvement on the classic linear theory, as clearly visible in Fig. 12.3 which shows that the zonal medians (i.e. the median values at each latitude) of the speeds predicted by the new extended theory match the observed speeds significantly better (outside the $\pm 10^\circ$ band) than those predicted by the classic linear theory.

The development of an accurate theory can also increase our understanding of the processes and dynamics that occur within the oceans. The fact that waves were seen to propagate faster than predicted has led to the discovery of the importance of contributions from sources such as the ocean's background mean flow, local bathymetry and external forcing. The inclusion of a mean background baroclinic flow in the planetary wave theory (Killworth and Blundell, 1999) led to an increase in the predicted wave speeds, reinforcing the idea that the ocean's local properties factor greatly into how the waves propagate. This opens-up a new avenue for thought, as one has to consider the climate scenario in which we live.

With predictions for a heating world and the implications that might have on the oceanic internal structure, the possibility for changes in the properties of planetary waves and eddies is a real one. How those changes might impact on the manner in which momentum and information are distributed across the oceans, begs for further research. In light of this, Fyfe and Saenko (2007) modelled the alteration in stratification of the ocean's upper layers, based on the changes predicted by all the IPCC emissions scenarios. Using only linear theory, the authors found that the heating of the ocean's upper layers would induce a wave speed-up that begins to show at the lower latitudes by the end of the twentieth century, extending to the higher latitudes as time progresses.

The model runs showed a 20–40% increase across all the model scenarios and in particular a 35% speed increase for scenario A2 by the end of the twenty-first century (all compared to pre-industrial era speeds). These results reinforce the notion that the expected changes in ocean properties will change planetary wave speeds, decreasing the ocean's response time to external forcing. This has the potential to change ocean dynamics, such as the time-set of the ENSO, ocean-gyre circulation and western boundary currents, thus impacting on climate. A proper understanding of planetary wave speeds seems therefore to be fundamental to our knowledge of the oceans/climate system and the way we model it.

The coexistence of multiple altimeters with different spatial and temporal sampling patterns (due to the different orbital configuration of the satellite platforms, which results in different orbit inclinations and orbital repeat cycles) has prompted for a merging of the data, based on optimal interpolation techniques (Le Traon et al., 1998; Ducet et al., 2000) in order to increase the resolution of the SSH fields. This merging has improved our view of the mesoscale, allowing a much better resolution of those scales typical of oceanic eddies (for a review see Le Traon and Morrow, 2001). Chelton et al. (2007) have investigated the mesoscale variability of the global ocean using the improved fields (namely, merged TOPEX/Poseidon and ERS-1 and ERS-2 satellite datasets distributed by AVISO), finding that a significant fraction of that variability is accounted for by eddies, mostly non-linear, with amplitudes of 5–25 cm and diameters of 100–200 km. This study and other recent studies on eddies are reviewed by Fu (Chapter 9) in this same volume. Ongoing research is attempting to decompose the westward propagating energy into spectral "macro-components" that can be unambiguously mapped into different processes, and its early results confirm the co-existence of eddies and planetary waves over most of the ocean, with linear waves larger within 20–30° of the equator and non-linear eddies prevailing outside of that band (Matthew Thomas, Personal communication).

12.4 Current Research and Open Questions

Current research on westward propagating features is now focusing on a few questions opened by altimetric observations, alone or in combination with other satellite datasets. In this section we review two classes of "open questions": those that only concerns the physics of the propagating features, and those that instead concern the features' impact on the biology.

12.4.1 Waveguides and Normal Modes

One important dynamical problem is why there appear to be distinctive *waveguides* of enhanced westward propagation in the oceans, as noted for instance at 33–34°N in the Atlantic (Cipollini et al., 1997; Cromwell, 2001). Simulations done with a ray tracing approach (Killworth and Blundell, 2004, 2005) do yield zonal waveguides of enhanced propagation energy due to the convergence of many rays, but sometimes at different latitudes from those observed in the data.

Another significant problem for ocean dynamicists is to ascertain the occurrence and relative importance of the different normal modes of propagation. The barotropic (i.e. depth-independent) mode of planetary waves is believed to propagate too fast to be properly resolved by altimetry, especially when the data are gridded on 10- or 7-day time steps; however Fu (2004) has been able to demonstrate a significant presence of "fast" westward-propagating barotropic energy in some oceanic basins by using TOPEX/Poseidon data gridded on a 3-day orbital sub-cycle. This suggests that at least part of the barotropic energy could be mapped by adopting spatially coarse grids with time resolution of the order of 1 day or less, something that might become feasible with a constellation of small number (<10) of altimeters.

Maharaj et al. (2007) have looked at the significance of the different baroclinic (i.e. variable with depth) modes in the South Pacific by splitting the energy in wavenumber-frequency spectra of SSH anomalies on the basis of "spectral boundaries". These boundaries are dictated by the dispersion curves for the different modes predicted by the various theories of planetary wave propagation. The adoption of Killworth and Blundell's extended theory in place of the classical linear theory results in explaining up to 60% more of the variance in the observed power spectral energy as planetary waves.

As far as the relative importance of the different modes, Maharaj et al. (2007) found that mode 1 is by far the most important, and that mode 2 is significant in places, while modes 3 and 4 are negligible, as shown in Fig. 12.4. The dominance of the first baroclinic mode is evident also in a modelling study carried out over the north Atlantic by Lecointre et al. (2008), and based on the ATL6-ERS26 1/6° simulation (Penduff et al., 2004) performed during the French CLIPPER project. In addition, Lecointre et al. (2008) found a puzzling result that calls for further investigation: while at the surface the model wave speeds agree reasonably well with their counterparts observed in altimetry, below the surface the westward propagating disturbances in the model exhibit a systematic deceleration with increasing depth, by a factor that appears to vary geographically.

This questions the usual normal mode assumption that the speed of propagation of the disturbances is independent of depth. A crucial contribution to a better understanding and full 3-D characterization of the modal structure of planetary waves (and eddies) is expected from the integration of altimetric data and vertical profiles of temperature and salinity (hence density) from the ARGO floats (Gould et al.,

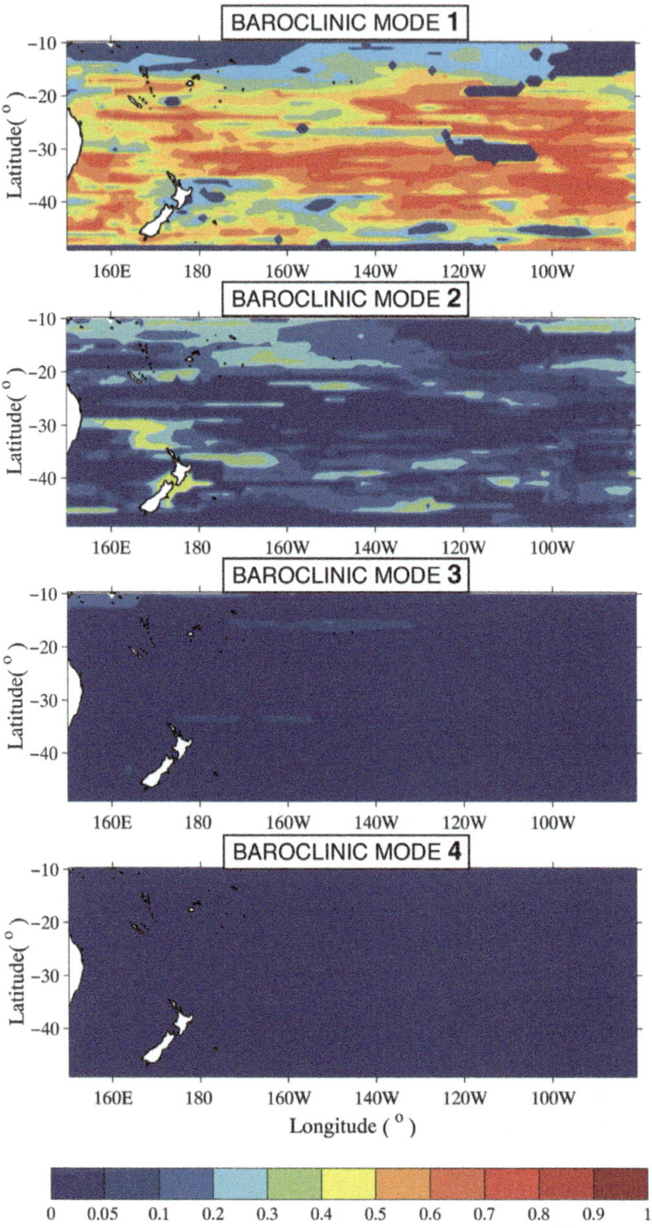

Fig. 12.4 Proportion of the total westward variance in the wavenumber/frequency spectra of SSH over the South Pacific that is assigned to each of the first four baroclinic modes of planetary wave propagation, on the basis of Killworth and Blundell's extended theory predictions. Figure from Maharaj et al. (2007)

2004), whose concentration has been rising in recent years and is approaching the desired level of one float for every 3° × 3° box.

12.4.2 Westward Propagation in Temperature and Ocean Colour

The existence of a signature of westward propagating features in satellite-derived Sea Surface Temperature (SST) data has been known for some time. Hill et al. (2000) were the first to show, using SST data from the ATSR infrared radiometer on board ERS-1, that this signature was almost ubiquitous and at speeds close to those expected for planetary waves, therefore strongly supporting the hypothesis that, in addition to eddies, planetary waves are also visible in these data. This SST signature is important as influential for the processes of ocean-atmosphere interaction.

An even bigger surprise came to the scientific community when, at the beginning of this century, a couple of studies (Cipollini et al., 2001; Uz et al., 2001) showed global, unambiguous evidence of wave-like, westward propagating signals in longitude-time plots of chlorophyll concentrations from ocean colour satellites. Figure 12.5 shows an example of this at around 32°N in the North Atlantic. This prompted several questions on which mechanisms provoke this signature, and more importantly whether its presence indicates a net effect on primary production and ultimately on the carbon budget. A number of studies have investigated these issues in recent years, but this new field of research on the biological effect of planetary waves has not been completely explored and more surprises could be around the corner.

The possible mechanisms involved in the generation of the ocean colour signature of planetary waves can be horizontal – like horizontal advection of meridional (i.e. north–south) gradients of phytoplankton – and/or vertical – like vertical advection of phytoplankton or even upwelling of nutrient due to the passage of the wave, that in turn stimulates growth: this latter mechanism has been dubbed *rototiller effect* (Siegel, 2001). It has also been suggested that the features could be due to convergence/divergence of particles at the surface, with the waves acting as a "hay rake" (Dandonneau et al., 2003) but there does not seem to be widespread consensus on this (Killworth, 2004). While little or no impact on production is to be expected from horizontal and surface mechanisms, the vertical ones are more interesting for their potential effects on the carbon cycle, so it is crucial to ascertain whether and where they occur.

To date, the most comprehensive attempt to model the signals due to all the different processes by which planetary waves would impact on the ocean colour field is the one by Killworth et al. (2004). In parallel to their process modelling, Killworth et al. (2004) performed a global cross-spectral analysis of satellite-derived SSH and chlorophyll which allowed them to estimate which processes were taking place in the real ocean, through the comparison of the observed cross-spectral amplitudes and phases with those predicted for the various processes. Their conclusion is that horizontal advection seems to be the dominant mechanism, but vertical mechanisms cannot be completely ruled out, due to both phase ambiguities between different

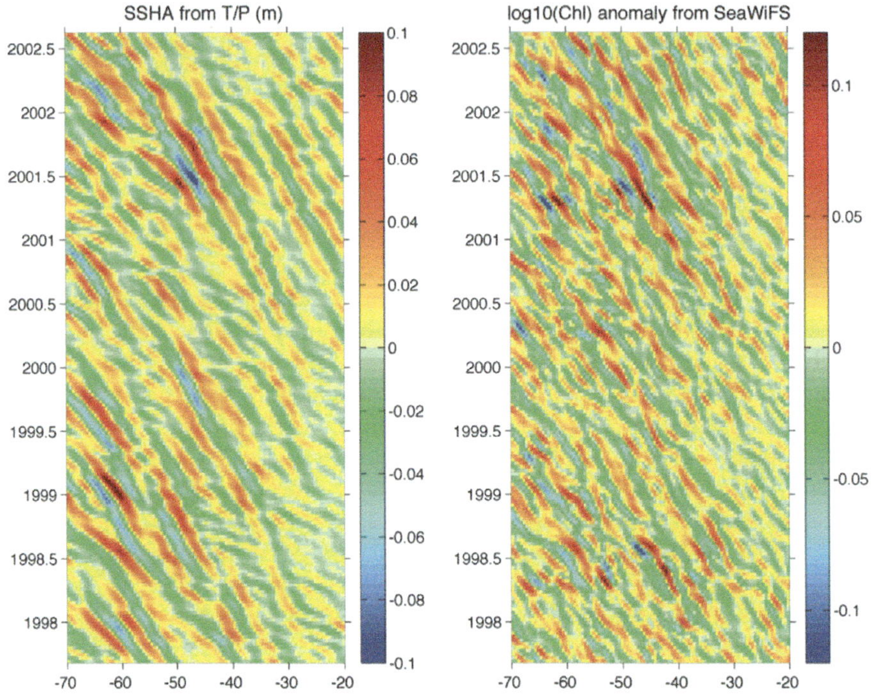

Fig. 12.5 Longitude/time plots of SSH Anomaly (SSHA, *left*) from TOPEX/Poseidon and log10 of the Chlorophyll concentration anomaly from SeaWiFS at 32°N in the Atlantic, clearly showing the signature of similar westward propagating features in the two datasets. The two datasets have been bandpass filtered to retain wavelengths expected for eddies and planetary waves in the westward-propagating quadrants

mechanisms, and the fact that the predicted amplitude for the horizontal advection case is in places lower than the signal observed in the real data.

Charria et al. (2006) have attempted to further quantify the contribution of the different mechanisms with a statistical decomposition of the observed wave signal in ocean colour in the North Atlantic, based on Killworth et al. (2004) models. Their results are obviously strongly dependent on both the process modelling adopted and the statistical assumptions in the decomposition, but nevertheless show a strong prevalence of horizontal advection south of 28°N, while polewards of 28°N horizontal advection and upwelling each contribute approximately half of the observed signal.

The contribution of uplifting is everywhere much smaller than the other two. More recently, Charria et al. (2008) have used a 3-D coupled physical-biogeochemical model to look for the direct influence of planetary waves on primary production, and found some significant local effects, namely increases (generally associated with the chlorophyll wave crest) and decreases (generally associated with the chlorophyll wave trough) in primary production of about ±20% of the estimated background primary production. The symmetric increase/decrease suggests a net weak effect over the basin, but the question is still open and will undoubtedly be the subject of further studies. These will have to make the most of the intrinsic synergies between biogeochemical modelling and satellite observations, but, as ever, altimetry is expected to be the reference for the detection and characterization of the propagating features.

12.5 Conclusions

The clarity with which eddies and planetary waves manifest themselves in altimetric SSH fields has allowed for an unprecedented progress in our knowledge of these important features of ocean dynamics. Brilliant scientific advances – like the characterization of planetary waves that has prompted physical oceanographers to revisit the theoretical framework – have been accompanied by puzzling discoveries (like the occurrence of planetary wave signals in ocean colour fields) that deserve further research. This chapter has reviewed some of the findings and looked briefly at several of the questions still open on planetary waves and eddies.

A very important possibility that opens now, by virtue of the altimetric record now approaching 18 year long, is to detect decadal-scale changes and trends in the occurrence and qualities of eddies and planetary waves, changes that in some cases might be related to climate change. This possibility is certainly going to be explored in detail in the immediate future, while forthcoming altimetric missions hold a lot of promise to extend and further improve the formidable altimetric record that is the foundation for this branch of research.

Acknowledgements We are grateful to Jeff Blundell for his advice and for providing the theoretical speeds from the extended theory. The altimeter products were produced by SSALTO/DUACS and distributed by AVISO with support from CNES via the web site www.aviso.oceanobs.com/duacs/

References

Anderson DLT, Gill AE (1975) Spin-up of a stratified ocean, with application to upwelling. Deep Sea Res 22:583–596

Antonov JI, Locarnini RA, Boyer TP, Mishonov AV, Garcia HE (2006) World Ocean Atlas 2005, Volume 2: salinity. In: Levitus S (ed.) NOAA Atlas NESDIS 62, US Government Printing Office, Washington, DC, 182pp

Charria G, Dadou I, Cipollini P, Drévillon M, De Mey P, Garçon V (2006) Understanding the influence of Rossby waves on surface chlorophyll concentrations in the North Atlantic Ocean. J Mar Res 64:43–71

Charria G, Dadou I, Cipollini P, Drévillon M, Garçon V (2008) Influence of Rossby waves on primary production from a coupled physical bio-geo-chemical model in the North Atlantic Ocean. Ocean Sci 4:199–213

Chelton DB, Schlax MG (1996) Global observations of oceanic Rossby waves. Science 272: 234–238

Chelton DB, Schlax MG, Samelson RM, de Szoeke RA (2007) Global observations of large oceanic eddies. Geophys Res Lett 34:L15606

Cipollini P, Challenor PG, Colombo S (2006a) A method for tracking individual planetary waves in remotely sensed data. IEEE Trans Geosci Rem Sens 44:159–166

Cipollini P, Cromwell D, Challenor PG, Raffaglio S (2001) Rossby waves detected in global ocean colour data. Geophys Res Lett 28:323–326

Cipollini P, Cromwell D, Jones MS, Quartly GD, Challenor PG (1997) Concurrent altimeter and infrared observations of Rossby wave propagation near 34 N in the Northeast Atlantic. Geophys Res Lett 24:889–892

Cipollini P, Quartly GD, Challenor PG, Cromwell D, Robinson IS (2006b) Remote sensing of extra-equatorial planetary waves. In: Rencz AB, Gower JFR (eds.) Manual of Remote Sensing, Volume 6: Remote Sensing of Marine Environment, American Society for Photogrammetry and Remote Sensing, Bethesda MD, USA, pp. 61–84

Cromwell D (2001) Sea surface height observations of the 34°N 'waveguide' in the North Atlantic. Geophys Res Lett 28:3705–3708

Dandonneau Y, Vega A, Loisel H, du Penhoat Y, Menkes C (2003) Oceanic Rossby waves acting as a "hay rake" for ecosystem floating byproducts. Science 302:1548–1551

de la Rosa S, Cipollini P, Snaith HM (2007) An application of the radon transform to study planetary waves in the Indian Ocean. ESA SP-636. Envisat Symposium 2007, Montreux

Ducet N, Le Traon P-Y, Reverdin G (2000) Global high-resolution mapping of ocean circulation from the combination of T/P and ERS-1/2. J Geophys Res 105:477–498

Fu L-L (2004) Latitudinal and frequency characteristics of the westward propagation of large-scale oceanic variability. J Phys Oceanogr 34:1907–1921

Fu L-L, Chelton DB (2001) Large-scale ocean circulation. In: Fu L-L, Cazenave A (eds.) Satellite Altimetry and Earth Sciences: A Handbook of Techniques and Applications, Academic Press, San Diego, pp. 133–169

Fu L-L, Qiu B (2002) Low-frequency variability of the North Pacific Ocean: the roles of boundary- and wind-driven baroclinic Rossby waves. J Geophys Res 107(C12):3220, doi:10.1029/2001JC001131

Fyfe JC, Saenko OA (2007) Anthropogenic speed-up of oceanic planetary waves. Geophys Res Lett 34:L10706, doi:10.1029/2007GL029859

Gill AE (1982) Atmosphere-Ocean Dynamics, Academic Press, New York

Gould J, Roemmich D, Wijffels S, Freeland H, Ignaszewsky M, Jianping X, Pouliquen S, Desaubies Y, Send U, Radhakrishnan K, Takeuchi K, Kim K, Danchenkov M, Sutton P, King B, Owens B, Riser S (2004) Argo profiling floats bring new era of in situ ocean observations. Eos Trans 85:185–190

Hill KL, Robinson IS, Cipollini P (2000) Propagation characteristics of extratropical planetary waves observed in the ATSR global sea surface temperature record. J Geophys Res 105: 21927–21945

Hirschi JJ-M, Killworth PD, Blundell JR (2007) Subannual, seasonal, and interannual variability of the North Atlantic meridional overturning circulation. J Phys Oceanogr 37:1246–1265

Jacobson AR, Spiesberger JL (1998) Observations of El Nino-Southern oscillation induced Rossby waves in the Northeast Pacific using in situ data. J Geophys Res 103:24585–24596

Killworth PD (2004) Comment on "Oceanic Rossby waves acting as a 'hay rake' for ecosystem floating byproducts". Science 304:390

Killworth PD, Blundell JR (1999) The effect of bottom topography on the speed of long extratropical planetary waves. J Phys Oceanogr 29:2689–2710

Killworth PD, Blundell JR (2003a) Long extratropical planetary wave propagation in the presence of slowly varying mean flow and bottom topography. Part I: The local problem. J Phys Oceanogr 33:784–801

Killworth PD, Blundell JR (2003b) Long extratropical planetary wave propagation in the presence of slowly varying mean flow and bottom topography. Part II: Ray propagation and comparison with observations. J Phys Oceanogr 33:802–821

Killworth PD, Blundell JR (2004) The dispersion relation for planetary waves in the presence of mean flow and topography. Part I: Analytical theory and one-dimensional examples. J Phys Oceanogr 34:2692–2711

Killworth PD, Blundell JR (2005) The dispersion relation for planetary waves in the presence of mean flow and topography. Part II: Two-dimensional examples and global results. J Phys Oceanogr 35:2110–2133

Killworth PD, Chelton DB, de Szoeke RA (1997) The speed of observed and theoretical long extratropical planetary waves. J Phys Oceanogr 27:1946–1966

Killworth PD, Cipollini P, Uz BM, Blundell JR (2004) Physical and biological mechanisms for planetary waves observed in satellite-derived chlorophyll. J Geophys Res 109:C07002

Killworth PD, John HS (2001) Rossby waves. In: Steele JH, Thorpe SA, Turekian KK (eds.) Encyclopedia of Ocean Sciences, Academic Press, Oxford, pp. 2434–2443

Le Traon P-Y, Morrow R (2001) Ocean currents and eddies. In: Fu LL, Cazenave A (eds.) Satellite Altimetry and Earth Sciences: A Handbook of Techniques and Applications, Academic Press, San Diego, pp. 171–210

Le Traon P-Y, Nadal F, Ducet N (1998) An improved mapping method of multisatellite altimeter data. J Atmosph Ocean Technol 15:522–533

Lecointre A, Penduff T, Cipollini P, Tailleux R, Barnier B (2008) Depth dependence of westward-propagating North Atlantic features diagnosed from altimetry and a numerical 1/6° model. Ocean Sci 4:99–113

Locarnini RA, Mishonov AV, Antonov JI, Boyer TP, Garcia HE (2006) World Ocean Atlas 2005, Volume 1: temperature. In: Levitus S (ed.) NOAA Atlas NESDIS 61, US Government Printing Office, Washington, DC, 182 pp

Maharaj AM, Cipollini P, Holbrook NJ, Killworth PD, Blundell JR (2007) An evaluation of the classical and extended Rossby wave theories in explaining spectral estimates of the first few baroclinic modes in the South Pacific Ocean. Ocean Dyn 57:173–187

Pedlosky J (1987) Geophysical Fluid Dynamics, Springer-Verlag, New York

Penduff T, Barnier B, Dewar WK, O'Brien JJ (2004) Dynamical response of the oceanic eddy field to the North Atlantic oscillation: a model-data comparison. J Phys Oceanogr 34:2615–2629

Polito PS, Liu TW (2003) Global characterization of Rossby waves at several spectral bands. J Geophys Res 108(C1):3018

Siegel DA (2001) The Rossby rototiller. Nature 409:576–577

Tailleux R, McWilliams JC (2000) Acceleration, creation, and depletion of wind-driven, Baroclinic Rossby waves over an Ocean Ridge. J Phys Oceanogr 30:2186–2213

Tailleux R, McWilliams JC (2001) The effect of bottom pressure decoupling on the speed of extratropical, Baroclinic Rossby waves. J Phys Oceanogr 31:1461–1476

Uz BM, Yoder JA, Osychny V (2001) Pumping of nutrients to ocean surface waters by the action of propagating planetary waves. Nature 409:597–600

Chapter 13
Sea Surface Temperature Measurements from Thermal Infrared Satellite Instruments: Status and Outlook

Craig J. Donlon

13.1 Introduction

Thermal Infrared (TIR) sensors have been deployed on earth observing satellites for over 30 years providing measurements of Sea Surface Temperature (SST), clouds and many other products. Developed initially for meteorology and now used widely by the oceanographic and climate communities, TIR derived SST measurements are available in an operational context in Near Real Time (NRT) from a wide variety of satellite missions. TIR sensors have a characteristically high spatial resolution of 0.5–1.1 km (at nadir) with quasi global coverage on a daily basis (using two operational wide swath TIR missions). TIR sensors are typically calibrated using on-board reference blackbody systems alone (e.g. Corlett et al., 2006) or a combination of blackbody and deep-space "cold" views to an accuracy of 0.1–0.2 K (e.g. Robinson, 2004). On-board calibration is sometimes supplemented with vicarious calibration adjustments implicit in some Level-2 SST retrieval algorithms that compensate for the atmospheric attenuation of water leaving radiances using in-situ SST measurements (e.g. Kilpatrick et al., 2001; Zhang et al., 2009). Other approaches to atmospheric correction rely on the use of radiative transfer modes to derive look-up-tables that can be applied to brightness temperature measurements using a suitable SST retrieval algorithm (e.g. Merchant and Le Borgne, 2004).

This approach has the benefit of releasing in-situ observations for use in on-going verification and validation work and for a more detailed investigation of sensor and algorithm biases, essential activities for the production of fundamental climate data records (e.g. Merchant et al., 2008b). Most importantly, well defined and error quantified measurements of SST are required for climate time series (in the form of Fundamental Climate Data Records, or FCDR) that can be analyzed to reveal the role of the ocean in short and long term climate variability.

This chapter first presents a summary of key TIR satellite sensors from 2000 to 2020. In Section 13.3 it outlines the primary on-going challenges and issues

C.J. Donlon (✉)
ESTEC (EOP-SME), European Space Agency, Noordwijk, 2201 AZ, The Netherlands
e-mail: Craig.Donlon@esa.int

V. Barale et al. (eds.), *Oceanography from Space*,
DOI 10.1007/978-90-481-8681-5_13, © Springer Science+Business Media B.V. 2010

associated with the use of TIR data for accurate retrieval of SST. In Section 13.4 impact of the Group for High Resolution SST (GHRSST, Donlon et al., 2007) will be reviewed. Finally, conclusions are presented and a forward perspective for the coming decade is provided.

13.2 Key TIR Satellite Sensors Since 2000

Development in the definition, availability, future planning and service provision of TIR satellite sensors and data has matured significantly in the last 10 years. According to the Committee for Earth Observation Satellites (CEOS) on-line database (CEOS, 2008, 2009) over 20 satellite missions capable of measuring SST in a variety of orbits (polar, low inclination and geostationary) have been launched since 1999. The tables reported in Appendix list the main TIR sensors and their basic characteristics for missions operating from 2000 and up to 2020. It is interesting to note the transition of the (A)ATSR instrument series to the Sea and Land Surface Temperature Radiometer (SLSTR) carried by the Sentinel-3 operational mission. The NOAA AVHRR/3 series, a traditional workhorse TIR sensor, will end with NOAA-19, to be replaced with the new NPOESS/NPP VIIRS with enhanced capability. New geostationary imager capability has emerged in the last 10 years in Europe, with the MSG SEVIRI instrument now providing high quality operational SST. Also, the development of new capability in China through the FY-satellite series is noteworthy.

The accuracy that can be obtained for SST derived from TIR data is now at the limit of the capability of available operational in-situ infrastructure (~0.1–0.2 K). Comparisons between the AATSR and drifting-buoy measurements made by the UK Met Office have shown that AATSR is capable of achieving biases in Global SST which typically <0.15 K (O'Carroll et al., 2008). Such error analyses show clearly that AATSR SST data can act as a "benchmark" of accuracy, against which data from other sources can be bias-corrected. This approach has been adopted at operational centres (e.g. Stark et al., 2007).

In summary, Appendix shows that TIR satellite sensors have matured (research instruments are now flown on operational missions), advanced (Sentinel-3 SLSTR has a much wider swath, ~1,400 km, compared to the ENVISAT 512 km swath of AATSR) and both polar and geostationary missions are being sustained until 2020. This is considerable progress since the *Oceans From Space* meeting in 2000.

13.3 On-Going Challenges and Issues

13.3.1 Data Access

Wide and open access in near real time to many TIR satellite SST data products has been established in an operational-like manner using existing data user-driven distribution protocols, tools and services coordinated by the GHRSST project (Donlon et al., 2007). This is a significant development since *Oceans From Space* in 2000 and has led to increased scrutiny, research, development and operational uptake of TIR data. Over 26 Gb of data are provided in NRT every day by GHRSST Services, and over 25,500 international users have accessed GHRSST products. This framework needs to be maintained and evolve as new satellite TIR instruments come on line in the coming decade.

13.3.2 Cloud Flagging of SST Derived from TIR Data

The SST fields obtained TIR sensors are corrupted by clouds, with the temperature of cloud contaminated pixels generally colder than the actual SST. Inclusion of contaminated pixels in final products renders data inaccurate and difficult to use. For these reasons, flagging of cloud contaminated pixels in SST fields has received a great deal of attention over the past 30 years. Despite the effort devoted to such algorithms, significant problems and challenges remain. For applications in which the absolute accuracy of the retrieved SST values is central to their use, it is important to exclude any pixel that is even slightly cloud contaminated.

In contrast, applications in which the location of oceanographic features is important make use of the relative accuracy of adjacent SST values and have some tolerance to cloud contamination. Most cloud screening algorithms are sensitive to large gradients in the retrieved fields and pixels in a high gradient region are generally flagged as cloud contaminated. Approaches make use of the structural characteristics of fronts to either reset the quality mask for those pixels that are believed to be frontal pixels that were falsely flagged as clouds (Cayula and Cornillon, 1996) or add a new flag. One advantage of this test is that it can be applied after the SST retrieval and quality fields have been obtained.

Development of cloud screening algorithms has focused on applications for which the absolute accuracy of the SST value is paramount and typically makes full use of both visible and TIR data available from the sensor in the day time. Only the IR channels are available at night further complicating cloud detection. Algorithms rely on differences in emissivity, reflectivity, temperature and spatial structure between the ocean surface and clouds. Some work well in identifying cloud-contaminated pixels under most open ocean conditions. However, because screening is based on thresholds associated with specific parameters and the underlying distributions are in most cases continuous, there will be ambiguity when one or more of the parameter values is close to a threshold value. The problem is therefore intrinsically probabilistic, with a trade-off between false alarms and hits, a balance that depends critically on the user's application.

Many SST fields are now provided with a separate "quality" field, which is often derived from the cloud screening portion of the retrieval algorithm. This field allows users to mask SST values based on the quality threshold that meets their specific needs. Quality fields are derived differently by different data providers with different meanings that are not always described in sufficient detail making it difficult for the user to apply them consistently. This challenge requires careful attention in the future. In addition to providing quality fields with the SST data, there is a trend toward increasing use of simulations in near-real time from national weather programs to inform the discrimination – either by dynamically calculating thresholds or as input to a probabilistic calculations (Merchant et al., 2005). Further development of this approach is expected in the future.

Cloud screening of TIR satellite data is remains a significant challenge and more effort is required to develop effective systems to minimize the data loss due to inappropriate cloud screening and the increase in error where clouds are not properly detected.

13.3.3 Improved Treatment of Atmospheric Aerosol Contamination

The performance of TIR derived SST retrievals is degraded in the presence of atmospheric aerosols (e.g., Saharan dust, volcanic eruptions). This has been a particular problem for the Meteosat-8 SEVIRI instrument SST retrieval. During the initial phase of operations the occurrence of Saharan dust outbreaks lead to SST bias errors of ~1 K. These problems have been mitigated to a certain degree by upgrading the MSG algorithms to include a Saharan dust index scheme (Merchant et al., 2006) and the use of ENVISAT AATSR data to derive a bias correction for the aerosol (and other) contaminated data. There are several aspects to improving atmospheric aerosol detection and flagging algorithms that will provide increased sensitivity and performance:

1. in strong SST gradient regions,
2. when sub-pixel clouds and optically thin cirrus are present,
3. when only limited instrument channels are available,
4. when aggregated data are used (e.g. AVHRR GAC),
5. when multi-angle view data are available,
6. based on multi-satellite synergy (e.g. use of geostationary data, (A)ATSR, and passive microwave sensors),
7. based on probabilistic techniques,
8. based on improved conventional threshold, histogram and spatial coherence techniques.

It is expected that significant progress will be made in the next decade on these issues as climate quality SST data sets are derived for a variety of TIR sensors.

13.3.4 Improving Current and Future SST Measurements Through Better Uncertainty and Error Estimation

A key user request from all user communities (and in particular the SST community) is the provision of uncertainty estimates to be attached to each pixel in SST products. A framework has emerged from the GHRSST activity called Single Sensor Error

Statistics (SSES) designed to take into account uncertainties for specific instrument/platforms (Donlon et al., 2007). Bias and uncertainty estimates are generally derived from near contemporaneous match ups between satellite and in-situ SST measurements which are periodically analysed to provide SSES. The EUMETSAT OSI-SAF has developed a statistical method to derive SSES bias and standard deviation estimates by associating a confidence level assigned to the retrieved SST estimate. The confidence levels are based on tests to the reliability of the cloud mask and the SST algorithm conditions. Regional (and seasonal) characteristics need to be accounted for in this scheme although it is successfully used in operations.[1]

An alternative approach called the Hypercube has also been developed based on a match-up data base for the Aqua and Terra MODIS sensors. In this case, the MDB includes near-contemporaneous, co-located satellite brightness temperatures, in-situ buoy and radiometer SST, auxiliary data from model or satellite observed fields, and the satellite viewing geometry. A series of quality tests is applied during processing of the MODIS data to identify cloud and dust aerosol contaminated retrievals and assign pixels to one of four different quality levels with quality 0 being the best quality possible. The relative immunity of the MODIS 3.95 and 4.05 μm bands to both water vapour and aerosols as compared to the increased sensitivity to both in the MODIS 11 and 12 μm bands is used to identify aerosol data. After eliminating records with quality levels greater than 1, each match-up database is partitioned into a multi-dimensional array with the following 7 dimensions: time by season (4 values), latitude bands (5 steps in 20° increments from 60°S to 60°N), surface temperature (8 increments in 5° steps), satellite zenith angle (4 increments), brightness temperature difference as a proxy for water vapour (4 intervals for 4 μm and 3 intervals for 11–12 μm SST), retrieved satellite SST quality level (2 intervals) and day/night selection (2 intervals). The bias (satellite-in-situ) and standard deviation are then computed for each element to define a hypercube look up table (LUT). The LUT is then used during satellite data processing to predict the SSES bias and standard deviation of the SST retrieval. The hypercube approach provides more control over the specification of uncertainty estimates and is being actively developed within the framework of GHRSST.

Finally, it is important to recognize that more work is required to ensure that uncertainty values, where possible, are traceable to accepted international reference standards and SI units. Satellite TIR instruments and ground truth instrumentation should also be traceable to the same reference standards. More effort is required in this area.

13.3.5 New SST Retrieval Techniques Using TIR Data

A single-view TIR imager with channels at roughly 3.7, 11 and 12 μm can demonstrate global 1 km accuracy approaching 0.3 K at night-time (i.e., when all three

---

[1] See http://www.osi-saf.org

channels are used), although instability of calibration, cloud-detection failures, and episodes of atmospheric aerosol can each degrade this potential significantly. With only the 11 and 12 μm channels available (as in day-time), coefficient-based retrievals have been limited to accuracies of about 0.4 K. Recent work on METOP-A AVHRR data shows that optimal estimation techniques can drive accuracy down to ~0.3 K for SST estimates where the retrieval cost is low (Merchant et al., 2008a). All these quoted errors have a random element, but are in large part correlated on the synoptic scales of the atmosphere. Further research is required to develop and refine SST optimal estimation techniques for TIR sensors that maximize the error reduction from having multiple complementary observing systems in space.

13.3.6 Improving SST Provision in the High Latitude Regions

Accurate retrieval of SST at high latitudes using TIR satellite sensors requires that (a) the discrimination between ice-free and ice-covered water at the resolution, temporal and spatial, of the SST retrieval schemes is well known; and (b) the atmospheric attenuation on the infrared radiation as it propagates from the sea surface to the satellite radiometer is determined. For infrared SST retrievals, during the day, reflected sunlight provides a powerful mechanism for identifying open, cloud-free water.

During the polar night the problem of identifying ice becomes more difficult. A simple temperature threshold test might be adequate to identify pack ice but this would not be sufficient in the more complex marginal ice zone. Surface temperature retrievals below −1.8°C, the freezing point of seawater, can be classified as ice cover. However, this is prone to error as (a) there is noise in the satellite-derived surface temperature, so that ice-free retrievals could fall below the threshold, and ice-covered pixels fall above the threshold; and (b) when melting, sea ice, especially if covered by snow, may remain frozen at temperatures above the threshold. More effort should be given to define and implement ice masking procedures and techniques in Polar Regions for TIR satellite observations.

Considering the impact of atmospheric attenuation on the water leaving signal it is clear that the polar atmosphere is generally very dry and cold, and is thus an extreme in terms of the climatological distribution of atmospheric properties. It represents an anomalous set of conditions for routine SST atmospheric correction algorithms optimized for the global range of atmospheric variability (e.g. Walton et al., 1998; May et al., 1998). It is to be expected that systemic retrieval errors in the derived SSTs will result: bias errors, usually result in warm SST errors that can be greater than 1 K (Vincent et al., 2008b). Loss of the correlation between the brightness temperatures measured at 10.5 and 11.5 μm with the atmospheric water vapour that occurs in very dry atmospheres and Vincent et al. (2008a, b) show using AVHRR brightness temperature data collocated with ship-based radiometric skin SST measurements that a simple, single channel retrieval algorithm can produce improved accuracy in the measurement of skin SST and Ice Surface Temperature. Single-channel algorithms appear to be better suited to the problem than current multi-channel approaches. Satellite SST data providers using infrared systems should review the performance of their atmospheric correction algorithms in polar atmospheres and take steps to develop more appropriate algorithms for these regions.

13.4 Principles and Lessons Learned from the GHRSST International Framework

The GHRSST project (Donlon et al., 2007[2]) was a significant contribution to progress in SST over the last decade as it nurtured a community of scientists from the scientific and operational agencies and institutions. GHRSST established a set of user requirements for all GHRSST activities from a bottom up collaborative and open discussion in five areas: (1) scientific development and applications, (2) operational agency requirements, (3) SST product specifications, (4) programmatic organization of an international SST service and (5) developing and sharing scientific techniques and insight to improve data products and exploit the observing system.

These requirements formed an essential part of the GHRSST evolution and were critical to establishing a framework and a work plan. A consensus GHRSST Data Processing specification (GDS) was developed that described how satellite data providers should process satellite data streams; a common format and content of data products; the basic approaches to providing uncertainty estimates and auxiliary data sets that should be included in products to help users interpret the SST measurements. GHRSST also conducted scientific research and developed a data management framework, including long term stewardship of all products.

GHRSST realised the benefits of creating modular data processing architectures in which many partners around the world can contribute to improve global monitoring. The GHRSST service has also encouraged the development of new SST monitoring and forecasting initiatives by operational agencies that use the new data products. In particular the end-to-end service reveals how a system which enables the complementary use of data from different sources reinforces the importance of each, as it leads to new records of SST with enhanced accuracy and improved spatial and temporal resolution. A full discussion of GHRSST success over the last 10 years is reported in Donlon, 2008; Donlon et al., 2009. The key developments include:

- International agreement on the definition of different SST parameters in the upper layer of the ocean that distinguish between measurements made by infrared radiometers, passive microwave radiometers, in-situ sub-surface observations and SST merged analysis outputs. These definitions have been registered in the Climate Forecast (CF) standard name table (Donlon, 2008).
- Diverse satellite SST data product formats and product content have been homogenised according to international consensus and user requirements to

[2]See http://www.ghrsst.org

include measurement uncertainty estimates for each derived SST value and supporting auxiliary data sets to facilitate their use by data assimilation systems.

- GHRSST advisory groups have conducted extensive research to ensure that SST diurnal variability (DV) is properly flagged within observational data; developed methods to correct for bias in different satellite data sets; provided uncertainty estimates on a measurement by measurement basis, developed high resolution sea ice data sets and accurate SST products in the marginal ice zone.
- New cost-effective approaches to an integrated and optimised SST measurement system have been developed and are used operationally, to reduce bias error in AVHRR data using targeted global deployment strategies for drifting buoys (Zhang et al., 2009).
- New SST analysis products using new methods to merge in-situ data with complementary microwave and infrared satellite data have been developed and implemented operationally.
- Inter-comparison frameworks – e.g., the GHRSST Multiproduct Ensemble (GMPE)[3] – have been developed at resolutions of 10 km or better for the global ocean and other regions of interest. An operational High Resolution Diagnostic Data Set (HR-DDS)[4] has been established for real time inter-comparisons and validation/verification of GHRSST products allowing real time monitoring of satellite and in-situ SST data streams.
- A delayed-mode intercomparison framework has been established in conjunction with the GCOS SST and Sea Ice Working Group to understand the links between the modern era satellite-based SST record and historical primarily ship-based SST reconstructions.[5]
- Methods to convert between radiometric "skin" SST and the SST at depths measured by ships and buoys have been developed (e.g., Donlon et al., 2002) that are now used by operational SST analysis systems (e.g., Stark et al., 2007).
- An internationally distributed suite of user focussed services are now provided in a sustained Regional/Global Task Sharing (R/GTS) framework that addresses international organisational challenges and recognises the implementing institutional capacities, capabilities and funding prospects. Long term stewardship, user support and help services including standards-based data management and interoperability have been developed that are manned and operated within the R/GTS on a daily basis.
- Methods to manage long-term SST data sets, for use in a reanalyses that considers SST data for the entire satellite era, have begun.

GHRSST has earned broad recognition as the international authority for modern-era SST activities because it has successfully built and nurtured a framework in which the exchange of satellite SST data has flourished and given new life to the study and application of high-resolution SST using TIR satellite and in-situ data.

[3] See http://ghrsst-pp.metoffice.com/pages/latest_analysis/sst_monitor/daily/ens/index.html

[4] See http://www.hrdds.net

[5] See http://ghrsst.nodc.noaa.gov

Applications have demonstrated positive impact in ocean and atmospheric forecasting systems and a new generation of data products and services to serve these and other users have been built and are operated on a day-to-day basis. The success of GHRSST stems from the Agencies and Offices that have supported the activities of the Pilot Project allowing a dedicated group of scientists and operational entities to successfully work together and bridge the gap between operations and science. All good operational systems are underpinned by excellent science and GHRSST has endeavoured to provide a forum in which operational systems and scientists can meet and discuss problems and solutions to address the real-world challenges associated with the application of high-resolution SST data sets.

13.5 Conclusion and Future Outlook

The future outlook for TIR sensors is very good. Over the last decade there have been many successes in terms of the TIR instruments that are flying in polar and geostationary orbits. Some systems are in the process of transitioning from research to operational systems (e.g. AATSR mobbing to the SLSTR on Sentinel-3). The accuracy of TIR retrievals from space is in some cases better than 0.2 K when using a dual view along track scanning approach. Accuracy of SST from geostationary TIR systems is less than this but still extremely useful especially when bias adjusted using other satellite or in-situ data.

Challenges remain for improved quality of SST products derived from TIR satellite data. These include better cloud clearing, better treatment of atmospheric aerosols that contaminate TIR data, better techniques for SST retrieval and better retrieval algorithms in the polar atmosphere. Better uncertainty estimates for SST data products derived from TIR data are essential and must be continually refined and updated based on the best tools and techniques.

The activities within the SST community over the last decade have transformed the measurement of SST using a complementary satellites and in-situ measurements working in synergy together. The establishment of a GHRSST framework for the exchange and management of international SST data has been successfully implemented and is operating on a daily basis. A thriving user community has developed in which integrated SST data sets are being used at scientific institution and operational agencies. Tools and data services have been developed and implemented to serve this user community. Through the activities of GHRSST many lessons have been learned that provide the basis for an optimal configuration for the SST observing system in the next 10 years.

The major challenges focus on augmenting and maintaining high quality SST measurements from both in-situ and satellite instruments, maintaining and developing the scientific and operational SST community, providing robust and sustained methods and tools that provide uncertainty and error estimates in a format that is easy to use by users and developing and maintaining an SST data stewardship and reanalysis program that is able to tackle the development and validation of SST climate data records.

Appendix

Key TIR sensors and their basic characteristics for missions operating from 2000 and up to 2020 (data obtained from CEOS 2009):

- Along Track Scanning Radiometer 2 (ATSR-2)
- Advanced along Track Scanning Radiometer (AATSR)
- Advanced Very High Resolution Radiometer 3 (AVHRR/3)
- Moderate Resolution Imaging Spectroradiometer (MODIS)
- Spinning Enhanced Visible and Infrared Imager (SEVIRI)
- Visible and Infrared Sounder (VIRS)
- Meteosat Third Generation (MTG)
- MTSAT Imager
- GOES Imager
- Visible/Infrared Imager Radiometer Suite (VIIRS)
- Sea and Land Surface Temperature Radiometer (SLSTR)
- Multispectral Visible and Infrared Scan Radiometer (10 channels)
- Visible and Infra-red Scan Radiometer (VIRR)

The following tables list each instruments' name, mission(s), spatial resolution, swath, wavebands, description, as in:

Instrument Mission(s) Spatial resolution Description	Swath width	Spectral bands

Along track scanning radiometer 2 (ATSR-2)		
ERS-2 (1995-04-21 2011-12-31)		
1.1 km	512 km	VIS – SWIR: 0.65 μm 0.85 μm 1.27 μm 1.6 μm SWIR-TIR: 1.6 μm 3.7 μm 11 μm 12 μm
Imaging Vis/IR radiometer exploiting different viewing conditions		

Advanced along track scanning radiometer (AATSR)

ENVISAT (2002-03-01 2013-12-31)		
1.1 km	512 km	VIS – NIR: 0.555 μm 0.659 μm 0.865 μm SWIR: 1.6 μm MWIR: 3.7 μm TIR: 10.85 μm 12 μm
Imaging Vis/IR radiometer exploiting different viewing conditions		

Advanced very high resolution radiometer 3 (AVHRR/3)

NOAA-12 (1991-05-14 2005-12-31)		
NOAA-14 (1994-12-30 2005-12-31)		
NOAA-15 (1998-05-01 2010-12-31)		
NOAA-16 2000-09-21 2012-12-31		
NOAA-17 (2002-06-24 2014-12-31)		
NOAA-18 (2005-05-20 2015-12-31)		
NOAA-19 (2009-02-04 2016-03-01)		
EUMETSAT Metop-A (2006-10-19 2011-11-01)		
Metop-B (2012-04-02 2017-05-01)		
Metop-C (2016-04-02 2021-12-01)		
1.1 km	~3,000 km ensures full global coverage twice daily	VIS: 0.58–0.68 μm NIR: 0.725–1.1 μm SWIR: 1.58–1.64 μm MWIR: 3.55–3.93 μm TIR: 10.3–11.3 μm 11.5–12.5 μm
Multi-purpose imaging Vis/IR radiometer Imaging multi-spectral radiometers (vis/IR)		

Moderate resolution imaging spectroradiometer (MODIS)

Terra (1999-12-18 2011-09-30) Aqua (2002-05-04 2011-09-30)		
250 m (day)	2,330 km	VIS – TIR:
1,000 m (night) SST: 1,000 m		36 bands in range 0.4–14.4 μm
Medium-resolution spectro-radiometer		

Spinning enhanced visible and infrared imager (SEVIRI)

Meteosat-8 (2002-08-13 2011-06-30)		
Meteosat 9 (2005-12-21 2014-06-30)		
HRV = 1 km all others = 3 km (spatial sampling distance at SSP)	Full earth disk	HRV: ~0.48–0.91 μm VIS: 0.6 μm 0.8 μm NIR: 1.6 μm IR: 3.9 μm 6.3 μm 7.3 μm 8.7 μm 9.7 μm 10.8 μm 12.0 μm 13.4 μm
Multi-purpose imaging Vis/IR radiometer, in geostationary orbit Imaging multi-spectral radiometers (vis/IR)		

Visible and infrared sounder (VIRS)

Tropical rainfall mapping mission (TRMM) (1997-11-27 2011-09-30)		
2 km	720 km	VIS: 0.63 μm SWIR – MWIR: 1.60 μm 3.75 μm TIR: 10.8 μm 12.0 μm
NASA/JAXA Imaging multi-spectral radiometers (vis/IR) Multi-purpose imaging Vis/IR radiometer		

Meteosat third generation (MTG)

MTG Imager-1 (2016-12-15 2025-06-15)		
MTG Imager-2 (2021-06-15 2029-12-15)		
MTG Imager-3 (2025-01-15 2033-07-15)		
MTG Imager-4 (2029-06-15 2037-12-15)		
VIS/SWIR:	Full earth disk	VIS:
0.5, 1.0 km		0.4 μm
IR:		0.5 μm
2.0 km		0.6 μm
		0.8 μm
		0.9 μm
		NIR:
		1.3 μm
		1.6 μm
		2.2 μm
		3.8 μm
		6.3 μm
		7.3 μm
		8.7 μm
		9.7 μm
		10.5 μm
		12.3 μm
		13.3 μm
Multi-purpose imaging Vis/IR radiometer, in geostationary orbit		
Imaging multi-spectral radiometers (vis/IR)		

MTSAT imager

MTSAT-1, 2 and 3		
VIS: 1 km TIR: 4 km	Full earth disk (every hour)	VIS – SWIR: 0.55 – 0.80 μm MWIR – TIR: 3.5 – 4 μm 6.5 – 7 μm 10.3 – 11.3 μm 11.5 – 12.5 μm
Imaging multi-spectral radiometers (vis/IR)		
Multi-purpose imaging Vis/IR radiometer		

GOES imager

GOES-10 GOES-11 GOES-12 GOES-8 GOES-9 GOES-14 GOES-P GOES-13		
10 km	Full earth disk	GOES 8 – 11 VIS: (1 channel, 8 detectors) IR: (4 channels) 3.9 μm 6.7 μm 10.7 μm 12 μm GOES 12 – Q VIS: (1 channel, 8 detectors) IR: (4 channels) 3.9 μm 6.7 μm 10.7 μm 13.3 μm
Imaging multi-spectral radiometers (vis/IR) Multi-purpose imaging Vis/IR radiometer		

Visible/Infrared imager radiometer suite (VIIRS)

NPP NPOESS preparatory project (2010-06-02 2015-06-02) NPOESS-1 (2013-01-31 2020-01-01) NPOESS-2 (2016-01-31 2022-01-01) NPOESS-3 (2018-01-31 2025-01-01) NPOESS-4 (2020-01-31 2027-01-01)		
400 m–1.6 km	3,000 km	VIS – TIR: 22 channels range 0.4–12.5 μm
NASA/NOAA and USA DoD Imaging multi-spectral radiometers (vis/IR) Multi-purpose imaging Vis/IR radiometer		

Sea and land surface temperature radiometer (SLSTR)			
Sentinel-3A (2012-10-01 2019-10-01)			
Sentinel-3B (2015-10-01 2022-10-01)			
VNIR/SWIR:	Near-nadir view:	S1	0.555 μm
500 m	1,400 km	S2	0.659 μm
TIR:	Backward view:	S3	0.865 μm
1 km	750 km	S4	1.375 μm
		S5	1.61 μm
		S6	2.25 μm
		S7	3.74 μm
		S8	10.95 μm
		S9	12 μm
ESA/EC			
Imaging multi-spectral radiometers (vis/IR)			
Multi-channel/direction/polarisation radiometer			

Multispectral visible and infrared scan radiometer (10 channels)		
FY-1C and 1D		
1.1 km	3,200 km	VIS:
		0.43–0.48 μm
		0.48–0.53 μm
		0.53–0.58 μm
		0.58–0.68 μm
		NIR:
		0.84–0.89 μm
		NIR–SWIR:
		0.90–0.965 μm
		1.58–1.68 μm
		3.55–3.93 μm
		TIR:
		10.3–11.3 μm
		11.5–12.5 μm
Chinese Space Agency		
Imaging multi-spectral radiometers (vis/IR)		
Multi-purpose imaging Vis/IR radiometer		

Visible and infra-red scan radiometer (VIRR)		
FY-3A		
FY-3B		
FY-3C		
FY-3D		
FY-3E		
FY-3F		
FY-3G		
1.1 km	2,800 km	10 channels range 0.43–10.5 μm
Chinese space agency		
Visible and infra-red scan radiometer		
Multi-purpose imaging Vis/IR radiometer		

References

Cayula JF, Cornillon P (1996) Cloud detection from a sequence of SST images. Rem Sens Environ 55:80–88

CEOS (2008) The Earth Observation Handbook: Climate Change Special Edition 2008. Available at http://www.eohandbook.com/

CEOS (2009) The Earth Observation Handbook Online Database. Available at shttp://database.eohandbook.com/

Corlett GK, Barton IJ, Donlon CJ, Edwards MC, Good SA, Horrocks LA, Llewellyn-Jones DT, Merchant CJ, Minnett PJ, Nightingale TJ, Noyes EJ, O'Carroll AG, Remedios JJ, Robinson IS, Saunders RW, Watts JG (2006) The accuracy of SST retrievals from AATSR: an initial assessment through geophysical validation against in situ radiometers, buoys and other SST data sets. Adv Space Res 37(4):764–769

Donlon CJ (2008) The next generation of multi-sensor merged sea surface temperature data sets for Europe. In Barale V, Gade M (eds.) Remote sensing of the European Seas, Springer, Heidelberg, pp. 177–188

Donlon CJ, Casey KS, Robinson IS, Gentemann CL, Reynolds RW, Barton I, Arino O, Stark J, Rayner N, LeBorgne P, Poulter D, Vazquez-Cuervo J, Armstrong E, Beggs H, Llewellyn Jones D, Minnett PJ, Merchant CJ, Evans R (2009) The GODAE high resolution sea surface temperature pilot project (GHRSST-PP). Oceanography 22(3):34–45

Donlon CJ, Minnett PJ, Gentemann C, Nightingale TJ, Barton IJ, Ward B, Murray J (2002) Toward improved validation of satellite sea surface skin temperature measurements for climate research. J Climate 15:353–369

Donlon CJ, Robinson I, Casey KS, Vazquez-Cuervo J, Armstrong E, Arino O, Gentemann C, May D, LeBorgne P, Piollé J, Barton I, Beggs H, Poulter DJS, Merchant CJ, Bingham A, Heinz S, Harris A, Wick G, Emery B, Minnett P, Evans R, Llewellyn-Jones D, Mutlow C, Reynolds R, Kawamura H, Rayner N (2007) The global ocean data assimilation experiment (GODAE) high resolution sea surface temperature pilot project (GHRSST-PP). Bull Amer Meteor Soc 88(8):1197–1213, doi:10.1175/BAMS-88-8-1197

Kilpatrick KA, Podesta GP, Evans R (2001) Overview of the NOAA/NASA advanced very high resolution radiometer pathfinder algorithm for sea surface temperature and associated matchup database. J Geophys Res 106(C5):9179–9197

May DA, Parmeter MM, Olszewski DS, McKenzie BD (1998) Operational processing of satellite sea surface temperature retrievals at the naval oceanographic office. Bull Am Met Soc 79: 397–407

Merchant CJ, Embury O, Le Borgne P, Bellec B (2006) Saharan dust in nighttime thermal imagery: detection and reduction of related biases in retrieved sea surface temperature. Remote Sens Environ 104(1):15–30

Merchant CJ, Harris AR, Maturi E, MacCallum S (2005) Probabilistic physically-based cloud screening of satellite infra-red imagery for operational sea surface temperature retrieval, quart. J Royal Met Soc 131:2735–2755

Merchant CJ, Le Borgne P (2004) Retrieval of sea surface temperature from space based on modeling of infrared radiative transfer: capabilities and limitations. J Atmos Ocean Technol 22(11):1734–1746, doi:10.1175/JTECH1667.1

Merchant CJ, Le Borgne P, Marsouin A, Roquet H (2008a) Optimal estimation of sea surface temperature from split-window observations. Rem Sens Env 112(5):2469–2484, doi:10.1016/j.rse.2007.11.011

Merchant CJ, Llewellyn-Jones D, Saunders RW, Rayner NA, Kent EC, Old CP, Berry D, Birks AR, Blackmore T, Corlett GK, Embury O, Jay VL, Kennedy J, Mutlow CT, Nightingale TJ, Ocarroll AG, Pritchard MJ, Remedios JJ, Tett S (2008b) Deriving a sea surface temperature record suitable for climate change research from the along-track scanning radiometers. Adv Sp Res 41(1):1–11, doi:10.1016/j.asr.2007.07.041

O'Carroll AG, Eyre JR, Saunders RW (2008) Three-way error analysis between AATSR, AMSR-E, and in situ sea surface temperature observations. J Atmos Oceanic Technol 25:1197–1207

Robinson IS (2004) Measuring the oceans from space: the principles and methods of satellite oceanography, Springer/Praxis, Berlin, Germany, ISBN 3-540-42647-7, p. 670

Stark JD, Donlon CJ, Martin MJ, McCulloch ME (2007) OSTIA: an operational, high resolution, real time, global sea surface temperature analysis system. Oceans '07 IEEE Aberdeen, Conference Proceedings, Marine Challenges: Coastline to Deep Sea, Aberdeen, Scotland

Vincent RF, Marsden RF, Minnett PJ, Buckley JR (2008a) Arctic waters and marginal ice zones: Part 2 – An investigation of arctic atmospheric infrared absorption for AVHRR sea surface temperature estimates. J Geophys Res 113:C08044, doi:10.1029/2007JC004354

Vincent RF, Marsden RF, Minnett PJ, Creber KAM, Buckley JR (2008b) Arctic waters and marginal ice zones: a composite arctic sea surface temperature algorithm using satellite thermal data. J Geophys Res 113:C04021, doi:10.1029/2007JC004353

Walton CC, Pichel WG, Sapper JF, May DA (1998) The development and operational application of nonlinear algorithms for the measurement of sea surface temperatures with the NOAA polar-orbiting environmental satellites. J Geophys Res 103:27999–28012

Zhang HM, Reynolds RW, Lumpkin R, Molinari R, Arzayus K, Johnson M, Smith TM (2009) An integrated global observing system for sea surface temperature using satellites and in situ data: research to operations. Bull Amer Meteor Soc 90:31–38

Chapter 14
The Validation of Sea Surface Temperature Retrievals from Spaceborne Infrared Radiometers

Peter J. Minnett

14.1 Introduction

There are uncertainties associated with all measurements, and the magnitude of the uncertainties imposes restrictions on how the measurements should be applied or interpreted. The uncertainties can result from a variety of causes that relate to the nature of the variable being measured, and how the measurements are being made. Furthermore the techniques employed to assess the magnitude and characteristics of the uncertainties are also prone to error.

In the case of Sea Surface Temperature (SST) derived from measurements taken by infrared radiometers on earth-observing satellites, the sources of uncertainties can be divided into those that result from the characteristics of the radiometer, and how well the measurements are calibrated, and those that arise from imperfections in the atmospheric correction algorithm that is applied to remove the effects of the intervening atmosphere, including identifying the effects of clouds and aerosols. Typically, for want of a better defensible approach, the uncertainties that arise from the method of validating the satellite-derived SSTs are attributed to the satellite data.

In this chapter we mention the sources of uncertainties in the SST retrieval from the on-orbit measurements, and focus on how the uncertainties can be determined. We do not consider directly the physics of the measurement, nor the particulars of the atmospheric correction algorithms or approaches to cloud screening; these have recently been reviewed elsewhere (Minnett and Barton, 2010), and some aspects are discussed elsewhere in this volume. But first we have to establish what is meant by "SST" and what are desired, or at least acceptable, levels of uncertainties.

14.2 What Is Sea-Surface Temperature?

The radiance measured in space by infrared radiometers has its origin in the skin layer of the ocean and not in the body of the water below, the "bulk temperature" of

P.J. Minnett (✉)
Meteorology and Physical Oceanography, Rosenstiel School of Marine and Atmospheric Science, University of Miami, Miami, FL 33149-1098, USA
e-mail: pminnett@rsmas.miami.edu

V. Barale et al. (eds.), *Oceanography from Space*,
DOI 10.1007/978-90-481-8681-5_14,

which is measured by in-situ thermometers below the surface. The near-surface temperature gradients result from three distinct processes: the absorption of insolation, the heat exchange with the atmosphere and levels of subsurface turbulent mixing. In conditions of low wind speed, the heat generated in the upper ocean by the absorption of solar radiation is not well mixed through the surface layer, but causes thermal stratification with temperature differences between the uppermost layer of the ocean and the water below. There is a strong diurnal component to the magnitude of these temperature gradients, as well as a dependence on cloud cover, which modulates the insolation, and, importantly, wind speed which influences the turbulent mixing (e.g. Price et al., 1986; Fairall et al., 1996; Gentemann and Minnett, 2008).

The surface, skin layer of the ocean, much less than 1 mm thick (Hanafin, 2002; Hanafin and Minnett, 2001), is nearly always cooler than the underlying water because the heat flux is nearly always from the ocean to the atmosphere. The heat flow, supplying energy for both the turbulent and radiant heat loss to the atmosphere, is accomplished by molecular conduction through the aqueous side of the interface and this is associated with a temperature gradient in the surface skin layer. The relationship between skin and bulk SSTs just below the surface (at ~5 cm) is reasonably well behaved (Minnett et al., 2010). The relationship with deeper bulk temperature, at depths of a few meters where many bulk SST measurements are taken, is the same on average during the night, and during the day for wind speed conditions of >~6 m/s (Donlon et al., 2002). But under low winds the relationship is very variable – vertically, horizontally and temporally (Minnett, 2003; Ward, 2006). The difference between the skin temperature and that measured by a bulk, in-situ thermometer is very variable and highly dependent on the depth of the bulk measurement (Fig. 14.1). Use of the bulk temperature for satellite-validation introduces

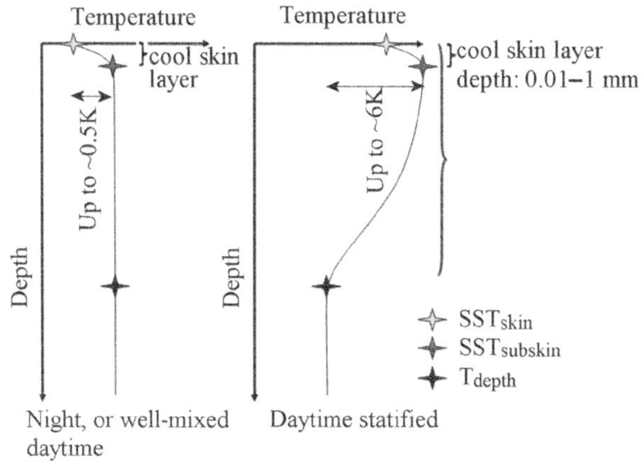

Fig. 14.1 Schematic representation of mean vertical profiles of near-surface temperature in the ocean. At *left* is the situation at night time, or daytime with good vertical mixing in the *upper layer*, and, at *right*, daytime during conditions conducive to the formation of a diurnal warm layer. The depth scale is non-linear: the skin layer is <1 mm in thickness, and the diurnal warm layer can extend through many meters (after Gentemann and Minnett, 2008)

these near-surface gradients into the error budget of the satellite retrieval and leads to an over-estimate of the uncertainties (Kearns et al., 2000). Physical models of the growth and decay of the diurnal thermocline (e.g. Woods and Barkmann, 1986; Price et al., 1986; Schiller and Godfrey, 2005; Gentemann et al., 2009a) require high temporal resolution forcing fields to produce reliable predictions, and this is a limitation on their use in relating bulk to skin temperatures for the validation of satellite-derived SSTs.

14.3 Required Accuracies

Some applications of satellite-derived SSTs require good precision, or relative accuracy, and the absolute accuracy is of a lesser importance. Examples include monitoring the positions and evolution of the surface expressions of thermal fronts in the ocean. However for many applications, it is the accuracy of the SSTs derived from satellite data that is of prime importance. The application with the most demanding accuracy requirement is "climate research" where a multi-decadal time series of global SSTs is required to detect small changes that are expected to reveal the response of the climate to changing forcing. Analysis of a time-series of SSTs to search for signatures of climate change will not lead to a convincing result if the uncertainties associated with the measurements are larger than the anticipated signal, which is likely to be <0.2 K/decade which requires 15–20 years of consistent and accurate SSTs with uncertainties <0.3 K.

Time series intended for use in Climate Research are referred to as "Climate Data Records" (CDRs), which has been defined as "a data set designed to enable study and assessment of long-term climate change, with 'long-term' meaning year-to-year and decade-to-decade change. Climate research often involves the detection of small changes against a background of intense, short-term variations" (NRC, 2000). To derive CDRs from satellite data "calibration and validation should be considered as a process that encompasses the entire system, from the sensor performance to the derivation of the data products." Furthermore, it is important to continue validation efforts over the lifetimes of the spacecraft sensors to ensure that the effects of degradation of the instruments in orbit are not misinterpreted as being caused by environmental signals (NRC, 2000). In generating time series of surface temperatures that span several satellite missions, the role of validation includes providing the necessary continuity in the derived fields.

An important aspect of the validation exercise is sampling the full ranges of orbital and atmospheric conditions. The orbital aspect is important as the thermal conditions on the spacecraft change markedly around the orbit, and these can propagate to the radiometers with the consequence of a changing thermal environment in and around the instrument. A prime example is the thermal shock experienced as the satellite enters and leaves the shadow of the earth (Brown et al., 1985).

An example of how knowledge of the uncertainties in the SST retrieval can be used is in the assimilation of SST retrievals in Numerical Weather Prediction

(NWP) where the sea surface is the bottom boundary condition for an atmospheric model (e.g. Chapter 15 by Beggs, this volume). A temperature measurement that is relatively inaccurate is therefore given less weight than one that has smaller uncertainties. A root-mean-square (rms) error of 0.3 K about a zero mean bias is the target accuracy for WMO observational requirements for global NWP applications (Eyre et al., 2009). A more relaxed requirement of 0.5 K is given as the "breakthrough" level which if achieved would result in a significant improvement in the targeted application. An rms uncertainty of 1 K is defined as the "threshold" accuracy, meaning that SSTs with greater uncertainties would not be of use in NWP applications.

14.4 Validation Techniques

The standard approach to establish the uncertainties in satellite-derived SSTs is to compare them with coincident measurements from independent sources. This is called "validation" as it leads to a verification of the in-flight calibration and the performance of the algorithms used to derive SST from the "top-of-atmosphere" brightness temperature measurements. The objectives of the validation exercise are to reveal the residual effects of instrumental artifacts in the raw measurements that have been imperfectly corrected, and of uncompensated effects of the intervening atmosphere. If patterns are identifiable in the uncertainties that reveal a systematic component to the sources of uncertainties, these may lead to improved correction algorithms. Included in such analyses is the quest for dependences on other relevant parameters that influence the satellite measurements, such as the water-vapor content of the atmosphere. If no clear patterns or dependences are found, the properties of the random or non-systematic uncertainties provide guidance on the averaging, spatial or temporal, that may be required to reach a specific level of accuracy required for a particular application.

Ideally, the reference measurements should be free of error and be taken at the same time and place as the satellite measurements, and have the same temporal and spatial sampling characteristics. But such measurements do not exist, so we have to endeavor to ensure that they are at least more accurate than the satellite retrievals, and they should share as many of the same characteristics as possible. This means that radiometric validation measurements are preferred over a sub-surface thermometric measurement, so that "like is compared with like" and additional sources of uncertainty caused by the near surface temperature gradients (Fig. 14.1) can be rendered negligible. Since the reference measurements are imperfect, having their own uncertainties, and the method of comparison introduces additional uncertainties, an error budget has to be constructed that takes into account the contributions from all sources.

There are several approaches to validating satellite-derived surface temperatures that use different instruments. Some are mounted on aircraft, others on ships or buoys.

14.4.1 SST Validation Using Radiometers

The validation of SSTs with infrared radiometers can be done using instruments mounted on ships (e.g. Kearns et al., 2000; Noyes et al., 2006) fixed platforms (McMillan et al., 2003) and aircraft (Smith et al., 1994). Aircraft have the advantage that they can be flown to areas free of clouds at the time of the satellite overpasses, but suffer from a serious disadvantage of being costly. Mounting radiometers on ships for long-term deployment is feasible, and the cruises of the ships, especially on transoceanic voyages, provide the opportunity of sampling a wide range of conditions. Cloud cover is an issue that leads to up to about 90% of possible overpasses being unsuitable for satellite SST validation (Kilpatrick et al., 2001), but the remaining ~10% can make a significant contribution. Radiometers mounted on fixed platforms require the advection of weather systems to provide a range of atmospheric conditions. We focus on the approach of ship-based measurements in the following discussion.

For the highest quality data to be used in the validation of satellite SSTs, the ship-based radiometers must be mounted on the ships so they have a clear view of the sea surface ahead of the ship's bow wave. Otherwise they do not take measurements of the skin SST undisturbed by the presence of the ship. Because the emissivity of the sea surface is not unity, a small component of the signal measured by the radiometer when it is directed at the sea surface is reflected sky radiance. To correct for this a measurement of the downwelling atmospheric radiance is required and thus, the validating instrument must be able to view the sky at the same angle to zenith as the sea view is inclined to nadir (Fig. 14.2). Depending on the size and layout of

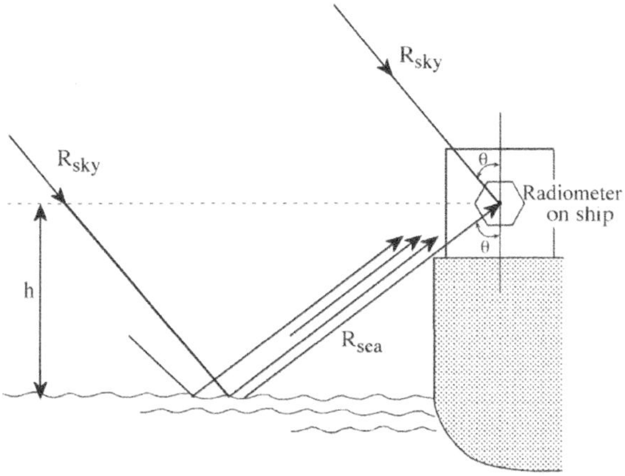

Fig. 14.2 The measurement geometry of a ship-board radiometer measuring the skin SST. R_{sea} is the emission from the sea surface that contains the information on the skin SST, and R_{sky} is the emission from the sky. The sky view is necessary to provide a correction for the small component in the sea-view measurements that is reflected infrared emission from the sky. The intersection of the radiometer field of view and the sea surface should be ahead of the bow wave to minimize the local influence of the ship

the ship, the radiometers are mounted at heights of a few meters to a few tens of meters above the sea surface, and even for those mounted at height the effects of the atmosphere below are negligible or easily corrected to good accuracy (Minnett et al., 2001; Smith et al., 1996). The radiometers on the ships, as those on aircraft, must be calibrated throughout the field deployment using internal calibration targets; and the calibration procedure should be checked using laboratory facilities before and after each deployment.

The ship-based instruments are of two types: filter radiometers where the band-pass (the relative spectral response) of the instrument is determined by an optical filter, and hyperspectral interferometers which measure a broad spectral range, a section of which is selected for the skin SST measurement. The filter radiometers developed for the validation of satellite-derived SSTs include the Infrared Scanning Autonomous Radiometer (ISAR; Donlon et al., 2008), the Calibrated InfraRed in-situ Measurement System (CIRIMS; Branch et al., 2008; Jessup and Branch, 2008), and the Scanning Infrared Sea Surface Temperature Radiometer (SISTeR; described in Barton et al., 2004). These are autonomous instruments that can be installed on ships to operate for weeks or months at a time with little or no operator intervention. For example, ISARs have been mounted on the M/V *Val de Loire* and M/V *Pride of Bilbao* for validation of SST in European waters (Corlett et al., 2006), and on M/V *Jingu Maru* and M/V *Andromeda Leader* of NYK Lines for trans-oceanic measurements (Fig. 14.3). A CIRIMS has been installed on the research vessels *Thomas G Thompson* and the *Ronald H Brown*, and the US Coast Guard icebreaker *Polar Sea* (Jessup and Branch, 2008). These radiometers, and others, are described elsewhere (Barton et al., 2004).

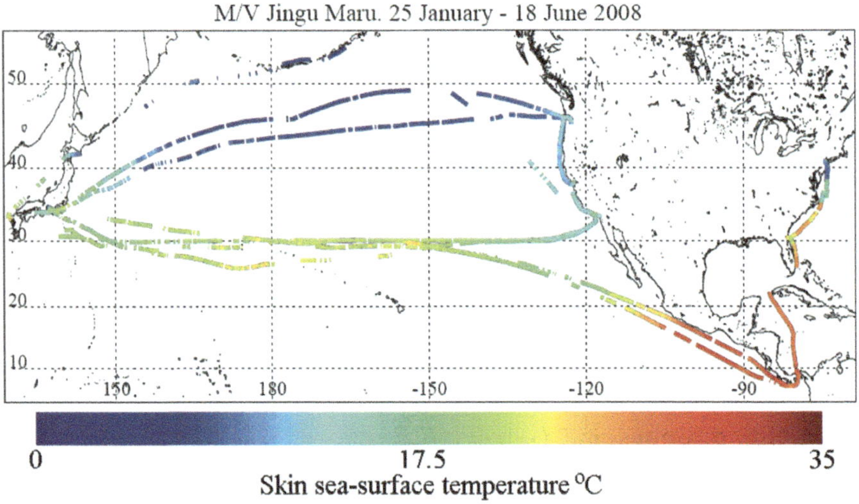

Fig. 14.3 Tracks of the M/V *Jingo Maru* between 25 January and 18 June 2008. The colors indicate the skin SST measured by an ISAR. Even though the cruise tracks are essentially zonal, they have sampled most of the SST temperature range in this period of about 6 months

In all of the ship-based radiometers, the data are recorded internally or on a local computer and transmitted to shore via satellite telemetry link. The ISAR and CIRIMS use commercial radiometers with bandpass filters covering the 9.6–11.5 μm wavelength range. The SISTeR has a filter wheel with three filters corresponding to the mid- and thermal infrared bands of the AVHRRs and (A)ATSRs. All three types of ship-board radiometers have two small internal blackbody calibration targets, for real-time calibration in the field.

The hyperspectral validation data are derived from the Marine-Atmospheric Emitted Radiance Interferometer (M-AERI; Minnett et al., 2001). This is a Fourier Transform InfraRed (FTIR) interferometric spectroradiometer that operates in the range of infrared wavelengths from ~3 to ~18 μm and measures spectra with a resolution of ~0.5/cm. It was developed specifically for the validation of MODIS skin SST retrievals and includes two very accurate internal blackbody cavities for calibration in the field. Two infrared detectors, needed to achieve this wide spectral range, are cooled to close to the boiling point of liquid nitrogen (~78 K) by a Stirling cycle mechanical cooler to reduce the noise equivalent temperature difference to levels well below 0.1 K. A gold-plated scan mirror, which is programmed to step through a pre-selected range of angles, directs the field of view from the interferometer to either of the internal blackbody calibration targets or to the environment from nadir to zenith. The interferometer integrates measurements over a pre-selected time interval, typically 60 s–100 s, to obtain a good signal-to-noise ratio. A typical cycle of measurements, including two view angles to the atmosphere, one to the ocean, and calibration measurements, takes about 5–10 min.

The absolute accuracy of the infrared spectra produced by the M-AERI is determined by the effectiveness of the blackbody cavities as calibration targets. The absolute accuracy of the spectral measurements of the M-AERI is better than 0.03 K (Minnett et al., 2001). The absolute uncertainties of the retrieved skin SST, determined by operating two M-AERI's side-by-side and by comparing M-AERI measurements with those from other well-calibrated radiometers, are less than 0.05 K (Barton et al., 2004; Minnett et al., 2001) which are sufficiently small to give confidence in the use of such data in the validation of satellite SST retrievals.

Since the launch of the NASA Earth Observing System satellite *Terra* in December 1999, over 40 M-AERI cruises have been undertaken on research vessels (Fig. 14.4). These and other radiometer cruises (e.g. Jessup and Branch, 2008) have spanned a wide range of climatological regimes, from polar regions to the tropics. In addition to sampling the full range of SST, from freezing to the high temperatures in the Red Sea and Tropical Western Pacific, validation data have been taken in a wide range of marine atmospheres. In addition, an M-AERI and a suite of atmospheric sensors have been deployed on the cruise ship *Explorer of the Seas*, operated by Royal Caribbean International. This ship was equipped at construction with scientific laboratories for oceanographic and atmospheric research (Williams et al., 2002). This vessel has been a valuable source of infrared radiometric measurements for the validation of satellite SST retrievals.

By taking validation data from moving ships, the integration time of the radiometric measurements along the shiptrack, or the averaging of rapidly sampled data,

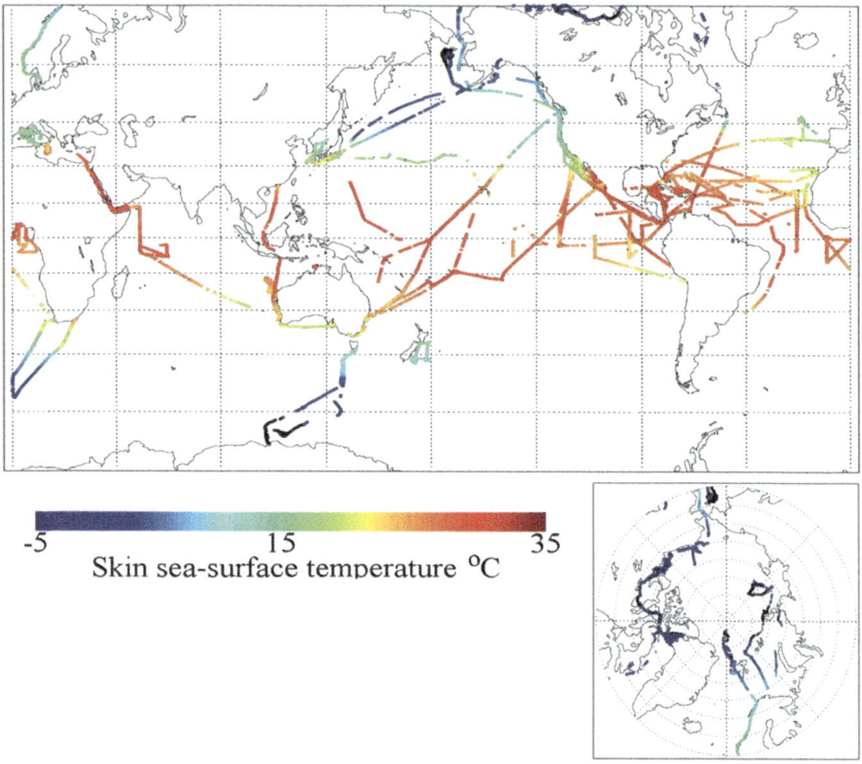

Fig. 14.4 Skin SST measured by the M-AERI on research vessels from 1998 to 2008. The coverage spans the global range of SST and atmospheric conditions

provides spatially-averaged information that is comparable to the km-scale infrared retrieval determined by the spacecraft radiometer footprint. This requires only the assumption that a one-dimensional average along the ships' tracks is a good approximation of the two-dimensional average in the spacecraft radiometer field of view. In most situations away from strong non-isotropic thermal features, such as the outcropping of strong fronts, this is believed to be a reasonable assumption.

The use of aircraft to provide SST validation data permits the deployment of radiometers that can be constructed to replicate closely the measurements of the satellite instruments. As with the spacecraft instrument, the aircraft radiometer, must have accurate internal calibration, but unlike the satellite sensor, the calibration procedure of the aircraft instrument can be verified before and after each flight. The flight speed of the aircraft permits measurements over distances that are comparable to the footprint of the satellite instruments, and the aircraft can be flown to areas that are cloud-free at the time of the satellite overpass. High-altitude transects underneath the satellite provide data to replicate the satellite measurement, but a source of uncertainty is in the effects of the atmosphere above the aircraft. This would provide information about the accuracy of the "top-of-atmosphere" brightness temperatures. Low-level flights provide a measurement of the skin SST, but here the source of uncertainty is in the effects of the atmosphere beneath the aircraft. The major drawback of using aircraft for surface temperature validation is cost.

14.4.2 SST Validation Using Buoys

While accurate infrared radiometric measurements of skin SST provide the most physically appropriate data for the validation of satellite-derived skin SSTs, they are relatively small in number compared to those provided by the network of drifting buoys (Fig. 14.5). The buoy measurements are telemetered via satellite along with a determination of the buoy position. In addition to the uncertainties introduced into the satellite SST validation by the near-surface temperature gradients (Fig. 14.1), the buoy data have uncertain calibration (Emery et al., 2001). The thermometers on the buoys are calibrated to 0.1 K accuracy prior to use, but after they have been deployed they are only very rarely recovered to assess calibration drift, sensor damage or contamination.

By comparing temperatures reported by pairs of buoys that drifted within 10 km of each other, an estimate of the rms errors in the measurement was found to be ~0.15 K (Emery et al., 2001). This value includes a contribution from the horizontal variability of the near-surface temperatures (Minnett, 1991), and Emery et al. (2001) report an increase in the rms difference to ~0.8 K if they allow separations of the buoy pairs to be up to 50 km. By comparing SSTs derived from two types of satellite radiometers and from buoys, both drifting and moored, the buoy temperature uncertainties, expressed as a standard deviation, was found to be 0.23 K

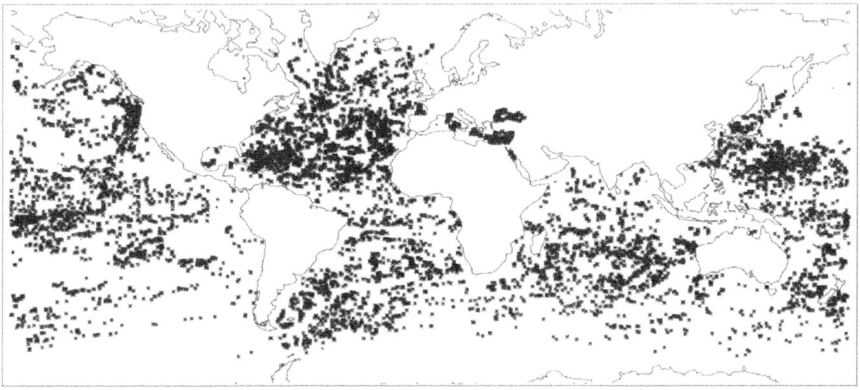

Fig. 14.5 The distribution of bulk SST measurements from drifting and moored buoys that matchup with high-confidence clear-sky SST retrievals from the MODIS on *Aqua*. There are 12,536 such matchups for 2003. Generally, less than 10% of all matchups between buoys and satellite data pass the stringent cloud-screening tests (Kilpatrick et al., 2001; after Minnett and Barton, 2010)

(O'Carroll et al., 2008). The large numbers of the drifting buoy data set render it too valuable a resource for SST validation for it to be neglected, but the quality control of the buoy data sets has to be thorough.

Moored buoys in the tropical Pacific and Atlantic Oceans, and in US waters are also used as a source of validating data. They generally have a thermometer at a depth of about 3 m, and while these are recovered for post-deployment recalibration (Freitag et al., 1999) the disturbance to the upper ocean by the presence of the mooring introduces a source of uncertainty into the comparison that is difficult to quantify.

The distribution of collocated (within 0.1° of latitude and longitude) and coincident (within 30 min) measurements from the *Aqua* MODIS and drifting and moored buoys is shown in Fig. 14.5 for the year 2003 (Minnett and Barton, 2010). There are 12,536 match-ups in the plot and only those satellite retrievals with the best quality flag (i.e. confidently cloud-free and satellite zenith angle <45°) are shown, but even with this large number there are ocean areas that are poorly sampled.

14.4.3 Traceability to Temperature Standards

Given that CDRs of SST span several satellite missions, ensuring that the validating measurements are themselves accurate over the CDR period is of prime importance. Without this assurance, systematic changes in the characteristics of the data sets used to validate satellite SST could be misinterpreted as systematic changes in the upper ocean and the climate. The only way of ensuring stability in the calibration of the sensors used to provide validation data is to have a traceable calibration chain to a national SI temperature standard. Since buoys are seldom recovered for re-calibration, the key to the generation of an SST CDRs lies in the calibration of the ship-based radiometers. The path to national temperature standards, therefore, is through the calibration of the radiometers used to validate the satellite retrievals, and this requires, and provides, radiometric traceability to national standards. In the USA, the national reference standards are maintained by the National Institute of Standards and Technology (NIST).

As part of the pre-launch characterization of the satellite radiometers, they are carefully calibrated in thermal-vacuum chambers to replicate the conditions on orbit. The pre-launch calibration is traceable to national standards, but the satellite radiometers are not recovered at the end of the mission for recalibration and re-characterization. To ensure traceability to NIST standards, an infrared calibration facility has been set up at the Rosenstiel School of Marine and Atmospheric Science at the University of Miami. Three international workshops have been held at which many of the ship-board radiometers used to validate satellite-derived SSTs were calibrated using a water-bath blackbody calibration target, built to a NIST design (Fowler, 1995). This calibration target consists of a black-painted, thin-walled, hollow, tapered, copper cone surrounded by a water bath, the temperature of which can be very accurately controlled. The water bath temperature is monitored by two thermometers, which have calibrations traceable to NIST standards. The internal calibration of ship-board radiometers is assessed by pointing them into the cone. The radiation emerging from the cone depends not only on its temperature, as given by the thermometers in the water bath, but also on its emissivity. The emissivity was determined, and hence the calibration system characterized, by the NIST Transfer Radiometer (TXR; Rice and Johnson, 1998), which is the infrared radiometric standard for the NASA Earth Observing System program (Rice and Johnson, 1996). The TXR was also used to characterize the laboratory blackbody calibrators used elsewhere to check the internal calibration of the ship-deployed radiometers (Rice et al., 2004). The comparative performance of the radiometers in conditions such as they experience in the field has been determined by mounting them together on a short cruise of the R/V *F. G. Walton Smith* (Barton et al., 2004), or on a pier. The outcome of these exercises is that the radiometers with internal calibration and an effective correction for the reflected sky radiance are capable of measuring the skin SST with uncertainties <0.1 K. Thus they can be used to help generate CDRs of SST. Additional at-sea comparisons between radiometers of different design have taken place on a more ad-hoc basis, but with similar results (e.g. Branch et al., 2008).

14.5 Uncertainties

The uncertainties in the retrievals of SST from spacecraft radiometers are often expressed as a mean error, or bias, and a scatter, or standard deviation about the mean; this terminology assumes a Gaussian error distribution. However, the reduction of the uncertainty fields to a pair of numbers does not indicate the complexity of the information that is required for some applications, such as the assimilation of the data in NWP and Ocean Forecasting models (Donlon et al., 2007). Some of the uncertainties are caused by systematic dependences on some of the parameters that determine the conditions at the time of the satellite measurement, such as the satellite zenith angle (through the slant path length through the atmosphere and the surface emissivity), the atmospheric water vapor and temperature distribution, the characteristics of aerosols, and the surface wind speed. And, of course, the uncertainties also depend on the details of the processing algorithms, in particular the cloud screening approaches, and the effectiveness of the atmospheric correction algorithms. These algorithms are generally optimized in the sense of producing minimum errors on a global basis, and application of the resulting SST fields in regionally or seasonally constrained analyses can result in different error characteristics (Eugenio et al., 2005; Kumar et al., 2003; Minnett, 1990; Shenoi, 1999).

The following summary of accuracies of SST derived from various satellite radiometers is determined by comparison with independent measurements and generally the discrepancies are ascribed to the satellite measurement. Since the independent measurements have their own uncertainties, and the method of comparison introduces additional uncertainties, the SST retrievals from the satellite-based measurements are likely to be more accurate than indicated. The study of O'Carroll et al. (2008) involved collocated measurements of three different sources of SSTs provides sufficient information to enable the standard deviation of error in each observation type to be estimated, thus allowing a more appropriate apportioning of the uncertainties.

14.5.1 AVHRR

The statistics shown in Table 14.1 are for the AVHRR Pathfinder SST fields (Kilpatrick et al., 2001) and show a near-zero mean error and standard deviation of about half a degree. The coefficients in the atmospheric correction algorithms used in the Pathfinder project are derived by robust regression between the AVHRR brightness temperatures and in-situ measurements from drifting buoys, so the retrieved SSTs are a "bulk SST" and they are related to the skin SST, through the mean skin temperature differences occurring in the matchups between satellite and in-situ data.

Table 14.1 Uncertainties in the Pathfinder AVHRR SST retrievals, from cloud-free comparisons between all AVHRR's in the period 1985–1998 and drifting buoys (from Kilpatrick et al., 2001)

Data	Mean	St dev
All	0.02 K	0.53 K

Using a radiometric SST from the M-AERI (Minnett et al., 2001), reduces the error (Table 14.2), with the exception of polar regions (Kearns et al., 2000) where the multichannel atmospheric correction algorithm has been shown to be prone to larger uncertainties (Vincent et al., 2008).

Table 14.2 Uncertainties in the AVHRR SST retrievals vs. M-AERI (after Kearns et al., 2000)

				Mean (K)	St dev (K)	N
Data	Mean (K)	St dev (K)	N	Excluding arctic cruise		
All	0.14	0.36	299	0.07	0.31	219
Day	0.18	0.40	142	0.00	0.24	62
Night	0.10	0.33	157	0.10	0.33	157

14.5.2 MODIS

The statistics of the uncertainties in the MODIS (Esaias et al., 1998) SST retrievals from the instruments on *Terra* and *Aqua* are shown in Tables 14.3, 14.4, 14.5 and 14.6, derived from comparisons with buoys and radiometers. The periods covered are from the start of each of the missions to mid 2007, and the processing scheme is V5 for both instruments. The statistics are shown for daytime and nighttime retrievals using two channels in the 11–12 μm atmospheric transmission window,

Table 14.3 Uncertainties in the *Terra* MODIS SST retrievals vs. buoys

	Mean (K)	St dev (K)	N	Mean (K)	St dev (K)	N
Data	11 μm SST			4 μm SST		
All	−0.15	0.61	254,834			
Day	−0.11	0.64	149,893			
Night	−0.20	0.55	104,941	−0.18	0.47	114,562

Table 14.4 Uncertainties in the *Aqua* MODIS SST retrievals vs. buoys

	Mean (K)	St dev (K)	N	Mean (K)	St dev (K)	N
Data	11 μm SST			4 μm SST		
All	−0.17	0.49	243,826			
Day	−0.15	0.55	151,814			
Night	−0.22	0.48	92,012	−0.22	0.42	99,986

Table 14.5 Uncertainties in the *Terra* MODIS SST retrievals vs. M-AERI

	Mean (K)	St dev (K)	N	Mean (K)	St dev (K)	N
Data	11 μm SST			4 μm SST		
All	0.04	0.53	4,751			
Day	0.09	0.58	1,999			
Night	0.01	0.53	2,752	−0.02	0.43	3,056

Table 14.6 Uncertainties in the *Aqua* MODIS SST retrievals vs. M-AERI

	Mean (K)	St dev (K)	N	Mean	St dev (K)	N
Data	11 μm SST			4 μm SST		
All	0.00	0.56	2,093			
Day	0.04	0.59	832			
Night	−0.02	0.53	1,261	−0.06	0.45	1,399

and the nighttime retrievals using measurements of brightness temperatures in the 4 μm, atmospheric transmission window. The MODIS SST retrievals are of a skin temperature and the mean errors (biases) that result from the comparison with the buoys is a manifestation of the thermal skin effect, which is typically in the range of 0.15–0.20 K (Donlon et al., 2002). The bias errors approximate to zero in the comparisons with the radiometer data (which are far fewer in number than comparisons with buoys). The standard deviations of the uncertainties are higher during the day than at night, reflecting the larger contribution of the variability in diurnal heating between the surface and the depth of the buoy measurements by day (Gentemann and Minnett, 2008). The day-night differences in the scatter are much smaller in the radiometer comparisons. The nighttime 4 μm SST retrievals show markedly smaller scatter than the 11–12 μm SST, which reflects not only the radiometric advantage of measurements in the shorter wavelength atmospheric window, which is a consequence of being on the short-wavelength side of the peak of the Planck Function, but also the better instrument performance. The dual-sided paddle-wheel scan mirror has a multi-layer interference coating that introduces a dependence of the reflectivity on the scan angle for wavelengths greater than about 8 μm (Guenther et al., 2002), and imperfect corrections for this effect introduce an additional uncertainty in the 11–12 μm SST retrievals.

The uncertainty characteristics in the *Aqua* MODIS retrievals are better than for the *Terra* MODIS and this presumably results from improvements in the instrument construction and pre-launch characterization of the second instrument, as some of the lessons learned in the development of the *Terra* MODIS could be applied to the MODIS on *Aqua* (Xiong et al., 2008, 2009).

14.5.3 AATSR

The advantage of an atmospheric correction algorithm based on dual views through different atmospheric path lengths, as used by the (A)ATSRs (Prata et al., 1990; Závody et al., 1995), is demonstrated in the uncertainties (Table 14.7). The disadvantage of the narrow swath (~500 km) is apparent in the relatively small number of comparisons. The nighttime algorithm uses all three infrared channels ($\lambda = 3.7$, 11 and 12 μm) at both views, while the daytime algorithm uses both views of the two longer wavelength channels. The buoy comparisons are taken from the 12-month period from 19th August 2002, and comprise about 30 per day, with collocation and coincidence criteria of 10 min of latitude and longitude, and 3 h (Corlett et al., 2006). Given that the AATSR SST retrieval is a skin temperature, the smaller bias errors derived from the comparison to buoys to that with radiometers is curious and probably indicates a warm bias in the SSTs, which is apparent in the daytime radiometer comparisons. The smaller nighttime bias in the radiometer comparison, along with the smaller standard deviation compared to the daytime retrievals, indicates the benefit of including the 3.7 μm measurements in the SST retrieval.

Table 14.7 Uncertainties in the *Envisat* AATSR SST retrievals (from information in Corlett et al., 2006)

	Mean (K)	St dev (K)	N	Mean	St dev (K)	N
Data	Buoy comparisons			Radiometer comparisons		
Day	0.02	0.39	~5,500	0.11	0.33	18
Night	0.04	0.28	~5,500	−0.06	0.20	12

Table 14.8 Uncertainties in the *Envisat* AATSR SST retrievals vs. M-AERI skin SST in the Caribbean Sea, measured from the *Explorer of the Seas* (from information in Noyes et al., 2006)

Data	Mean (K)	St dev (K)	N
Day	0.16	0.36	32
Night	0.04	0.26	84

AATSR skin SST validation using M-AERI data from the *Explorer of the Seas* in the Caribbean Sea (Noyes et al., 2006) supports the evidence of a small warm bias in the retrievals with a clear day-night difference (Table 14.8). The standard deviations revealed in all of the AATSR validation are notably smaller than for any other infrared radiometer.

14.5.4 Geostationary Meteorological Satellites

In principle, the SSTs retrieved from infrared radiometers on geostationary satellites should be comparable to those on polar-orbiters, given that the atmospheric corrections are essentially the same. However, the standard deviations of the SST uncertainties, when compared to temperatures measured from buoys, are in general greater than for those derived from radiometers on polar-orbiting satellites (Wick et al., 2002). For the GOES SST retrievals, part of the elevated scatter (0.6–0.9 K) is caused by errors that are systematic in the time of day (Wick et al., 2002), and are therefore influenced by instrumental artifacts with a diurnal character.

14.6 Future Sensors

At the time of writing, two new series of satellite infrared radiometers capable of SST retrievals are being developed for launch within the next several years. These are VIIRS (Visible/Infrared Imager Radiometer Suite[1]) which will be part of the payload on the US National Polar-orbiting Operational Environmental Satellite System (NPOESS), with the first VIIRS to fly on the NASA NPP (NPOESS Preparatory Program) satellite, and the SLSTR (Sea and Land Surface Temperature Radiometer) on the ESA Sentinal-3 series of satellites (Aguirre et al., 2007). VIIRS is a linear scanning radiometer with 22 spectral bands, which are a subset of the MODIS bands plus a sensitive panchromatic visible band. It has four bands suitable for SST retrieval, two in each of the thermal- and mid-infrared atmospheric transmission windows. SLSTR is a derivative of the ATSR series, having nine spectral bands of which three (two in the thermal infrared and one in the mid-infrared) are for SST measurements, and two views through different atmospheric path lengths. It

[1] See http://jointmission.gsfc.nasa.gov/science/VIIRScontent.REM.html and http://npoess.noaa.gov/index.php?pg=viirs

is anticipated that these instruments will continue the SST time series into the next decade, perhaps longer for VIIRS, and thus will require as extensive validation as their heritage instruments if they are to contribute to the SST CDR.

14.7 Conclusions

While there have been significant improvements in the demonstrated accuracies of SSTs derived from infrared radiometers on satellites in recent years, the most stringent requirements have not yet been met. These improvements have resulted from better atmospheric correction algorithms, and more accurate cloud and aerosol screening. They have also benefited from improved validation methods including the increasingly more widespread use of ship-bard radiometers. The usual approach is to attribute all of the uncertainties derived from the validation exercise to the satellite retrieval, despite there being additional sources of error. Thus we are still in the position of believing that the satellite SST retrievals are more accurate than demonstrated, but are not able to quantify with confidence the true accuracies. Innovative analysis approaches (e.g. O'Carroll et al., 2008) indicate a possible way forwards, and improved understanding of the physics controlling the skin layer and diurnal heating will lead to better corrections for these effects (e.g. Donlon et al., 2002; Gentemann et al., 2009a).

An important aspect of recent developments in this area is the increasing coordination between researchers in the field, and the GHRSST project (formerly the GODAE High Resolution SST Pilot Project; now the Group for High Resolution SST; Donlon et al., 2007) and the national components (e.g. Gentemann et al., 2009b) have played a significant role in this. Also the establishment of a methodology to cooperatively refer radiometer calibration to a NIST radiometric standard (Rice et al., 2004; Barton et al., 2004) has also laid the foundation of deriving a Climate Data Record for SST.

Acknowledgements The support of NASA over many years is gratefully acknowledged as it has facilitated much of the work that is included here. Discussions with many colleagues have helped clarify the issues involved in the physics of the measurements and in the approaches to SST validation.

References

Aguirre M, Berruti B, Bezy J-L, Drinkwater M, Heliere F, Ulf Klein, Mavrocordatos C, Silvestrin P, Greco B, Benveniste J (2007) Sentinel-3. The ocean and medium-resolution land mission for GMES operational services. ESA Bull 131:24–29

Barton IJ, Minnett PJ, Donlon CJ, Hook SJ, Jessup AT, Maillet KA, Nightingale TJ (2004) The Miami2001 infrared radiometer calibration and inter-comparison: 2. Ship comparisons. J Atmos Ocean Tech 21:268–283

Branch R, Jessup AT, Minnett PJ, Key EL (2008) Comparisons of shipboard infrared sea surface skin temperature measurements from the CIRIMS and the M-AERI. J Atmos Ocean Tech 25:589–606

Brown OB, Brown JW, Evans RH (1985) Calibration of advanced very high resolution radiometer infrared observations. J Geophys Res 90:11667–11677

Corlett GK, Barton IJ, Donlon CJ, Edwards MC, Good SA, Horrocks LA, Llewellyn-Jones DT, Merchant CJ, Minnett PJ, Nightingale TJ, Noyes EJ, O'Carroll AG, Remedios JJ, Robinson IS, Saunders RW, Watts JG (2006) The accuracy of SST retrievals from AATSR: an initial assessment through geophysical validation against in situ radiometers, buoys and other SST data sets. Adv Space Res 37:764–769

Donlon C, Robinson I, Casey KS, Vazquez-Cuervo J, Armstrong E, Arino O, Gentemann C, May D, LeBorgne P, Piollé J, Barton I, Beggs H, Poulter DJS, Merchant CJ, Bingham A, Heinz S, Harris A, Wick G, Emery B, Minnett P, Evans R, Llewellyn-Jones D, Mutlow C, Reynolds RW, Kawamura H, Rayner N (2007) The global ocean data assimilation experiment high-resolution sea surface temperature pilot project. Bull Am Met Soc 88:1197–1213

Donlon C, Robinson IS, Reynolds M, Wimmer W, Fisher G, Edwards R, Nightingale TJ (2008) An infrared sea surface temperature autonomous radiometer (ISAR) for deployment aboard volunteer observing ships (VOS). J Atmos Ocean Tech Res 25:93–113

Donlon CJ, Minnett PJ, Gentemann C, Nightingale TJ, Barton IJ, Ward B, Murray J (2002) Toward improved validation of satellite sea surface skin temperature measurements for climate research. J Clim 15:353–369

Emery WJ, Baldwin DJ, Schlüssel P, Reynolds RW (2001) Accuracy of in situ sea surface temperatures used to calibrate infrared satellite measurements. J Geophys Res 106:2387–2405

Esaias WE, Abbott MR, Barton I, Brown OB, Campbell JW, Carder KL, Clark DK, Evans RH, Hoge FE, Gordon HR, Balch WM, Letelier R, Minnett PJ (1998) An overview of MODIS capabilities for ocean science observations. IEEE Trans Geosci Rem Sens 36: 1250–1265

Eugenio F, Marcello J, Hernández-Guerra A, Rovaris E (2005) Regional optimization of an atmospheric correction algorithm for the retrieval of sea surface temperature from the Canary Islands-Azores-Gibraltar area using NOAA/AVHRR data. Int J Rem Sens 26:1799–1814

Eyre J, Anderson E, Charpentier E, Ferrantier L, Ferranti L, Lafeuille J, Ondras M, Pailleux J, Rabier F, Riishojgaard LP (2009) Requirements of Numerical Weather Prediction for Observations of the Ocean, Venice, Italy

Fairall C, Bradley E, Godfrey J, Wick G, Edson J, Young G (1996) Cool-skin and warm-layer effects on sea surface temperature. J Geophys Res 101:1295–1308

Fowler JB (1995) A third generation water bath based blackbody source. J Res Natl Inst Stand Technol 100:591–599

Freitag HP, McCarty ME, Nosse C, Lukas R, McPhaden MJ, Cronin MF (1999) COARE Seacat data: calibrations and quality control procedures. NOAA Tech. Memo. ERL PMEL-115, National Oceanic and Atmospheric Adminstration, Silver Spring, MD, USA, p. 41

Gentemann CL, Minnett PJ (2008) Radiometric measurements of ocean surface thermal variability. J Geophys Res 113:C08017

Gentemann CL, Minnett PJ, Sienkiewicz J, DeMaria M, Cummings J, Jin Y, Doyle JD, Gramer L, Barron CN, Casey KS, Donlon CJ (2009b) MISST: the multi-sensor improved sea surface temperature project. Oceanography 22:76–87

Gentemann CL, Minnett PJ, Ward B (2009a) Profiles of ocean surface heating (POSH): a new model of upper ocean diurnal thermal variability. J Geophys Res 114:C07017

Guenther B, Xiong X, Salomonson VV, Barnes WL, Young J (2002) On-orbit performance of the earth observing system moderate resolution imaging spectroradiometer; first year of data. Rem Sens Environ 83:16–30

Hanafin JA (2002) On sea surface properties and characteristics in the infrared. Ph.D. Thesis, University of Miami, Miami, p. 111

Hanafin JA, Minnett PJ (2001) Profiling temperature in the sea surface skin layer using FTIR measurements. In: Donelan MA, Drennan WM, Saltzmann ES, Wanninkhof R (eds.) Gas Transfer at Water Surfaces, American Geophysical Union Monograph, pp. 161–166

Jessup AT, Branch R (2008) Integrated ocean skin and bulk temperature measurements using the calibrated infrared in situ measurement system (CIRIMS) and through-hull ports. J Atmos Oceanic Tech 25:579–597

Kearns EJ, Hanafin JA, Evans RH, Minnett PJ, Brown OB (2000) An independent assessment of Pathfinder AVHRR sea surface temperature accuracy using the Marine-atmosphere emitted radiance interferometer (M-AERI). Bull Am Met Soc 81:1525–1536

Kilpatrick KA, Podestá GP, Evans RH (2001) Overview of the NOAA/NASA Pathfinder algorithm for sea surface temperature and associated matchup database. J Geophys Res 106: 9179–9198

Kumar A, Minnett PJ, Podesta G, Evans RH (2003) Error characteristics of the atmospheric correction algorithms used in retrieval of sea surface temperatures from infrared satellite measurements; global and regional aspects. J Atmos Sci 60:575–585

McMillan W, Hoff R, Strow L, Comer J, Lightner K, Maddy E, McCann K, McCourt M, Rutledge K (2003) ABOVE: The BBAERI AIRS Ocean Validation Experiment, Optical Remote Sensing. OSA Technical Digest. Optical Society of America, Quebéc City, Canada.

Minnett PJ (1990) The regional optimization of infrared measurements of sea-surface temperature from space. J Geophys Res 95:13497–13510

Minnett PJ (1991) Consequences of sea surface temperature variability on the validation and applications of satellite measurements. J Geophys Res 96:18475–18489

Minnett PJ (2003) Radiometric measurements of the sea-surface skin temperature – the competing roles of the diurnal thermocline and the cool skin. Int J Rem Sens 24:5033–5047

Minnett PJ, Barton IJ (2010) Remote sensing of the earth's surface temperature. In: Zhang ZM, Tsai BK, Machin G (eds.) Radiometric Temperature Measurements and Applications, Academic Press/Elsevier, New York, pp. 333–391

Minnett PJ, Knuteson RO, Best FA, Osborne BJ, Hanafin JA, Brown OB (2001) The marine-atmospheric emitted radiance interferometer (M-AERI), a high-accuracy, sea-going infrared spectroradiometer. J Atmos Oceanic Tech 18:994–1013

Minnett PJ, Smith M, Ward B (2010) Measurements of the oceanic thermal skin effect. Deep Sea Res II (in review)

Noyes EJ, Minnett PJ, Remedios JJ, Corlett GK, Good SA, Llewellyn-Jones DT (2006) The accuracy of the AATSR sea surface temperatures in the Caribbean. Rem Sens Environ 101:38–51

NRC (2000) Issues in the Integration of Research and Operational Satellite Systems for Climate Research: II. Implementation. National Academy of Sciences, Washington, DC

O'Carroll AG, Eyre JR, Saunders RW (2008) Three-way error analysis between AATSR, AMSR-E, and in situ sea surface temperature observations. J Atmos Oceanic Tech 25:1197–1207

Prata AJ, Cechet RP, Barton IJ, Llewellyn-Jones DT (1990) The along-track scanning radiometer for ERS-1 – scan geometry and data simulation. IEEE Trans Geosci Rem Sens 28:3–13

Price JF, Weller RA, Pinkel R (1986) Diurnal cycling: observations and models of the upper ocean response to diurnal heating, cooling and wind mixing. J Geophys Res 91:8411–8427

Rice JP, Butler JJ, Johnson BC, Minnett PJ, Maillet KA, Nightingale TJ, Hook SJ, Abtahi A, Donlon CJ, Barton IJ (2004) The Miami2001 infrared radiometer calibration and intercomparison: 1. Laboratory characterization of blackbody targets. J Atmos Oceanic Tech 21:258–267

Rice JP, Johnson BC (1996) A NIST thermal infrared transfer standard radiometer for the EOS program. Earth Obs 8:31

Rice JP, Johnson BC (1998) The NIST EOS thermal-infrared transfer radiometer. Metrologia 35:505–509

Schiller A, Godfrey JS (2005) A diagnostic model of the diurnal cycle of sea surface temperature for use in coupled ocean-atmosphere models. J Geophys Res 110:C11014

Shenoi SC (1999) On the suitability of global algorithms for the retrieval of SST from the north Indian ocean using NOAA/AVHRR data. Int J Rem Sens 20:11–29

Smith AH, Saunders RW, Závody AM (1994) The validation of ATSR using aircraft radiometer data over the Tropical Atlantic. J Atmos Oceanic Tech 11:789–800

Smith WL, Knuteson RO, Revercomb HE, Feltz W, Howell HB, Menzel WP, Nalli N, Brown O, Brown J, Minnett P, McKeown W (1996) Observations of the infrared radiative properties of the ocean – implications for the measurement of sea surface temperature via satellite remote sensing. Bull Am Met Soc 77:41–51

Vincent RF, Marsden RF, Minnett PJ, Creber KAM, Buckley JR (2008) Arctic waters and marginal ice zones: a composite arctic sea surface temperature algorithm using satellite thermal data. J Geophys Res 113:C04021

Ward B (2006) Near-surface ocean temperature. J Geophys Res 111:C02005

Wick GA, Bates JJ, Scott DJ (2002) Satellite and skin-layer effects on the accuracy of sea surface temperature measurements from the GOES satellites. J Atmos Ocean Tech 19:1834–1848

Williams E, Prager E, Wilson D (2002) Research combines with public outreach on a cruise ship. EOS 83:590–596

Woods JD, Barkmann W (1986) The response of the upper ocean to solar heating I: The mixed layer. Quart J Roy Meteor Soc 112:1–27

Xiong X, Kwo-Fu C, Aisheng W, Barnes WL, Guenther B, Salomonson VV (2008) Multiyear on-orbit calibration and performance of terra MODIS thermal emissive bands. IEEE Trans Geosci Rem Sens 46:1790–1803

Xiong X, Wenny BN, Aisheng W, Barnes WL, Salomonson VV (2009) Aqua MODIS thermal emissive band on-orbit calibration, characterization, and performance. Geosci Rem Sens IEEE Trans 47:803–814

Závody AM, Mutlow CT, Llewellyn-Jones DT (1995) A radiative transfer model for sea-surface temperature retrieval for the along track scanning radiometer. J Geophys Res 100:937–952

Chapter 15
Use of TIR from Space in Operational Systems

Helen M. Beggs

15.1 Introduction

Data from the thermal-infrared (TIR) channels of scanners flown on polar-orbiting and geostationary meteorological satellites are routinely used for the determination of Sea Surface Temperature (SST) at depths less than 0.1 mm (Robinson, 1985). There are many operational meteorological, oceanographic and ecosystem management systems that use SST data from these TIR satellite sensors. The aim of this chapter will be to review how satellite infrared SST (IR SST) observations are currently used in these operational systems, their particular requirements for SST and recommendations for future developments.

In the absence of cloud, thermal-infrared radiometers such as AVHRR, AATSR and MODIS on polar-orbiting satellites provide twice-daily repeat measurements of SST at the highest spatial resolution currently available from a satellite (1 km). The most accurate satellite TIR sensor is the dual-view AATSR on EnviSat, providing improved atmospheric correction resulting in accuracy <0.2°C (O'Carroll et al., 2008). However, this sensor has limited coverage compared with the less accurate (~0.3–0.5°C) AVHRR or MODIS, due to much narrower swath width.

Geostationary satellites (e.g. USA's GOES, Europe's MSG and Japan's MTSAT series) carry radiometers with IR window channels similar to AVHRR. Their horizontal resolution is coarser (3–5 km) but they sample at higher temporal resolution (generally half-hourly). Within the modern-era SST record (commencing with TIR from AVHRR in 1981) there are a wide variety of global, spatially-mapped SST products available for research and operational applications. The different types are:

H.M. Beggs (✉)
Centre for Australian Weather and Climate Research, Australian Bureau of Meteorology, Melbourne VIC 3001, Australia
e-mail: H.Beggs@bom.gov.au

V. Barale et al. (eds.), *Oceanography from Space*,
DOI 10.1007/978-90-481-8681-5_15, © Springer Science+Business Media B.V. 2010

- Level 1 (L1): Geolocated radiances (brightness temperatures) from individual sensors
- Level 2 (L2): Geolocated sea surface temperatures from individual satellite sensors over a single swath (orbit)
- Level 3 (L3): Composite, gridded, SST products created by averaging (but not interpolating) L2 data over space and time (may be produced from a single sensor or multiple sensors over a single orbit to several days)
- Level 4 (L4): Gridded, gap-free, SST analyses created by interpolating multiple observation sources
- Dynamically consistent SST outputs from ocean global circulation models that assimilate satellite SST observations at levels 2, 3 or 4

The following Section 15.2 describes a new format for level 2 satellite SST products developed by the Group for High-Resolution SST (GHRSST) and now used in various operational systems. Sections 15.3 and 15.4 summarise several operational level 3 (composite) and level 4 (analysis) products using IR SST data streams as inputs. Sections 15.5, 15.6, 15.7, 15.8 and 15.9 provide an overview of each of the types of operational systems currently using satellite IR SST level 2, 3 or 4 products, how the infrared SST products are used in each system and their particular requirements. Section 15.10 summarises goal requirements for remotely sensed infrared SST from a review of the latest conference proceedings and scientific literature. Section 15.11 presents recommendations from both IR SST product producers and users for future developments in TIR observations and processing that will benefit operational systems.

15.2 GHRSST Level 2 SST Products

In 2000, the GODAE High-Resolution SST Pilot Project (GHRSST-PP: Donlon et al., 2007), was initiated to develop and demonstrate a system that could deliver high-resolution, global-coverage SST data products operationally in near-real time according to GODAE specifications. GHRSST-PP (now the Group for High-Resolution SST – GHRSST[1]) provides two major types of near-real-time SST products (L2 pre-processing and L4 analysis products). GHRSST-L2P products are baseline measurement data sets. SST data generated from different sensors are made available to users in a common, self-describing format (netCDF) together with ancillary information to assist with their interpretation and use by data assimilation systems (Donlon et al., 2009b). For every level 2 file of input data, GHRSST produces a matching "L2P" (level 2 pre-processed) product (listed in Table 15.1) that contains identical SST values in the same geographical layout as those in the source L2 product.

[1] http://www.ghrsst.org

Table 15.1 GHRSST-L2P (Level 2 Preprocessed) SST products available in real-time and delayed mode from GHRSST[2]

L2P Product	Instrument type	Satellite orbit	Resolution	Time series	Data source agency
ATS_NR_2P	AATSR/infrared	Envisat/polar	1 km	Mar 2005 present	ESA
SEVIRI_SST	SEVIRI/infrared	MSG/geostationary	0.1°	Feb 2005 present	EUMETSAT
AMSRE	AMSR-E/microwave	Aqua/polar	25 km	Jun 2002 present	RSS
AVHRR16_G	AVHRR/infrared	NOAA-16/polar	8.8 km × 4.4 km	Feb 2005 Aug 2006	NAVOCEANO
AVHRR16_L LAC	AVHRR/infrared	NOAA-16/polar	2.2 km	Feb 2005 Oct 2005	NAVOCEANO
AVHRR17_G	AVHRR/infrared	NOAA-17/polar	8.8 km × 4.4 km	Feb 2005 present	NAVOCEANO
AVHRR17_L LAC	AVHRR/infrared	NOAA-17/polar	2.2 km	Feb 2005 present	NAVOCEANO
AVHRR17_L HRPT	AVHRR/infrared	NOAA-17/polar	1.1 km	Sep 2008 present	PML
AVHRR18_G	AVHRR/infrared	NOAA-18/polar	8.8 km × 4.4 km	Jan 2006 present	NAVOCEANO
AVHRR18_L LAC	AVHRR/infrared	NOAA-18/polar	2.2 km	Jan 2006 present	NAVOCEANO
AVHRR18_L HRPT	AVHRR/infrared	NOAA-18/polar	1.1 km	Sep 2008 present	PML
AVHRR MTA_G	AVHRR/infrared	MetOp-A/polar	9 km	Sep 2007 present	ESA
NAR16_SST	AVHRR/infrared	NOAA-16/polar	2 km	Feb 2005 Nov 2005	NAVOCEANO
NAR17_SST	AVHRR/infrared	NOAA-17/polar	2 km	Feb 2005 present	NAVOCEANO
NAR18_SST	AVHRR/infrared	NOAA-18/polar	2 km	Nov 2008 present	NAVOCEANO
MODIS_T	MODIS/infrared	Terra/polar	1 km	Oct 2006 present	NASA/OBPG/RSMAS/JPL
MODIS_A	MODIS/infrared	Aqua/polar	1 km	Jul 2006 present	NASA/OBPG/RSMAS/JPL
GOES11	GOES/infrared	GOES-11/geostationary	6 km/half-hour	Dec 2006 present	NOAA
GOES12	GOES/infrared	GOES-12/geostationary	6 km/half-hour	Dec 2006 present	NOAA
TMI	TMI/microwave	TRMM/equatorial	25 km	Jan 1998 present	RSS

[2] Source: http://ghrsst.jpl.nasa.gov/GHRSST_product_table.html accessed 24 October 2009.

On a pixel-by-pixel basis, each SST retrieval is augmented with estimates of the bias error and standard deviation error typically derived from statistical databases of in-situ and satellite data. Other ancillary fields for each pixel are surface wind speed, aerosol optical depth, sea ice concentration, time of measurement, difference from previous reference SST analysis field, and quality control flags, including a cloud proximity confidence flag. The standard deviation errors associated with the highest quality SST observations (denoted by a proximity_confidence flag of 5) are typically in the range 0.3–0.5°C for the various satellite infrared SST GHRSST-L2P products currently available (Table 15.1).

High Resolution Picture Transmission (HRPT) AVHRR data has been received from NOAA's operational polar orbiting satellites at local L-band receiving stations by various operational meteorological and oceanographic agencies since 1981. This data is used as a real-time source of 1.1 km resolution SST into operational systems.

Figure 15.1a–d show plots of the new Integrated Marine Observing System[3] (IMOS) 1 km resolution, HRPT AVHRR L2P SST for various values of cloud proximity flag, defined in this product as a function of distance to cloud, satellite zenith angle and day/night (Beggs et al., 2009c). Figure 15.1e–h present values of the derived SST standard deviation calculated for each cloud proximity flag value (Beggs et al., 2009c). These figures illustrate one of the particular strengths of the GHRSST L2P format for level 2 satellite SST data that users of L2P products can trade off between spatial coverage and accuracy, depending on the requirements of their application. Less spatial coverage in this IMOS AVHRR L2P product (Fig. 15.1) corresponds to less chance of cloud contamination and lower estimated errors.

The design of the L2P format was mainly driven by the requirements of the new "foundation" SST analyses for the "best" cloud-free SST with sufficient auxiliary data to enable conversion of SST observations within or immediately below the cool skin to foundation SST. By definition, the "foundation" SST (*SSTfnd*) product provides an SST that is free of any diurnal variations (daytime warming or nocturnal cooling) and is considered equivalent to the "subskin" ocean temperature (at the base of the conductive laminar sub-layer of the surface ocean) in the absence of any diurnal signal (Donlon et al., 2007). The L2P format has proven to be sufficiently universal that GHRSST L2P files from a range of IR and MW sensors (AATSR, MODIS, AVHRR, AMSR-E, TMI, SEVIRI and GOES Imagers) are now used in some operational systems for ocean forecasting and seasonal prediction.

By accessing L2P products in netCDF format via OPeNDAP[4] servers both in real-time[5] and delayed mode[6] it is possible to download only the variables or domain that the application requires, therefore saving both on communication costs and storage requirements. The consistent formats for GHRSST products also facilitate addition of new data streams into operational systems.

---

[3] http://www.imos.org.au

[4] http://www.opendap.org

[5] http://dods.jpl.nasa.gov

[6] http://data.nodc.noaa.gov/ghrsst

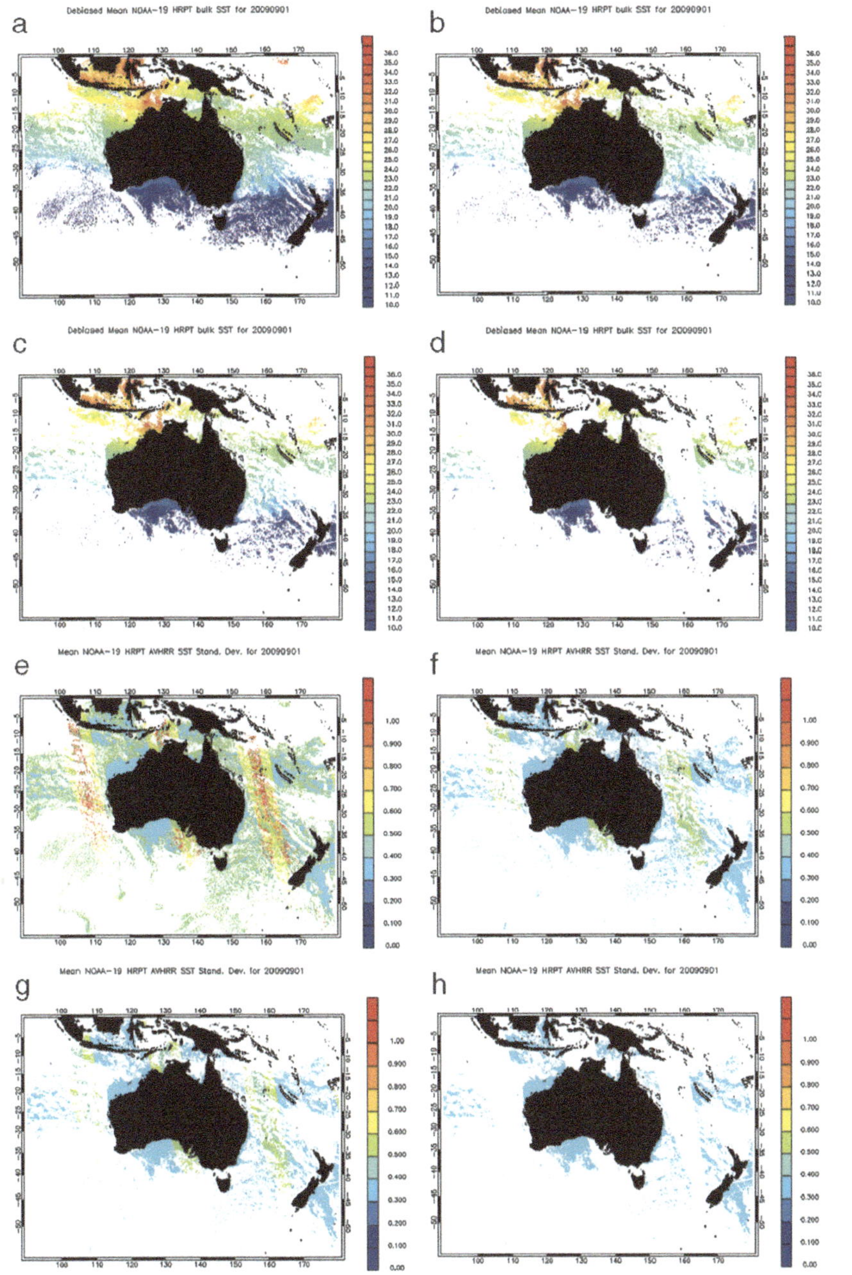

Fig. 15.1 Example GHRSST-L2P data set (from Bureau of Meteorology and IMOS) derived using HRPT AVHRR SST data collected from all ascending orbits of the NOAA-19 satellite during 1 September 2009 over the region 90°E–180°E, 0°N–55°S. SST is plotted in C for cloud proximity confidence flags (**a**) ≥1, (**b**) ≥3, (**c**) ≥4 and (**d**) 5. The corresponding single sensor error statistic (SSES) standard deviation error is plotted in C for cloud proximity confidence flags (**e**) ≥1, (**f**) ≥3, (**g**) ≥4 and (**h**) 5

15.3 Level 3 Composite SST Products

Level 2 (single orbit) SST products can be combined to produce either a single-sensor or multiple-sensor, gridded, "level 3" SST product. Table 15.2 gives a few examples of operational level 3 composite SST products formed from level 2 files of IR SST (GOES-11, GOES-12, MSG-9, MTSAT-1R, MODIS and AVHRR) that are used in operational applications.

These applications include fisheries management and protection of endangered species (e.g. TurtleWatch, Howell et al., 2008), hindcasting and nowcasting of coral bleaching and weather forecasting.

15.4 Level 4 SST Analyses

Gap-free, "level 4", analyses of satellite level 2 and in-situ SST are produced operationally at various oceanographic and meteorological agencies around the world using the method of optimal interpolation (OI) (e.g. Smith et al., 1991). These SST analyses provide sea surface boundary forcing in atmospheric and ocean models, and are used as a constraint to observations in modelling of the upper ocean circulation. Accurate SST fields are essential in computing different climate variables, such as heat transfer in ocean-atmosphere coupled systems and surface pCO_2 (Vinogradova et al., 2009).

Over the last 10 years or so, progress in the real-time analysis of satellite SST has been made possible by the following developments:

- Dramatic improvement in the quality and quantity of IR SST retrievals from AVHRR instruments aboard POES satellites
- Arrival of new sources of highly accurate IR SST (e.g. AATSR, errors from 0.16°C (O'Carroll et al., 2008))
- Arrival of new sources of accurate microwave SST (e.g. AMSR-E)
- Progress in facilitating access and exchange of data through international cooperation such as implemented by GHRSST-PP.

Infrared SST observations from AATSR, AVHRR and MODIS provide accurate estimates of SST at high resolution (~1 km) within the ocean cool skin (~10–20 μm depth; *SSTskin*), but only over cloud-free regions. Microwave observations from the AMSR-E instrument on Aqua provide SST observations in the subskin region (~1 mm depth; *SSTsubskin*), at coarser resolution (~25 km) but with the advantage of being able to measure SST through cloud (although not rain). In addition, IR SST observations from geostationary satellites provide up to half-hourly observations of *SSTskin*at ~5 km resolution in cloud-free regions, with the advantage of frequent temporal sampling enabling the measurement of SST over ocean regions experiencing ephemeral cloud. A summary of the various definitions of SST is presented in Table 15.3.

Table 15.2 Examples of operational level 3 composite SST products using IR SST level 2 files as inputs

Level 3 product	Spatial resolution	Temporal resolution (h)	Max latency (days)	Input products	References
Coastwatch POES SST	1 and 4 km	~6 (4 products/day)	~1/4	HRPT AVHRR L2 from all POES satellites	http://coastwatch.noaa.gov/poes_sst_overview.html
NOAA/ NESDIS GOES SST	0.05°	1, 3 and 24	1/24, 1/8 and 1	GOES-11 and -12 MTSAT-1R and MSG in L2P	Maturi et al. (2008) and http://www.coastwatch.noaa.gov
MODIS/AMSR-E SST (weighted average based on latency)	0.01°	~6 (4 products/day)	7	MODIS L2P (Aqua/Terra) AMSR-E L2P (Aqua) OSTIA L4	Jedlovec et al. (2009)
BoM/IMOS mosaic SST (weighted average based on latency)	0.01°	24	14	HRPT AVHRR L2 from all POES satellites	Rea (2004) and http://www.marine.csiro.au/remotesensing/imos/
NOAA/ NESDIS nighttime AVHRR SST	0.5 and 1	24	1	GAC AVHRR L2 from all POES satellites	http://www.osdpd.noaa.gov/ml/ocean/sst.html

Table 15.3 Definitions of SST product types (classified by depth) used in this chapter, following the GHRSST convention (Donlon et al., 2007)

SST product type	Nominal depth	SST sensors supplying product type
SST skin	~10–20 μm	In-situ radiometers, (A)ATSR, MODIS, SEVIRI, GOES, MTSAT-1R
SST subskin	~1 mm	AMSR-E, TMI
SST depth	Specified depth (~1 m for buoys)	Moorings, drifters, ships, Argo, XBT, CTD, AVHRR (calibrated using moorings/drifters to buoy depths)
SST fnd	1 mm to ~10 m (depending on solar, ocean, meteorological conditions)	Moorings, drifters, ships, Argo, XBT, CTD (depending on solar, ocean and meteorological conditions); all satellite IR and MW sensors using diurnal variation models to convert from skin or subskin to foundation SST
SST blend	10 μm to ~ 10 m	All

Combination of data from infrared and microwave sensors from polar-orbiting satellites and infrared sensors on geostationary satellites, often with in-situ measurements from drifting buoys and ships, now results in high-quality SST level 4 analyses with high spatial resolution, ranging from 0.25° (e.g. Operational MGDSST – Guan and Kawamura, 2004) down to 0.01° (experimental G1SST – Chao et al., 2009).

Table 15.4 presents a list of several of the operational, global, daily SST analyses that use IR SST along with the types of SST inputs used. Some analyses, such as OSTIA (Stark et al., 2007) and ODYSSEA (Autret and Piollé, 2007), use the highly accurate, dual-view, AATSR IR SST data in conjunction with in-situ SST to correct for biases in the other satellite SST data streams. Others use a combination of drifting buoy and ship SST observations to correct for regional and seasonal biases (e.g. MGDSST, CMC, AVHRR_OI, AVHRR_AMSR_OI). However, issues remain with the calibration of the IR and MW SST data streams, particularly at high latitudes north of 50°N and south of 50°S (Reynolds et al., 2010) where available in-situ SST observations are still relatively sparse.

Two operational analyses in Table 15.4 (OSTIA and GAMSSA – Zhong and Beggs, 2008) convert the radiometric *SSTskin* from IR satellite sensors and *SSTsubskin* from microwave sensors (AMSR-E and TMI) to an estimate of the foundation SST (see Section 15.2) by applying a simple empirical algorithm (Donlon et al., 2002) that filters out day-time input observations for surface winds <6 m/s. Comparisons between OSTIA and GAMSSA *SSTfnd* and independent buoy SST observations from the following day (0.4 and 0.5°C, respectively), indicate that filtering the input observations in this manner reduces the standard deviation errors compared with those obtained from a global *SSTblend* analysis such as NCDC AVHRR_AMSR_OI (0.6°C, Beggs et al., 2009b).

Table 15.4 Examples of operational, global, daily SST analysis (Level 4) products using IR SST level 2 files as inputs. SST analyses where all inputs are adjusted to a "foundation" (pre-dawn) SST are denoted as *SSTfnd*, following the GHRSST notation (Donlon et al., 2007), with those making no adjustment to SST inputs for the cool-skin and diurnal warm layer are denoted as *SSTblend*

Level 4 product	Spatial res.	SST type	Input products	Mean global error estimate (°C)	References
Met Office OSTIA	0.05°	*SSTfnd*	In-situ, L2P products from: LAC/GAC AVHRR, AATSR, AMSR-E, TMI, SEVIRI	0.44 Beggs et al. (2009b)	Stark et al. (2007)
Ifremer ODYSSEA	0.1°	*SSTfnd*	L2P products from: FRAC/HRPT/LAC/GAC AVHRR, AATSR, AMSR-E, TMI, GOES-11, GOES-12, SEVIRI	0.58 Beggs et al. (2009b)	Autret and Piollé (2007)
NOAA POES/GOES	0.1°	*SSTblend*	L2P products from: GAC AVHRR (POES satellites), GOES-11, GOES-12		Maturi et al. (2009)
NCODA	10 km	*SSTblend*	In-situ, L2 products from: GAC AVHRR, AMSR-E, GOES-11, GOES-12		Cummings (2005)
CMC	0.2°	*SSTblend*	L2 products from AATSR, AMSR-E and L2P products from GAC AVHRR	0.38 Brasnett (2008)	Brasnett (2008) and Brasnett (2009)
JMA MGDSST	0.25°	*SSTblend*	In-situ, L2 products from: HRPT/GAC AVHRR, AMSR-E	0.53 Brasnett (2008)	Guan and Kawamura (2004)
NCDC AVHRR_OI	0.25°	*SSTblend*	In-situ (buoys, ships), L2 products from Pathfinder AVHRR and L2P products from GAC AVHRR		Reynolds et al. (2007)
NCDC AVHRR AMSR_OI	0.25°	*SSTblend*	In-situ (buoys, ships), L2 products from Pathfinder AVHRR and AMSR-E, and L2P products from GAC AVHRR	0.64 Beggs et al. (2009b) 0.59 Brasnett (2008)	Reynolds et al. (2007)
GAMSSA	0.25°	*SSTfnd*	In-situ, L2 product from AATSR, L2P products from GAC AVHRR and AMSR-E	0.50 Beggs et al. (2009b)	Zhong and Beggs (2008)
NCEP RTG	0.5°	*SSTblend*	In-situ (ships, buoys), L2 products from GAC AVHRR	0.57 Brasnett (2008)	Thiebaux et al. (2003)

Some newer experimental analyses such as the 6-hourly, 4 km resolution, *SSTblend* Analysis over the Gulf of Maine, US (Vinogradova et al., 2009) combine hourly IR SST from geostationary satellites with daily multi-sensor SST analyses to attempt to resolve the diurnal cycle. Other methods being trialled for operational implementation in Numerical Weather Prediction (NWP) data assimilation systems (see Section 15.7) have been hourly skin SST analyses derived from foundation SST analyses combined with hourly *SSTskin* – *SSTfnd* estimates (Beggs et al., 2009a). These *SSTskin* – *SSTfnd* values can be calculated in real-time by inputting NWP forecast surface winds into an empirical model based on geostationary SEVIRI satellite IR *SSTskin* observations and AMSR-E wind observations (Gentemann et al., 2003).

15.4.1 TIR Requirements

It is very useful for operational SST analysis (level 4) producers to be able to obtain level 2 satellite SST products in a common format with associated error estimates and quality flags. The GHRSST L2P format products (Section 15.2) are very effective as they contain all the variables in one file required to use the SST data in an OI analysis system. These include error estimates (bias and standard deviation) for each SST value along with quality control and cloud proximity flags, surface wind and surface solar insolation (from either observations or NWP models), aerosol optical depth and time of observation.

Most of the operational SST analysis systems listed in Table 15.4 use some or solely GHRSST L2P IR SST products and several (OSTIA, ODYSSEA, NCODA, GAMSSA, AVHRR_OI and AVHRR_AMSR_OI) are available in the GHRSST-L4 netCDF format incorporating useful ancillary data fields such as analysis error, sea ice concentration and land mask.

With the future operational implementation of current experimental ultra-high-resolution SST analyses such as G1SST, there will be an increasing requirement from operational SST analysis systems for timely, accurate, 1 km resolution, IR SST level 2 products with reliable error statistics and effective cloud masking.

15.5 Ocean Models

Operational numerical ocean models providing forecasts of currents, temperature and salinity fields are used for a variety of applications including coral reef management, tide predictions, marine sanctuary and estuary management, diving operations, naval applications, oil and chemical spill drift forecasts, search and rescue operations, offshore oil drilling operations, cable laying and ship routing. Today, a dozen numerical ocean modelling systems routinely operate in the nine countries that participated in the Global Ocean Data Assimilation Experiment (GODAE) (Dombrowsky et al., 2009). They range from regional high resolution systems that include tides, to global eddy-resolving systems that provide estimates of the ocean state, updated regularly (from daily to monthly), providing forecasts from a few days to 1 month ahead (Dombrowsky et al., 2009).

15.5.1 Use of TIR Products

SST strongly co-varies with the ocean temperature over the mixed layer depth (50–100 m) and complements altimetry data in multi-variate ocean analyses (Brassington, 2009). Short-range ocean forecasting systems such as the UK Met Office Forecasting Ocean Assimilator Model (FOAM) (Martin et al., 2009) and BLUElink Ocean Model, Analysis and Prediction System (OceanMAPS: Brassington et al., 2007) assimilate level 2 SST data. The FOAM model is run daily in the operational suite at the UK Met Office at global 1/4° resolution with nested 1/12° resolution regional configurations. The OceanMAPS model is run twice-weekly at the Australian Bureau of Meteorology at global 0.1–2° horizontal resolution with nested 0.1° resolution over the Australian region (Brassington et al., 2007).

For satellite SST, FOAM uses only GHRSST-L2P files in its data assimilation. FOAM assimilates infrared SST data obtained from AATSR and AVHRR (from NOAA and MetOp-A satellites) in addition to microwave *SSTsubskin* data from AMSR-E. The satellite observations are combined into one observation ("super-obbed") by calculating the median of all observations of a particular type within a 13 km radius. The model counterparts of these observations are calculated during a 1-day model run in a first-guess-at-appropriate-time scheme (Martin et al., 2009). The observations undergo a bias correction using AATSR and in-situ data as reference data. The bias-corrected observations and their model counterparts are then used in an optimal interpolation type scheme to produce 2 dimensional SST increments (Martin et al., 2009).

OceanMAPS uses the BLUElink Ocean Data Assimilation System (BODAS – Oke et al., 2008) as its (re-)analysis component. The operational BODAS system assimilates AMSR-E *SSTsubskin* retrievals from the Aqua satellite and IR SST from AVHRR. It has been demonstrated that the addition of infrared GAC AVHRR L2P *SSTdepth* files from NAVOCEANO (Table 15.1) results in BODAS analyses closer to buoy observations compared with assimilation of the AMSR-E and GAC AVHRR SST data streams separately (Andreu-Burillo et al., 2010).

Other ocean modelling systems (such as TOPAZ, NMEFC, Mercator and MOVE/MRI.COM) assimilate gap-free (level 4) SST analyses rather than level 2 products (Dombrowsky et al., 2009).

15.5.2 TIR Requirements

Ocean models require satellite SST for assimilation that is reliable, timely and accurate. Spatial resolution of the IR SST assimilated into ocean models is not currently an issue due to the relative coarser horizontal resolution of the global and regional ocean models (>3 km – Dombrowsky et al., 2009). However, future ultra-high-resolution regional ocean models may require higher resolution (1 km) SST, particularly within 50 km of coastlines. In addition to assimilation, regional coastal models in particular (e.g. Baird et al., 2010) require accurate, ultra high-resolution (1 km), level 3 or level 4 SST products for validation.

SST in coastal and inland regions have a large variability due to the diurnal cycle of solar radiation, which enhances surface characteristics of the land and sea and forces land-air-sea interactions, i.e. land-sea breezes. Typically, coastal ocean modelling systems have a requirement for ultra-high resolution SST data sets (1 km spatial resolution and < 6 h temporal resolution), with good accuracy (<0.1°C) and hourly temporal coverage.[7] Overall, for global, regional and coastal ocean modelling maintenance of high resolution SST is critical as well as the ability to observe the diurnal cycle (Brassington, 2009).

15.6 Fisheries Management and Protection of Endangered Species

The prediction of fish habitat using SST (among other data) is currently being developed and used for fishery management. The near real-time mapping of top-predators' habitat is being used (e.g. Southern Bluefin Tuna, Hobday and Hartmann, 2006) or proposed (e.g. Atlantic Bluefin Tuna, Druon, 2010) for reducing discard and preserving the resource and fishery respectively. SST traces efficiently (with chlorophyll-a data) the biomass-rich oceanic fronts which are used to identify the feeding habitat of most pelagic species. The monthly accumulation of heat in the surface layer which can be retrieved from SST may also trace the spawning habitat of some fish species (Atlantic Bluefin Tuna, Druon, 2010). The potential of using SST with other satellite and ocean model products for improving scientific knowledge on most fish habitat and fishery management is high with the processing of both the multi-annual and near-real-time products.

Historically, assessment of the sustainable exploitation of a commercial fish stock has been largely based on data from the fishing fleet. Recently, international agencies have requested a more ecological approach to managing fisheries and other marine resources. In the US the "Pelagic Habitat Analysis Module" (PHAM) has been developed to improve stock assessment by integrating classical fisheries data with satellite SST and ocean colour maps, outputs from a Global Circulation Model and statistical algorithms to map habitat of pelagic species (Kiefer et al., 2009). Such information along with information on recruitment can then be introduced into existing stock assessment models (Kiefer et al., 2009).

Remotely sensed SST can also be used in management systems to protect endangered marine species, by identifying regions where the species is more likely to

[7] http://www.wmo.int/pages/prog/sat/documents/SOG-08_Ocean.doc

inhabit and thereby restricting fishing activities in these areas. The US National Marine Fisheries Service has set up the "TurtleWatch" tool, a near real-time map based on operational longline fishery characteristics, bycatch information, loggerhead sea turtle satellite tracks and satellite SST that identifies the fishing exclusion zones for shallow nets to avoid turtle entrapment (Howell et al., 2008).

15.6.1 Use of TIR

The US National Marine Fisheries Service use the NOAA/NESDIS 0.05° resolution GOES IR SST composite L3 product (Table 15.2) in support of coastal applications, particularly studies related to commercial fisheries management and protection of endangered species (Maturi et al., 2008). Another IR SST product used in the protection of endangered species is the AVHRR Pathfinder V4 SST. The TurtleWatch product uses a daily, 3 day temporal mean image of the Pathfinder SST (Howell et al., 2008).

15.6.2 TIR Requirements

Scientists involved in the stewardship of living marine resources require the characterisation of local error for each SST product, providing much needed reassurance for data users who are sceptical about the quality of satellite-based SST measurements (Foley, 2009). Standardised formats for SST products (such as those designed by GHRSST) allows for uniform access using tools compliant with those recommended by the US Integrated Ocean Observing System and its international counterparts (Foley, 2009).

15.7 Coral Bleaching Nowcasting and Forecasting Systems

Coral bleaching results from the loss of symbiotic algae, known as zooxantheallae, from coral tissues during times of stress, often due to temperatures higher than the coral colony's tolerance level (Glynn, 1993). NOAA's Coral Reef Watch Program's satellite data[8] provide current reef environmental conditions to quickly identify areas at risk for coral bleaching. Continuous monitoring of sea surface temperature at global scales provides researchers and stakeholders with tools to understand and better manage the complex interactions leading to coral bleaching. When bleaching conditions occur, these tools can be used to trigger bleaching response plans and support appropriate management decisions.

---

[8] http://coralreefwatch.noaa.gov/satellite/index.html

GBRMPA's ReefTemp[9] is a mapping product that provides information on coral bleaching risk for the Great Barrier Reef (GBR) region. It produces high-resolution now-casts of bleaching risk and provides an improved ability to monitor heat stress in the Great Barrier Reef.

Seasonal forecasts from coupled ocean-atmosphere models can be used to predict anomalous SST several months in advance. The Australian Bureau of Meteorology has implemented operational seasonal forecasts of SST anomalies[10] over the Great Barrier Reef to aid in the management of the Great Barrier Reef Marine Park using the operational POAMA seasonal dynamical prediction model (Spillman and Alves, 2009). In addition to SST anomalies, the system calculates a GBR index of the areal average of monthly SST anomalies within the GBR study region as a useful indicator of the likelihood of coral bleaching occurring.

15.7.1 Use of TIR

The NOAA Coral Reef Watch system uses the operational NOAA/NESDIS 0.5° nighttime AVHRR SST composites (Table 15.2) for near real-time monitoring of thermal stress that can cause coral bleaching (Liu et al., 2009). The NOAA Reynolds and Smith (Reynolds and Smith, 1994) weekly, 1°, OI v1 SST analyses are used for producing the bleaching outlook. NOAA is transitioning to using the recently implemented operational NOAA/NESDIS AVHRR Clear-Sky Processor for Oceans (ACSPO) SST and POES-GOES Blended SST data (Table 15.4) for Coral Reef Watch (Liu et al., 2009).

Australia's ReefTemp uses the BoM/IMOS daily, 0.01° resolution, Mosaic AVHRR SST product (Table 15.2) and a climatology produced by CSIRO Marine and Atmospheric Research from HRPT AVHRR SST data (Griffin et al., 2004) to produce its daily nowcasts of Degree Heating Weeks and Heating Rate to aid in the management of the Great Barrier Reef (Maynard et al., 2008).

The BoM GBR SST Anomaly products use the weekly, 1°, Reynolds OI.v2 SST (Reynolds and Smith, 1994; Reynolds et al., 2002).

15.7.2 TIR Requirements

Reef forecasting systems require highly robust instrumentation producing climate quality SST at 2–10 m depth for long periods, and reporting that data in near real-time (Gramer et al., 2009; Liu et al., 2009). These systems require higher spatial and temporal resolution than SST analysis and composite products currently used in NOAA's Reef Watch, with better coverage over shallow near-shore waters necessary to effectively model differences in bleaching response between distinct sub-regions of a reef system (Gramer et al., 2009 and references therein). Infrared SST 1 km resolution SST observations from AVHRR and MODIS sensors on polar-orbiters

[9] http://www.cmar.csiro.au/remotesensing/reeftemp/web/ReefTemp_application.htm

[10] http://www.bom.gov.au/oceanography/oceantemp/GBR_SST.shtml

are particularly useful for extreme shallow water coverage and to measure SST close to coasts. Issues remain, however, for the most effective way to cloud-clear and calibrate close to coasts.

Geostationary satellites, in clear-sky areas, provide a measure of the diurnal variation more effectively than SST products derived from polar-orbiters, which only observe at the same time or two each day. Observations of diurnal variation of SST are necessary to accurately monitor thermal stress on coral (Maturi et al., 2008). Current techniques use a single nighttime image to represent the SST for each day. Given that the coral is sensitive to absolute maximums, as well as thermal exposure (over time), a bleaching prediction technique based on full diurnal observations of SST (Leichter et al., 2006) is superior to the currently used polar orbiter techniques, which are being updated just once per day (Maturi et al., 2008).

15.8 Numerical Weather Prediction

Numerical weather prediction (NWP) uses current weather conditions as input into mathematical models of the atmosphere to predict the weather. Sea surface temperature and sea-ice affect the behaviour of the overlying atmosphere and vice versa. Consequently, NWP systems need to be regularly updated with the latest SST and sea-ice observations to ensure an accurate forecast. Daily analyses of both SST and sea-ice extent and concentration are required by many operational NWP systems. SST affects the formation and subsequent evolution of tropical cyclones, convection and thunderstorms, cyclogenesis, sea fog and sea breezes. It can also help upper air forecasters at the World Aviation Forecast Centre to monitor areas more likely to develop Cumulonimbus activity which can produce significant threat to aircraft.

15.8.1 Use of TIR

Numerical Weather Prediction models commonly use analyses (level 4) of SST as a boundary condition over the ocean. The US National Weather Service's Ocean Prediction Centre and the Tropical Prediction Centre assimilate the GOES SST level 3 products (Table 15.2) into their operational forecast models.

Another use of SST analyses in NWP is for the quality control of satellite atmospheric sounder channels that peak near the ocean surface. Currently, foundation or blended SST analyses are used, but in 2010 the Australian Bureau of Meteorology plans to trial the use of real-time, skin SST analyses produced from a combination of foundation SST analyses and a diurnal warming model using NWP forecast surface winds as inputs (Beggs et al., 2009a). In regions of the ocean experiencing high insolation and low winds, *SSTskin* can experience daily variations of up to 6 or more Kelvin. The presence of cloud can also result in anomalously cool *SSTskin* measurements from infrared sensors on satellites, and therefore a real-time estimate of *SSTskin* can aid in determining if atmospheric sounder profile data close to the ocean surface is affected by cloud.

15.8.2 TIR Requirements

The World Meteorological Organisation (WMO) goals for spatial resolution and accuracy of the SST analyses used as a boundary condition for global NWP models are 5 km and 0.3°C (Eyre et al., 2009). However, accurate ultra-high-resolution (1 km) SST is becoming more important for regional weather forecasting as regional NWP models are produced at higher spatial resolutions. Chelton et al. (2007) showed that using coarse resolution SST analyses as the boundary condition for regional weather forecast models does not properly portray the fluxes of heat and moisture from the ocean that drive the formation of low level clouds and precipitation over the ocean. High resolution SST data products (particularly those combining both IR and MW inputs) preserve SST gradients more effectively compared to analysis systems that rely on cloudy infrared and limited in-situ sources. A detailed analysis of SST and its diurnal cycle is therefore sometimes needed locally, in the case of important precipitation events which are very dependent on evaporation over the ocean.[11]

15.9 Seasonal and Interannual Forecasting

Several operational centres routinely issue seasonal forecasts of the Earth's climate using coupled ocean-atmosphere models, which require near-real-time knowledge of the state of the global ocean (Balmaseda et al., 2009). Seasonal forecasting systems are based on coupled ocean-atmosphere general circulation models that predict SSTs and their impact on atmospheric circulation. The aim of seasonal forecasts is to predict climate anomalies (e.g. temperature, rainfall, frequency of tropical cyclones) for the forthcoming seasons (Balmaseda et al., 2009). The strongest relationship between SST patterns and seasonal weather trends are found in tropical regions. The horizontal resolution of operational seasonal prediction models are typically of the order of 1° due to computational constraints (Balmaseda et al., 2008).

15.9.1 Use of TIR

Global SST analysis products are commonly used to initialize operational seasonal forecast models. Several operational seasonal prediction models use the global, weekly, 1° resolution, Reynolds OI v2 SST analysis of in-situ and GAC AVHRR SST data (Reynolds and Smith, 1994; Reynolds et al., 2002), either by itself or combined with the monthly, 1° resolution, in-situ SST and sea-ice HadISST analysis (Balmaseda et al., 2008).

---

[11] http://www.wmo.int/pages/prog/sat/documents/SoG-Regional-NWP.doc

15.9.2 TIR Requirements

The quality of the coupled models used for seasonal and interannual forecasting has reached the stage where they are now sensitive to the quality of the initial conditions. It is therefore important to have SST analyses that are unbiased, accurate and stable over time-scales of years to weeks, especially in the tropics. Ships and moored and drifting buoys provide observations of good temporal frequency and acceptable accuracy, but coverage is marginal or worse over large areas of the Earth. There is a WMO goal requirement for high quality, fast delivery bulk SST (ideally with accuracy <0.1°C on the 100 km spatial scale and <0.25°C on the 10 km spatial scale, available within 24 h.[12]

15.10 Summary of TIR Requirements

Operational systems require timely, accurate, reliable and robust SST level 2, 3 and 4 products in consistent, well-described formats. The specific requirements of each class of operational system are given in Table 15.5.

Table 15.5 Summary of operational applications and their goal TIR requirements

Application	Horizontal resolution (km)	Temporal resolution (h)	Timeliness (h)	Accuracy (°C)
Numerical weather prediction				
–Global	5	24	3	0.3
–Regional	1	<6	<1	0.3
Seasonal and	10	24	24	<0.1
interannual forecasting	100	24	24	<0.25
Ocean modelling and marine services				
–Coastal ocean	<1	<6	1	<0.1
–Open ocean	5–1	<6	1	<0.1
Coral reef management	<1	1	24	<0.3
systems		24		

15.11 The Way Forward

A review of the IR SST requirements from operational systems pointed to several areas requiring future work and improvements in order to reach the goal requirements summarised in Section 15.10. Current high-resolution, global, daily SST

[12] http://www.wmo.int/pages/prog/sat/documents/SOG-05_SIA.doc

analyses range in global mean RMS error from 0.4 to 0.6°C (Table 15.4) when compared with independent observations from Argo floats or drifting buoys. In order to meet the WMO goals for SST product accuracy for various operational systems of 0.1–0.3°C, it is essential that calibration and validation of satellite TIR over the ocean be improved. In particular, calibration of all satellite SST data needs to be improved over the Southern and Arctic Oceans (Reynolds et al., 2010; Donlon et al., 2009a). More in-situ SST observations at high latitudes are required to drive these improvements both for calibration and validation.

One important issue limiting satellite SST accuracy is that drifting and moored buoys are generally used to calibrate SST, with their associated representativeness and instrument errors and limited precision. Sea surface temperature measurements from buoys are recorded at a precision of 0.1°C on the WMO Global Telecommunications System (GTS) with an instrument error of around 0.3°C for point measurements at 0.5–2 m depth, whereas TIR sensors on satellites measure the skin SST over an ocean area of between 1 km × 1 km and 6 km × 6 km depending on sensor. A possible solution is to use physical retrieval methods such as those applied to AATSR IR SST (Merchant and Le Borgne, 2004) and more recently to AVHRR IR SST (Bogdanoff et al., 2009) to calibrate IR SST rather than regression against in-situ SST. In-situ measurements of skin SST can then be used to validate the satellite skin SST produced using this method. Since ship-borne radiometers are capable of routinely measuring skin SST to <0.1°C (Donlon et al., 2008), a global sustained array of these sensors on ships will be necessary to validate satellite IR SST to the goal accuracy of 0.1°C.

There is a clear need for global coverage (ideally 6 satellites in orbit) of at least hourly skin SST from geostationary satellites (Donlon et al., 2009a), and the calibration of SST from IR sensors on GOES and MTSAT-1R geostationary satellites in particular needs to be improved to approach the accuracy of AVHRR sensors (Beggs et al., 2009b). Physical retrieval calibration methods should be investigated for application to IR sensors on geostationary satellites.

Not only should the calibration of all single view satellite IR SST sensors be improved, but also methods for determining the single pixel SST bias and standard deviation. Operational systems require SST inputs with reliable error estimates per pixel determined by the producer, thereby removing the necessity for user-designed bias-correction schemes depending on either in-situ SST or other satellite IR SST data such as from AATSR. Logically, the IR SST data producers should be more expert in the error characteristics of their data stream than the operational users. Were operational systems to use consistent error estimates for each input SST data stream, then this would significantly aid SST inter-comparison efforts such as the GHRSST Multi-Product Ensemble and Intercomparison Project.[13]

The processing of the level 2 IR SST should also be improved, with a systematic review of cloud screening and ice masking methods being undertaken to potentially improve the quality of satellite IR SST data sets (Donlon et al., 2009a). Following

[13] http://www.ghrsst.org/Todays-global-SST.html

Table 15.6 Summary of future planned thermal infrared satellite instruments[14]

Sensor	Satellite orbit	Agency	Expected launch/activation date	Spatial resolution at nadir (km)
Sea and land surface temperature radiometer SLSTR/ATSR-4	Sentinel-3/polar	ESA	2012	1
Second generation global imager (SGLI)	GCOM-C1/polar	JAXA	2014	0.5 and 1
Visible infrared imager radiometer suite (VIIRS)	NPOESS/polar	NOAA	2011	~0.4
Infrared imager	MTSAT-2/geostationary at 145°E	JMA	2010	~5
Meteorological imager	COMS/geostationary at 128°E	KMA	2009/10 2nd satellite 2014	4
Infrared imager	GOES-13/geostationary at 105°W	NOAA	2006 (on-orbit storage)	~5
Infrared imager	GOES-14/geostationary	NOAA	2009 (on-orbit storage)	~5

the success of the AATSR IR sensor on EnviSat for producing highly accurate SST, a capability for satellite IR dual view along track scanning radiometry should be sustained in an operational context with redundancy (Donlon et al., 2009a).

Finally, there is an increasing need for 1 km resolution, accurate, IR SST level 2 products in GHRSST-L2P format for input into ever higher resolution SST analyses and coastal ocean models. Table 15.6 summarises future planned thermal infrared sensors on meteorological satellites capable of measuring sea surface temperatures. These sensors should ensure the continuity of IR SST into the future.

Acknowledgements Very helpful comments from Dr Claire Spillman and Ms Xinmei Huang from the Australian Bureau of Meteorology are gratefully acknowledged. The HRPT AVHRR L2P data presented was provided by the Integrated Marine Observing System (IMOS) – an initiative of the Australian Government being conducted as part of the National Collaborative Research Infrastructure Strategy.

References

Andreu-Burillo I, Brassington G, Oke P, Beggs H (2010) Including a new data stream in BLUElink ocean data assimilation system. Aust Meteorol Oceanogr J 59:77–86

[14]http://www.ghrsst.org/New-Satellite-Programs-and-GHRSST.html – accessed 18/11/2009.

Autret E, Piollé J-F (2007) Deliverable D2.2.10 implementation of a global SST analysis WP 02 Task 2.2, Météo-France, France, 29 pp. See: http://www.mersea.eu.org/Satellite/MERSEA-WP02-IFR-STR-001-1A.doc

Baird M, MacDonald H, Roughan M, Oke P (2010) Downscaling an eddy-resolving global ocean model for the continental shelf off South-East Australia. Ocean Mod (in press)

Balmaseda MA, Alves O, Arribas A, Awaji T, Behringer D, Ferry T, Fjii Y, Lee T, Rienecker M, Rosati T, Stammer D (2008) Ocean initialization for seasonal forecasting. Invited paper presented at the Final GODAE Symposium, Nice, France, 12–15 November 2008. See: http://www.godae.org/5.5-MB-abstract.html

Balmaseda MA, Alves OJ, Arribas A, Awaji T, Behringer DW, Ferry N, Fujii Y, Lee T, Rienecker M, Rosati T, Stammer D (2009) Ocean initialization for seasonal forecasts. Oceanography 22:154–159

Beggs H, Gentemann C, Steinle P (2009a) Real-time skin sea surface temperature analyses for quality control of data assimilated into NWP models, extended abstract. The Fifth WMO International Symposium on Data Assimilation of Observations in Meteorology, Oceanography and Hydrology, Melbourne, Australia, 5–9 October 2009. See: http://cawcr.gov.au/bmrc/ocean/BLUElink/SST/5WMO_DASymp_Extended_Abstract_Beggs_Sep2009.pdf

Beggs H, Majewski L, Andreu-Burillo I, Paltoglou G, Verein R (2009b) Report to GHRSST10 from Australia – BLUElink and IMOS. Submitted to the Proceedings of the 10th GHRSST Science Team Meeting, Santa Rosa, 1–5 June 2009. See: http://cawcr.gov.au/bmrc/ocean/BLUElink/SST/GHRSST10/GHRSST10_Meeting-Australian_Report_Beggs.pdf

Beggs H, Majewski L, Paltoglou G (2009c) New Australian High Resolution AVHRR SST Products from the Integrated Marine Observing System. GHRSST 2009 International Users Symposium Conference Proceedings, Santa Rosa, USA, 29–30 May 2009, pp. 29–32. See: http://www.ghrsst.org/modules/documents/documents/IGUS_Extended_Abstracts.pdf

Bogdanoff A, Clayson CA, Roberts B (2009) Calculation of sea surface temperature using a forward radiative transfer model approach. GHRSST 2009 International Users Symposium Conference Proceedings, Santa Rosa, USA, 29–30 May 2009, pp. 77–79. See: http://www.ghrsst.org/modules/documents/documents/IGUS_Extended_Abstracts.pdf

Brassington GB, Pugh T, Spillman C, Schulz E, Beggs H, Schiller A, Oke P (2007) BLUElink Development of operational oceanography and servicing in Australia. J Res Pract Inf Technol 39:151–164

Brassington GB (2009) Ocean Prediction Issues Related to Weather and Climate Prediction, Pre-CAS Public Web Consultation: Vision Paper (Agenda Item 8.5), WMO CAS XV, Seoul, Korea, 18–25 November 2009, 11 pp

Brasnett B (2008) The impact of satellite retrievals in a global sea-surface-temperature analysis. QJR Meteorol Soc 134:1745–1760

Brasnett B (2009) Recent developments in the Canadian Meteorological Centre Global SST Analysis. GHRSST 2009 International Users Symposium Conference Proceedings, Santa Rosa, USA, 29–30 May 2009, pp. 50–52. See: http://www.ghrsst.org/modules/ documents/documents/IGUS_Extended_Abstracts.pdf

Chao Y, Li Z, Farrara J, Huang P (2009) Blending sea surface temperatures from multiple satellites and in-situ observations for coastal oceans. J Atmos Ocean Tech 26:1435–1446

Chelton DB, Schlax MG, Samelson RM (2007) Summertime coupling between sea surface temperature and wind stress in the California current system. J Phys Oceanogr 37(3):495–517

Cummings JA (2005) Operational multivariate ocean data assimilation. QJR Meteorol Soc 131:3583–3604

Dombrowsky E, Bertino L, Brassington GB, Chassignet EP, Davidson F, Hurlburt HE, Kamachi M, Lee T, Martin MJ, Mei S, Tonani M (2009) GODAE systems in operation. Oceanography 22(3):80–95

Donlon CJ, Minnett P, Gentemann C, Nightingale TJ, Barton IJ, Ward B, Murray J (2002) Towards improved validation of satellite sea surface skin temperature measurements for climate research. J Clim 15(4):353–369

Donlon C, Robinson I, Casey K, Vasquez J, Armstrong E, Gentemann C, May D, LeBorgne P, Piolle J, Barton I, Beggs H, Poulter DJS, Merchant C, Bingham A, Heinz S, Harris A, Wick G, Emery B, Stuart-Menteth A, Minnett P, Evans B, Llewellyn-Jones D, Mutlow C, Reynolds R, Kawamura H, Rayner N (2007) The global ocean data assimilation project (GODAE) high resolution sea surface temperature pilot project (GHRSST-PP). Bull Am Met Soc 88: 1197–1213

Donlon C, Robinson IS, Reynolds M, Wimmer W, Fisher G, Edwards R, Nightingale TJ (2008) An infrared sea surface temperature autonomous radiometer (ISAR) for deployment aboard volunteer observing ships (VOS). J Atmos Oceanic Tech 25:93–113

Donlon CJ, Casey KS, Gentemann C, LeBorgne P, Robinson IS, Reynolds RW, Merchant C, Llewellyn-Jones D, Minnett PJ, Piolle JF, Cornillon P, Rayner N, Brandon T, Vazquez J, Armstrong E, Beggs H, Barton I, Wick G, Castro S, Hoeyer J, May D, Arino O, Poulter DJ, Evans R, Mutlow CT, Bingham AW, Harris A (2009a) Successes and Challenges for the Modern Sea Surface Temperature Observing System. Community Whitepaper to OceanObs09, Venice, Italy, 21–25 September 2009. See: http://www.oceanobs09.net/blog/

Donlon C, Casey K, Robinson I, Gentemann C, Reynolds R, Barton I, Arino O, Stark J, Rayner N, LeBorgne P, Poulter D, Vazquez-Cuervo J, Armstrong E, Beggs H, Llewellyn-Jones D, Minnett P, Merchant C, Evans R (2009b) The GODAE high-resolution sea surface temperature pilot project. Oceanography 22(3):34–45

Druon J (2010) Habitat mapping of the Atlantic bluefin tuna derived from satellite data: its potential as a tool for the sustainable management of pelagic fisheries. Mar Policy 34(2):293–297

Eyre J, Anderson E, Charpentier E, Ferrantier L, Ferranti L, Lafeuille J, Ondras M, Pailleux J, Rabier F, Riishojgaard LP (2009) Requirements of numerical weather prediction for observations of the ocean. Community WHITEPAPER to OceanObs09, Venice, Italy, 21–25 September 2009. See: http://www.oceanobs09.net/blog/

Foley D (2009) Applications of GHRSST data sets towards the stewardship of living marine resources: putting SST back in the saddle. GHRSST 2009 International Users Symposium Conference Proceedings, Santa Rosa, USA, 29–30 May 2009, pp. 92. See: http://www.ghrsst.org/modules/documents/documents/IGUS_Extended_Abstracts.pdf

Gentemann CL, Donlon CJ, Stuart-Menteth A, Wentz FJ (2003) Diurnal signals in satellite sea surface temperature measurements. Geophys Res Let 30:1140–1143

Glynn PW (1993) Coral reef bleaching: ecological perspectives. Coral Reefs 12:1–7

Gramer LJ, Hendee JC, Hu CM (2009) Integration of SST and other data for ecological forecasting on coral reefs. GHRSST 2009 International Users Symposium Conference Proceedings, Santa Rosa, USA, 29–30 May 2009, pp. 110–113. See: http://www.ghrsst.org/modules/documents/documents/IGUS_Extended_Abstracts.pdf

Griffin DA, Rathbone CE, Smith GP, Suber KD, Turner PJ (2004) A decade of SST Satellite Data. Final Report for the National Oceans Office, Contract NOOC2003/020, pp. 1–8. See: http://www.marine.csiro.au/remotesensing/oceancurrents/ten_years_of_SST.doc

Guan L, Kawamura H (2004) Merging satellite infrared and microwave SSTs: methodology and evaluation of the new SST. J Oceanogr 60(5):905

Hobday AJ, Hartmann K (2006) Near real-time spatial management based on habitat predictions for a longline bycatch species. Fish Manag Ecol 13(6):365–380

Howell EA, Kobayashi DR, Parker DM, Balazs GH, Polovina JJ (2008) TurtleWatch: a tool to aid in the bycatch reduction of loggerhead turtles (Caretta Caretta) in the Hawaii-based pelagic longline fishery. Endange Species Res 5:2–3. See: http://www.int-res.com/journals/esr/esr-specials/fisheries-bycatch-problems-and-solutions

Jedlovec G, LaFontaine F, Shafer J, Vazquez J, Armstrong E, Chin M (2009) An enhanced MODIS/AMSR-E SST composite product. GHRSST 2009 International Users Symposium Conference Proceedings, Santa Rosa, USA, 29–30 May 2009, pp. 47–49. See: http://www.ghrsst.org/modules/documents/documents/IGUS_Extended_Abstracts.pdf

Keifer D, Harrison D, Lam T, O'Brien F, Hinton M, Armstrong E (2009) A pelagic habitat analysis module for tuna of the Eastern Tropical Pacific. GHRSST 2009 International Users

Symposium Conference Proceedings, Santa Rosa, USA, 29–30 May 2009, pp. 125–127. See: http://www.ghrsst.org/modules/documents/documents/IGUS_Extended_Abstracts.pdf

Leichter JJ, Helmuth B, Fischer AM (2006) Variation beneath the surface: quantifying complex thermal environments on coral reefs in the Caribbean, Bahamas and Florida. J Mar Res 64: 563–588

Liu G, Christensen TRL, Eakin CM, Heron SF, Morgan JA, Parker BA, Skirving WJ, Strong AE (2009) NOAA Coral Reef Watch's application of satellite sea surface temperature data in operational near real-time global monitoring and experimental outlook of coral health and potential application of GHRSST. GHRSST 2009 International Users Symposium Conference Proceedings, Santa Rosa, USA, 29–30 May 2009, pp. 106–109. See: http://www.ghrsst.org/modules/documents/documents/IGUS_Extended_Abstracts.pdf

Martin M, Roberts-Jones J, Lea D, Stark J, Donlon C (2009) Use of SST and sea-ice data in operational analysis and assimilation systems at the UK Met Office. GHRSST 2009 International Users Symposium Conference Proceedings, Santa Rosa, USA, 29–30 May 2009, pp. 114–115. See: http://www.ghrsst.org/modules/documents/documents/IGUS_Extended_Abstracts.pdf

Maturi E, Harris A, Merchant C, Mittaz J, Potash B, Meng W, Sapper J (2008) NOAA's sea surface temperature products from operational geostationary satellites. Bull Am Met Soc 89(12): 1877–1888

Maturi E, Sapper J, Mittaz J, Harris A (2009) NOAA GHRSST data: GOES/MTSAT/MSG L2P and blended GOES/POES analysis. GHRSST 2009 International Users Symposium Conference Proceedings, Santa Rosa, USA, 29–30 May 2009, pp. 40–42. See: http://www.ghrsst.org/modules/documents/documents/IGUS_Extended_Abstracts.pdf

Maynard JA, Turner PJ, Anthony KRN, Baird AH, Berkelmans R, Eakin CM, Johnson J, Marshall PA, Packer GR, Rea A, Willis BL (2008) ReefTemp: an interactive monitoring system for coral bleaching using high-resolution SST and improved stress predictors. Geophys Res Let 35:L05603

Merchant CJ, Le Borgne P (2004) Retrieval of sea surface temperature from space, based on modeling of infrared radiative transfer: capabilities and limitations. J Atmos Oceanic Technol 21:1734–1746

O'Carroll AG, Eyre JR, Saunders RW (2008) Three-way error analysis between AATSR, AMSR-E and in situ sea surface temperature observations. J Atmos Oceanic Technol 25:1197–1207

Oke PR, Brassington GB, Griffin DA, Schiller A (2008) The BLUElink ocean data assimilation system (BODAS). Ocean Model 21:46–70

Rea A (2004) Recent Improvements to the NOAA AVHRR SST Product at the Australian Bureau of Meteorology. See: http://imos.org.au/srsdoc.html

Reynolds RW, Smith TM (1994) Improved global sea surface temperature analyses using optimum interpolation. J Clim 7:929–948

Reynolds RW, Rayner NA, Smith TM, Stokes DC, Wang W (2002) An improved in situ and satellite SST analysis for climate. J Clim 15:1609-1625.

Reynolds RW, Smith TM, Liu C, Chelton DB, Casey KS, Schlax MG (2007) Daily high-resolution blended analyses for sea surface temperature. J Clim 20:5473–5496

Reynolds RW, Gentemann CL, Corlett GK (2010) Evaluation of AATSR and TMI satellite SST data. J Clim 23:152–165

Robinson IS (1985) Satellite Oceanography – An Introduction for Oceanographers and Remote Sensing Scientists, Ellis Horwood, Chichester, 455 pp

Smith NR, Blomley JE, Meyers G (1991) A univariate statistical interpolation scheme for subsurface thermal analyses in the tropical oceans. Prog Oceanogr 28:219–256

Spillman CM, Alves O (2009) Dynamical seasonal prediction of summer sea surface temperatures in the Great Barrier Reef. Coral Reefs 28:197–206

Stark JD, Donlon CJ, Martin MJ, McCulloch ME (2007) OSTIA: an operational, high resolution, real time, global sea surface temperature analysis system. Oceans'07 IEEE Aberdeen, Conference Proceedings. Marine Challenges: Coastline to Deep Sea, Aberdeen,

Scotland, IEEE. See: http://ghrsst-pp.metoffice.com/pages/latest_analysis/docs/Stark_et_al_OSTIA_description_Oceans07.pdf
Thiébaux J, Rogers E, Wang W, Katz B (2003) A new high-resolution blended real-time global sea surface temperature analysis. Bull Am Meteorol Soc 84:645–656
Vinogradova NT, Zaccheo TS, Alcala CM, Vandemark D (2009) Operational high-resolution sea surface temperature product in the Gulf of Maine. J Oper Oceanogr 2(2):57–66
Zhong A, Beggs H (2008) Analysis and Prediction Operations Bulletin No. 77 – Operational Implementation of Global Australian Multi-Sensor Sea Surface Temperature Analysis, 2 October 2008. See: http://www.bom.gov.au/bmrc/ocean/BLUElink/SST/GAMSSA_BoM_Operational_Bulletin_77.pdf

Chapter 16
The Past, Present, and Future of the AVHRR Pathfinder SST Program

Kenneth S. Casey, Tess B. Brandon, Peter Cornillon, and Robert Evans

16.1 Introduction

With today's modern satellite sensors, many oceanic parameters are capable of being observed from space. Of these, only sea surface temperature (SST) from the Advanced Very High Resolution Radiometer (AVHRR) series on board the NOAA polar-orbiting satellites enjoys an unbroken, nearly 30-year long history of observation from the same class of instrument. These AVHRR observations date back to 1981, when the first five-channel AVHRR flew on board the NOAA-7 platform. The AVHRR series continues to fly today, and will continue perhaps another 10 years into the future on board the remaining NOAA polar orbiting satellites and the European METOP platforms.

Early in the life of this series of instruments, NASA Headquarters conceived the Pathfinder program to support the needs of the US Global Change Research Program (King and Greenstone, 1999). These Pathfinder efforts were to provide unprecedented access to large volumes of consistently processed satellite datasets, in advance of the launch of the NASA Earth Observing System (EOS) satellites. Climate-quality datasets would be generated and made available, providing valuable experience in reprocessing, archiving, and distributing large satellite data sets. Thus, these Pathfinder efforts would "find the path" forward in preparation for the ambitious EOS series of instruments and missions that were to follow.

In October of 1990, NOAA and NASA signed a Cooperative Agreement to commence work on three joint NOAA/NASA Pathfinder projects. These joint projects would use existing NOAA datasets, focusing on AVHRR Global Area Coverage (GAC) data, the Television and Infrared Observation Satellite (TIROS) Operational Vertical Sounder (TOVS) data, and the Geostationary Operational Environmental Satellite (GOES) data. Several months later, in 1991, a project involving Special Sensor Microwave/Imager (SSM/I) data, which was archived by NOAA under an agreement with the US Navy, became the fourth NOAA/NASA Pathfinder. In 1992

K.S. Casey (✉)
NOAA National Oceanographic Data Center, Silver Spring, MD 20910, USA
e-mail: Kenneth.Casey@noaa.gov

V. Barale et al. (eds.), *Oceanography from Space*,
DOI 10.1007/978-90-481-8681-5_16,

a NASA/EPA/USGS Landsat Pathfinder project was formed, as was the first NASA-only Pathfinder, using Scanning Multichannel Microwave Radiometer (SMMR) data. Funding for these activities came mainly from NASA.

Each of the Pathfinder projects would form a Science Working Group (SWG) that would determine the specific products to be created based on scientific need, identify community-consensus algorithms to generate those products, and make recommendations on product validation, storage, maintenance, and required data services. A NASA Distributed Active Archive Center (DAAC) would then make the data available, as would the EOS Data and Information System (EOSDIS).

Because the AVHRR data stream was perceived to be of value to several different areas in the Earth sciences, three product science working groups were formed under the AVHRR SWG umbrella: an atmosphere SWG, a land SWG and an oceans SWG (OSWG in the following). The official goals of the AVHRR Pathfinder projects, as presented on 08 January 1992 by Mary James of NASA Goddard Space Flight Center to a group that included the project chairpersons, were:

- To produce long term global datasets for research, modeling, and trend analysis. Datasets were to be consistently processed with the best available community consensus algorithms.
- To produce multiple geophysical products from a common input stream using integrated processing concepts.
- To assemble a consistent, low maintenance, readily accessible archive for AVHRR data with browse and on-line access capabilities.

In December of 1994 a Pathfinder NASA Research Announcement was released, marking the transition from the first phase to a second phase of the Pathfinder datasets. The NRA emphasized building long-term datasets and addressing consistency issues that arise when linking together multiple sensors over time. This second phase ended in 2001, which marked the formal end of NASA support for the NOAA/NASA Pathfinder efforts. However, because of the successes obtained and momentum built over the previous decade, some of these efforts continued operating under different mechanisms. For example, the AVHRR Pathfinder Atmosphere (PATMOS) project continued as a NOAA National Environmental Satellite Data and Information Service (NESDIS) effort (Stowe et al., 2002).

The OSWG met for the first time in February of 1991. As with the PATMOS project, the Pathfinder SST activities also continued with the goal of building on the successes of the previous decade. In the case of Pathfinder SST, the program continued through the efforts of the NOAA/NESDIS National Oceanographic Data Center (NODC), beginning in 2002. Details on the history, current status, and future directions of Pathfinder SSTs constitute the remainder of this chapter.

16.2 History of the Pathfinder SST Program

Reprocessing of AVHRR data to create long, accurate, and consistent SST records was one component of the overall NOAA/NASA Pathfinder Program. To remain consistent with terminology, the activities that took place between 1990 and 2001

Table 16.1 Summary of overall pathfinder SST program periods

The Pathfinder SST program		
NOAA/NASA AVHRR Oceans Pathfinder SST Project (Versions 1–4)	NOAA-Supported Pathfinder SST Program (Version 5)	Version 6 (in preparation)
1990–2001	2002–2008	2009–2010

are referred to as the "NOAA/NASA AVHRR Oceans Pathfinder SST Project." During that period, NASA was the primary sponsor of the effort. Beginning in 2002 NODC became involved and a variety of NOAA funding sources began supporting the efforts. Beginning in 2009, new NASA funding supported some specific aspects of the program. The overall set of activities, including both the early 1990–2000 and the 2002 and beyond periods is termed simply the "Pathfinder SST Program" (see Table 16.1).

The AVHRR Ocean Pathfinder SST Project was an outgrowth of the work of the NASA funded SST Archive Science Working Group (SASWG) formed in early 1987. The objective of this group was "to determine the needs of the scientific community for SST fields and the possibility of meeting these needs with existing sources of data." The SASWG met several times over the next 2 years and on 1 June 1989 delivered its final report detailing two useful classes of SST products and the steps necessary to produce these fields. One class of products, with fine spatial resolution and high relative accuracy, would address the needs of feature-related studies and the other, with high absolute accuracy, would address the needs of heat flux related studies. The SASWG final report identified the AVHRR GAC data stream as appropriate for the generation of both SST products outlined in the report.

Following the SASWG report, Stan Wilson (then of NASA Headquarters but currently with NOAA/NESDIS) assembled a subgroup of the SASWG plus several others within NASA with expertise in the processing of GAC data to explore the actual production of the desired fields. This group included a representative from the terrestrial science community in addition to those from the oceanographic community. Two observations emerged from this meeting: first, much of the required GAC data existed only at NOAA so a collaboration with NOAA would be required as part of any reprocessing effort and second, other Earth science communities could benefit from ready access to a complete global GAC data set. Based on these observations and the desire to learn how to handle large, multidisciplinary, open-ended data sets for the upcoming EOSDIS, Stan Wilson initiated a data transcription effort and a Cooperative Agreement with NOAA. The Cooperative Agreement, entitled "Early EOSDIS Pathfinder Data Set Activity", was signed on 15 October 1990. This agreement was the formal start of the Pathfinder Program, as described in Section 16.1.

However, despite the Cooperative Agreement, the available funding, and the desire of the NOAA staff (Bud Booth and Levin Lauriston) to support the nascent

Pathfinder group with the data transcription, little progress was made over the next 13 months. To determine the nature of the problem, in November 1990 Peter Cornillon and Otis Brown met with John Knauss, who was the NOAA Administrator at that time. In that meeting it became clear that the delays resulted form a misunderstanding within NOAA of a comment made by Knauss. When shown what NASA was doing with satellite data, Knauss made a statement to the effect, "Why can't NOAA do these things?" This statement was misinterpreted by NOAA staff to mean that the data were not to be provided to others and that NOAA was to undertake the processing steps. Knauss had not intended his statement to be interpreted this way and clarified the situation with his staff after the meeting with Cornillon and Brown. The data transcription began in earnest and progress continued steadily on the Pathfinder Program from that point forward.

The Pathfinder OSWG first met in February of 1991 at Goddard Space Flight Center in Greenbelt, Maryland, USA. Peter Cornillon of the University of Rhode Island chaired the OSWG, which included members from NOAA, NASA, and universities (see Appendix). The OSWG met approximately annually over the next 4 years continually evolving and improving the algorithms and reprocessing techniques. During this same period, the chairmen of the Pathfinder groups met approximately annually to coordinate the work of the different Pathfinder projects and subgroups met when needed to deal with data transcription and calibration issues. To give a sense of the technical challenges faced by the OSWG to simply manipulate the large volumes of data, Fig. 16.1 shows the first part of an e-mail sent to Cornillon on 22 February 1991 containing a draft report from the first OSWG meeting. Note the sender's concern that the text might be garbled because he had sent it at 9600 baud, which was considered to be very fast at the time. Much of the information emerging from the various groups was distributed by fax or over the Internet at extremely slow data rates by today's standards.

Sea surface temperature retrieval algorithms proposed by the OSWG were tested and implemented under the direction of Robert Evans at the University of Miami where the data were processed from Level 0 (L0)/Level 1b (L1b) through to Level 3 (L3) equal-area datasets on an integerized sinusoidal grid. These data were then sent to the NASA/Caltech JPL Physical Oceanography Distributed Active Archive Center (PO.DAAC) where they were converted to equal-angle, regularly gridded datasets in HDF4-Raster format and provided to users. The first four versions of Pathfinder SST were created at a reduced resolution of 9.28 km, due to computational resource limitations of the 1990s.

NASA funding for the initial effort ended in 2001–2002 and in 2002 NODC reinvigorated the Pathfinder SST activities with the goal of producing a Version 5 dataset at the full GAC resolution. Modest levels of NOAA funding from various sources including NODC base resources and the NOAA Coral Reef Conservation Program kept the Pathfinder SST Program alive from 2002 to 2006. During this period, the Version 5 data were produced at the University of Miami and then sent to NODC for final quality assurance, archiving, and provision to users. NODC provided the Version 5 data to the PO.DAAC as well, to maintain continuity for

```
From:    ST%"M.SCHWALLER@telemail" 22-FEB-1991 16:08
To:      PETE
Subj:    DRAFT SST REPORT

Date: Fri,, 22 Feb 91 13:48:00 EST
From: M.SCHWALLER (M.SCHWALLER@telemail)
Message-Id: (CGJB-4600-3717@telemail)
Subject: DRAFT SST REPORT
To: (pete)
Sender: Telemail-Daemon

Posted: Fri, Feb 22, 1991   1:48 PM EST              Msg: CGJB-4600-3717
From:    M.SCHWALLER
To:      P.CORNILLON
CC:      M.SCHWALLER
Subj:    DRAFT SST REPORT

Peter -- AsRather than write up the minutes in minute form, I decided to
write them as a draft report, since NASA/NOAA is expecting a report,
it made sense to me to skip the intermediate step.  I hope that makes
sense to you too.  Let me know if you want revisions, and when to send
a copy to the group.
PS I'm sending this at 9600, if it becomes jarbled I'll send it again
at a lower baud rate.
Matt Schwaller
))))))))))))))))))))))))))))))))))))))((((((((((((((((((((((((((((((((((((((
Z^B lp?/~SST PATHFINDER REPORT--DRAFT
BASED ON MINUTES FROM THE 1ST SST PATHFINDER MEETING

BACKGROUND
Previous studies (1) identified the need for accurate, long-term measurements of
global and regional Sea Surface Temperature (SST).  Such data sets are valuable
for process studies in oceanography, and are essential for global climate
studies and other investigations of global change.  NOAA and NASA recently made
the commitment to derive such SST data sets by reprocessing satellite data
currently available in Federal archives, and to provide the derived SSTs to the
scientific research community (2).  The agencies chartered an Oceans Science
Working Group (OSWG) to provide recommendations and an action plan for producing
a "SST Pathfinder Data Set" from NOAA AVHRR data (3).  The OSWG held its first
meeting at NASA/GSFC on February 14 and 15, 1989.  The following OSWG members
were present:

Peter Cornillon, chairman; Bob Evans, University of Miami; Dick Reynolds,
CAC/NMC; Richard Legeckis, NOAA/NESDIS; Gene Feldman, NASA/GSFC; Charles Walton,
NOAA/NESDIS


CHARTER OF THE OSWG
The terms of reference of the OSWG is actually quite general, and provides the
group with a lot of latitude in defining its goals.  A decision was made to
strengthen the terms of reference, to emphasize the group's focus on SST
products derived from AVHRR GAC data.  A new version of the OSWG Terms of
Reference is given in (3).
```

Fig. 16.1 Excerpt from a 22 February 1991 email to Peter Cornillon, demonstrating the slow data transmission rates available at the time. Despite the sender's concerns, the transmission was not garbled

users of earlier versions of the Pathfinder SST data who had grown accustomed to accessing it from the PO.DAAC systems. From 2007 to 2009, the NOAA Scientific Data Stewardship Program supported efforts to transition the reprocessing from the University of Miami to NODC, where it could be supported over the long term.

Beginning in 2009, new NASA funding augmented the continued NOAA support, with the goal of adding two new components to the Pathfinder SST Program. The first addition was the ability to generate error estimates, a critical component necessary to make the future Pathfinder Version 6 data compliant with the standards of the Group for High Resolution SST (GHRSST, Donlon et al., 2007). The second addition was to enable Pathfinder reprocessing of not just AVHRR GAC data, but also a large collection of 1 km resolution Local Area Coverage (LAC) and High Resolution Picture Transmission (HRPT) data from the AVHRR instruments. By the end of 2010, the GAC reprocessing system is expected to be functional at NODC and the LAC/HRPT at the University of Rhode Island. Long-term stewardship, archiving, and provision of the Version 6 data to users will take place at NODC.

16.3 Earlier Pathfinder SST Versions

Over its 20-year history, the Pathfinder SST Program has produced five distinct versions and is actively developing the sixth. The Pathfinder SST algorithm is based on the Non-Linear SST (NLSST) algorithm (Walton et al., 1998) and was originally chosen by the SWG because of the algorithm's adequate performance, operational nature, and its acceptance by the scientific user community (Kilpatrick et al., 2001). The basic form of the algorithm is

$$\mathrm{SST_{sat}} = a + bT_4 + c(T_4 - T_5)\mathrm{SST_{guess}} + d(T_4 - T_5)[\sec(\theta) - 1] \quad (16.1)$$

where

$\mathrm{SST_{sat}}$ = the satellite-derived SST estimate,
T_4 and T_5 = brightness temperatures in the 10.8 and 11.4 μm AVHRR bands (channels 4 and 5, respectively),
$\mathrm{SST_{guess}}$ = a first-guess SST value,
θ = the satellite zenith angle, and
a, b, c, and d = coefficients estimated from regression between collocated and coincident in–situ and satellite measurements.

Numerous modifications were made over time to improve the performance of the algorithm. A review of the algorithm evolution from Version 1 to 4 is provided below, based on information provided in an early version of the Pathfinder Users Manual (Vazquez et al., 1995) and Casey and Cornillon (1999). For more details, with an emphasis on Pathfinder Version 4, see Kilpatrick et al. (2001). Version 5 and Version 6 are discussed in more detail in the following sections.

16.3.1 Version 1

The first version of Pathfinder implemented several key enhancements to the NLSST algorithm. The form of the NLSST equation was slightly modified to include a time dependent term. Extra care was taken with the first steps in which the digital counts are converted to radiance and brightness temperatures. The first of these steps involves the linear transformation of counts to radiance based on the space-view and sensor base plate onboard calibration information. Next, a non-linear correction factor was applied based on pre-launch calibration data. Finally, lookup tables based on the sensor's operating temperature were used to convert the channel 4 and 5 radiances to brightness temperatures.

Pathfinder Version 1 also included improvements to the navigation of the satellite observations over previous AVHRR SST data products. These improvements focused on the clock drift and spacecraft and sensor attitude. Clock drifts caused uncertainties in the along-track position and were corrected using a database of satellite clock time and Earth time offsets determined by comparing precise time measurements recorded at the University of Miami Domestic Communication Satellite (DOMSAT, the communication satellite NOAA used to transfer AVHRR data) station with the time information in HRPT passes received at Wallops Island, VA, and Fairbanks, AK. Averaging over multiple geographic locations and times along the orbital track mitigated attitude errors.

Further improvements introduced in Pathfinder Version 1 were implemented during the next phase of reprocessing, in which SST values are calculated. Three sets of algorithm coefficients (a, b, c, and d in Equation 16.1) were generated using the satellite-in-situ matchup database: one for high atmospheric water vapor regimes where the $T_4 - T_5$ brightness temperature difference is greater than or equal to 1.8°C, one for low water vapor regimes where the $T_4 - T_5$ brightness temperature difference is less than 0.7°C, and one for moderate levels of water vapor where the $T_4 - T_5$ brightness temperature difference is greater than or equal to 0.7°C and less than 1.8°C. In Version 1, the coefficients were calculated on an annual basis, using 12 months of matchup data.

Finally, an extensive set of quality tests were implemented in Version 1, including checks for gross cloud contamination, uniformity tests, satellite zenith angle tests, stray sunlight tests, and a check against a reference SST field based on the Reynolds Optimally Interpolated SST (OISSTv1; Reynolds and Smith, 1994). Pixels on the edges of a scan line or in the first or last scan line of an orbital piece were excluded, and those subject to sun glint were also identified. Based on the combined results of these many tests, Version 1 Pathfinder SSTs were then assigned a quality level of between 0 (worst) and 4 (best). Pathfinder Version 1 data spanning 1987–1993 were released to the public beginning in 1995.

16.3.2 Version 2

A second version of Pathfinder was developed to better correct for temporal changes observed in the match-up statistics. The time-dependent term was removed and coefficients were generated on a monthly instead of annual basis. The coefficients were calculated using a temporally weighted 5-month running window and were found to better correct the temporal changes than the time-dependent term used in the NLSST equation in Version 1. The central month is given a 100% weight, the adjacent months an 80% weight, and the months at the ends of the window a 50% weight. This version of the Pathfinder data set was never released to the public.

16.3.3 Version 3

In the third version of Pathfinder SST, two sets of algorithm coefficients (a, b, c, and d in Equation 16.1) were generated using the satellite-in-situ matchup database instead of the three sets used in Version 1. The two sets were found to better reduce the overall bias between satellite and buoy SST. One set was for high atmospheric water vapor regimes where the $T_4 - T_5$ brightness temperature difference is greater than or equal to 0.7°C and the other was for low water vapor regimes where the

$T_4 - T_5$ brightness temperature difference is less than 0.7°C. To avoid discontinuities, in cases where the $T_4 - T_5$ brightness temperature difference was between 0.5 and 0.9°C the SST was calculated using a weighted combination of SST values determined using both sets of coefficients. Pathfinder Version 3 data were first released to the public in 1997 and spanned only 1991–1994. They were quickly superseded by Pathfinder Version 4, which was released only a short time later.

16.3.4 Version 4

In Version 4, a more robust algorithm coefficient scheme was introduced, and decision trees were implemented to better eliminate cloud-contaminated pixels. Quality levels were expanded to 0–7 to allow for greater user flexibility in determining trade-offs between data coverage and data quality. Version 4 data became available in 1997 and eventually covered 1985–2003. The Version 4 data marked the culmination of the NOAA/NASA AVHRR Oceans Pathfinder period.

16.4 Current State of Pathfinder SST: Version 5

Beginning in 2002, the Pathfinder SST Program entered a new phase targeted at producing an even more accurate, consistent, and finer resolution SST data set. By this time, the term "Climate Data Record" had emerged and the Pathfinder SST Program renewed its efforts toward delivering the SST climate data record. Improvements in spatial resolution, the land mask, and the way sea ice information was used in the quality level determinations were implemented. The 1° resolution OISSTv1 first-guess and reference SST field used in earlier Pathfinder versions was replaced with a newer version of the product, the OISSTv2 (Reynolds et al., 2002). For the earliest part of the AVHRR record from late 1981 to 1984, a new 25 km Daily Optimally Interpolated SST (DOISST, Reynolds et al., 2007) was used. A more quantitative analysis of the improvements achieved along with a description and evaluation of new SST climatologies derived from Pathfinder Version 5 is available in Casey et al. (2010).

By 2009, production, distribution, and long-term stewardship of the Pathfinder Version 5 SST data had become a joint University of Miami and NODC effort. The University of Miami processed the data from L0/L1b to L3, including a transformation from the integerized sinusoidal grid to the regularly gridded data in HDF4-Scientific Data Set format. These data were then transferred to NODC, where they were subjected to additional quality assurance involving browse image generation, visual inspection of every image, and comparison against the HadSST2 (Rayner et al., 2006) in-situ dataset. Any problems identified were then analyzed, and corrections implemented at the University of Miami where the data were regenerated before re-inspection at NODC. Once cleared through the quality assurance steps, the data were then archived at NODC and provided to users through its http, ftp, and OPeNDAP services. Access to the entire Pathfinder Version 5 collection

Fig. 16.2 Climatological mean SST for week 50, based on Pathfinder Version 5 data

is available on line.[1] An example image, generated from a climatological mean of Pathfinder Version 5 SST, is shown in Fig. 16.2.

The following steps provide the details on the current state of the Pathfinder processing system. Boldface items highlight the data and information provided to the processing system.

Step 1: Ingestion, calibration, and navigation of GAC data

1.1. Calibrate and convert AVHRR digital counts for channels 1–5 to radiances

1.1.1. Obtain *AVHRR channels 1–5 radiometer count data* from NOAA and University of Miami collections of GAC data.

1.1.2. For channels 1 and 2, use *pre-launch calibration coefficients* to perform a linear counts-to-radiance conversion, followed by a correction for temporal changes using *sensor decay rate data* and then a correction for inter-satellite differences using *inter-satellite standardization data to a NOAA-9 reference*, both of which use Libyan desert target area data.

1.1.3. For channels 3, 4, and 5 use both the above pre-launch calibration data and *onboard blackbody (space view and sensor base plate) data* to perform a non-linear counts-to-radiance conversion.

[1] At http://pathfinder.nodc.noaa.gov

1.2. Navigation, Clock, and Attitude Corrections

1.2.1. Apply satellite clock corrections using *Earth time offset data* based on University of Miami DOMSAT recorded HRPT files.
1.2.2. Apply attitude corrections made using *coastline comparison data*.
1.2.3. At this point, navigated, calibrated albedos/brightness temperatures are available for all five channels. Note that channels 1 and 2 are used in the Pathfinder Matchup Database decision trees (see Kilpatrick et al., 2001, Fig. 7) and channel 3 is used only in assignment of a quality indicator (see Step 2.4.1).

Step 2: SST Calculation

2.1. Convert channel 4 and 5 brightness temperatures to SST in °C using the Pathfinder algorithm (Equation 16.1), which requires a set of *monthly coefficients*.
2.2. These coefficients are derived using the *Pathfinder Buoy Matchup Database*, a set of in-situ SST observations and collocated AVHRR data. The in-situ data consist mainly of drifting buoys, but during 1981–1984 bias-corrected ship-based observations are also used.
2.3. The calculation of SST in Step 2.1 also requires a first-guess SST field. This first-guess field is the *OISSTv2* product for 1985–2009 and the *DOISST* for 1981–1984. Note that Versions 1–4 of Pathfinder used OISSTv1.
2.4. Quality Flag Assignment

2.4.1. A Channel 3, 4, and 5 brightness temperature test is performed. The brightness temperatures were calculated in Step 1.1.3.
2.4.2. The viewing angle is then evaluated using a *satellite zenith angle* check.
2.4.3. Next, a reference field comparison check is made against the OISSTv2/DOISST used in Step 2.3.
2.4.4. A stray sunlight test is then performed which requires information on whether the data in question are to left or right of nadir.
2.4.5. An edge test is performed next, which checks the location of the pixel within a scan line and the location of the scan line within the processing piece (a "piece" is a subset of an entire orbit file).
2.4.6. Then, a glint test is performed which requires a *glint index* calculated according to the Cox and Munk (1954) formulation.
2.4.7. Finally, these steps are all combined into an overall quality flag assignment for each pixel.

Step 3: Spatial Binning

3.1. The GAC pixels, converted to SST in Step 2, are then binned into an equal-area integerized sinusoidal grid.
3.2. To reduce discontinuities along the date line, a data-day based on a spatial data-day definition (Podestá, 1995) is used.
3.3. A *land mask* is applied to the data, identifying pixels that fall on land. In Pathfinder Version 5, an improved mask based on a 1 km resolution Moderate Resolution Imaging Spectroradiometer (MODIS) dataset (MOD12Q1) derived by the USGS Land Processes Distributed Active Archive Center is used.[2] In earlier versions of Pathfinder, the land mask was based on a Central Intelligence Agency database of navigation hazards, but no citation or further information is available.

Step 4: Temporal Binning

4.1. Step 4 begins by temporally binning the spatially binned pieces from Step 3 into a single daytime and single nighttime file for each day. In the case of overlapping satellite passes, only the highest quality pixels available for each are used. If there is more than one contributing pixel at the highest quality, these values are averaged. In Pathfinder Version 5, temporally binned files are also created for 5-, 7-, 8-day, monthly, and yearly periods.
4.2. An internal Pathfinder reference check comparison is then made to an internal 3-week Pathfinder comparison field. A *sea ice mask* based on *weekly SSM/I and sea ice information in the OISSTv2* is used to exclude pixels from the computation of the internal reference field. Note that sea ice information is not used in this way in the earlier versions of Pathfinder.
4.3. The SST, quality, and related fields are reformatted from equal-area to equal-angle for distribution and archiving in HDF4-Scientific Data Set format. Older Pathfinder versions were distributed in HDF4 Raster format.

Step 5: Quality Assurance, Archiving, and Distribution

5.1. Utilizing checksums to ensure file integrity, the HDF4-SDS files are then acquired by NODC from the University of Miami and browse graphics are generated.

---

[2] See http://www-modis.bu.edu/landcover/userguidelc/lc.html for more info

5.2. Each browse graphic is then visually inspected and problematic data are sent back to the University of Miami for correction and reprocessing.
5.3. After passing the visual inspection, an in-situ comparison against the HadSST2 data set (Rayner et al., 2006) is conducted. Again, any errors detected are investigated and, if the data are subjectively confirmed to be bad, they are returned to the University of Miami for correction and reprocessing.
5.4. After passing the visual inspection and in-situ comparisons, the Pathfinder data are formally archived by NODC.
5.5. The result of these steps is the Pathfinder Version 5 SST product. The data are distributed by NODC[3] and made available to the PO.DAAC for redistribution

Pathfinder Version 5 data were first released by NODC to the public in April of 2003. By December of 2003, data for 1985–2001 were available. Over the next several years, additional data were added incrementally. In April of 2009, Pathfinder data for 1981–1984 were made available for first time in the Pathfinder SST Program history. In prior versions, the lack of adequate in-situ matchups and suitable first-guess SST field made production of those early data from the NOAA-7 platform impossible. However, these limitations were overcome through the use of not just drifting buoys in the matchup database, but also by including bias-corrected ship observations. In addition, working iteratively, the DOISST, which itself uses Pathfinder SST data, was extended to include 1981–1984. The DOISST then served as the first guess and reference SST field for this period. By 2010, Pathfinder Version 5 data were available for 1981–2009.

16.5 Future Directions in Pathfinder SST: Version 6

Despite the improvements implemented over the years, problems still remain in Pathfinder Version 5. These errors are being corrected, and implemented in a new, Version 6 of Pathfinder. Version 6 will improve upon Version 5 in several notable ways. First, known errors in the land mask will be corrected, and Version 6 will utilize the DOISST throughout the record. Use of the finer resolution DOISST is resulting in improved ability of the quality procedures to properly identify cloudy pixels and to retain good pixels in the vicinity of coasts and strong gradient regions. Sea ice information will also be used to a greater extent in Version 6, and will be provided as part of the data set. Many spatial and seasonally varying biases evident in the Pathfinder Version 5 comparisons with HadSST2 will also be minimized through the use of new coefficients generated using a latitudinal band scheme.

[3] http://pathfinder.nodc.noaa.gov

Another problem in Version 5 manifests itself as a jump in SST every 18 lines in the regularly gridded data. While not clearly visible in a single image, edge detection algorithms that examine sequences of data identified this problem. The problem is related to the mapping procedures by which the equal-area integerized sinusoidal bins in which the data are generated are mapped into the equal-angle, uniform grid space in which the data are distributed. This problem has been resolved through the use of a refined mapping procedure that will be used in Version 6.

Significantly, Version 6 will also conform to the newest GHRSST data format, data content, and metadata requirements currently being developed, known as the GHRSST Data Specification Version 2 (GDS v2). The GDS requires each SST pixel value to have an associated bias and uncertainty error. These error estimates are being developed for Pathfinder Version 6 using an error hypercube approach which partitions the matchup database into a multi-dimensional lookup table. Pathfinder Version 6 will also be in netCDF and will contain Climate and Forecast metadata attributes. For the first time, Pathfinder SST data will be made available in not just collated L3 files, but also at the L2 swath and uncollated L3 processing levels.

Another addition to Version 6 will be to include a large collection of HRPT and LAC data collected at stations around the world. These full-resolution, 1 km data sets will provide key inputs to ultra-high resolution analysis systems and other fine scale applications. HRPT passes covering the east and west coasts of the United States, the western Pacific off of Japan, and the waters around Australia, have already been identified for inclusion into the Pathfinder Version 6 processing system and other HRPT station data are being actively sought.

16.6 Conclusion

For 20 years, the ongoing efforts of the AVHRR Pathfinder SST Program have contributed, and continue to contribute to a wide range of marine applications. To date, over 65,000 unique users have accessed AVHRR Pathfinder Version 5 SST data and applied them to a diverse set of applications and research topics. While it is difficult to accurately quantify the use of Pathfinder SST data in the scientific literature due to varying ways in which the data have been acknowledged or cited, searches of ScienceDirect and the Web of Science indicate that number to be in the hundreds and perhaps thousands. These studies include climate research areas such as SST trend analysis (e.g., Casey and Cornillon, 2001), numerical modelling applications (e.g., Shu et al., 2009), and the use of Pathfinder data as the basis for higher-level SST products (e.g., Reynolds et al., 2007; Marullo et al., 2007). A wide range of ecosystem-related studies have also relied on Pathfinder SST data (e.g., Somoza et al., 2008; Halpern et al., 2008).

Many other scientific studies, too numerous to summarize, have utilized Pathfinder SST across a wide range of disciplines and topics. Direct applications of the Pathfinder data to various societal benefits are also numerous. For example, Pathfinder data have been used to characterize humpback whale distributions to reduce ship collisions, improve marine protected area design and placement, plan optimal routing for yacht races, evaluate performance of ocean and numerical weather prediction models, and establish baselines for integrated ecosystem assessments and marine spatial planning projects.

The improvements coming in Version 6 will enable an even greater range of science and applications, enhance the compatibility of Pathfinder data with numerous other GHRSST-compliant data streams, and will result in a more accurate, consistent, and useful climate data record for SST.

Acknowledgements The authors wish to thank their respective institutions for support during the preparation of this manuscript and to Jorge Vazquez and Ed Armstrong of the NASA Physical Oceanography DAAC for their long-term efforts to support the Pathfinder SST community of users. Thanks also go to the many NASA and NOAA program and project managers who kept the AVHRR Pathfinder SST Program alive over the course of the last two decades, including Eric Lindstrom and John Bates. Finally, special thanks go to the last three Directors of the NOAA National Oceanographic Data Center, most recently Margarita Conkright Gregg and before her, Zdenka Willis and Lee Dantzler. All three deemed the Pathfinder effort sufficiently worthy to contribute significant NODC resources over the years. Without these contributions, the history of Pathfinder SST would have come to an end circa 2001 and would have fallen well short of making the most of the available AVHRR record.

Appendix: Original Members of the AVHRR Pathfinder Oceans SWG

Peter Cornillon, University of Rhode Island (Chair)
Robert Evans, University of Miami
Gene Feldman, NASA
Richard Legeckis, NOAA
Charles McClain, NASA
Richard W. Reynolds, NOAA
Charles Walton, NOAA

References

Casey KS, Cornillon P (1999) A comparison of satellite and in situ based sea surface temperature climatologies. J Clim 12:1848–1863

Casey KS, Cornillon P (2001) Global and regional sea surface temperature trends. J Geophys Res 14(18):3801–3818

Casey KS, Evans R, Kilpatrick K, Kearns EJ, Brandon TB, Reynolds RW, Vazquez-Cuervo J (2010) Toward the sea surface temperature climate data record from space: an improved advanced very high resolution radiometer pathfinder dataset. J Clim (in preparation)

Cox C, Munk W (1954) Measurement of the roughness of the sea surface from photographs of the sun's glitter. J Opt Soc Am 44(11):838–850

Donlon C, Robinson I, Casey KS, Vazquez-Cuervo J, Armstrong E, Arino O, Gentemann C, May D, LeBorgne P, Piollé JF, Barton I, Beggs H, Poulter DJS, Merchant CJ, Bingham A, Heinz S, Harris A, Wick G, Emery B, Minnett P, Evans R, Llewellyn-Jones D, Mutlow C, Reynolds RW, Kawamura H, Rayner N (2007) The global ocean data assimilation experiment high-resolution sea surface temperature pilot project. Bull Am Meteorol Soc 88(8):1197–1213, doi:10.1175/BAMS-88-8-1197

Halpern BS, Walbridge S, Selkoe KA, Kappel CV, Micheli F, D'Agrosa C, Bruno JF, Casey KS, Ebert C, Fox HE, Fujita R, Heinemann D, Lenihan HS, Madin EMP, Perry MT, Selig ER, Spalding M, Steneck R, Watson R (2008) A global map of human impact on marine ecosystems. Science 319(5865):948–952, doi:10.1126/science.1149345

Kilpatrick KA, Podestá GP, Evans R (2001) Overview of the NOAA/NASA advanced very high resolution radiometer pathfinder algorithm for sea surface temperature and associated matchup database. J Geophys Res 106(C5):9179–9197

King MD, Greenstone R (eds.) (1999) Earth Observing System (EOS) Reference Handbook, National Aeronautics and Space Administration, EOS Project Science Office, 355pp

Marullo S, Nardelli BB, Guarracino M, Santoleri R (2007). Observing the Mediterranean Sea from space: 21 years of Pathfinder-AVHRR sea surface temperatures (1985 to 2005): re-analysis and validation. Ocean Sci 2(2):299–310

Podestá GP (1995) Case studies for SeaWiFS calibration and validation, part 3, SeaWifs global fields: what's in a day? NASA Technical Memorandum No. 10456, 27pp

Rayner NA, Brohan P, Parker DE, Folland CF, Kennedy JJ, Vanicek M, Ansell T, Tett SFB (2006) Improved analyses of changes and uncertainties in sea surface temperature measured in situ since the mid-nineteenth century: the HadSST2 data set. J Clim 19:446–469

Reynolds RW, Rayner NA, Smith TM, Stokes DC, Wang W (2002) An improved in situ and satellite SST analysis for climate. J Clim 15:1609–1625

Reynolds RW, Smith TM (1994) Improved global sea surface temperature analysis using optimum interpolation. J Clim 7:929–948

Reynolds RW, Smith TM, Liu C, Chelton DB, Casey KS, Schlax MG (2007) Daily high-resolution-blended analyses for sea surface temperature. J Clim 20(22):5473–5496

Shu YQ, Zhu J, Wang DX, Yan CX, Xiao XJ (2009) Performance of four sea surface temperature assimilation schemes in the South China Sea. Continental Shelf Res 29(11–12):1489–1501.

Somoza RD, Kampel M, Rudorff FD, Sousa RB, Cobas S (2008) Lobster (Panulirus argus) captures and their relation with environmental variables obtained by orbital sensors for Cuban waters (1997–2005). Braz J Oceanogr 56(3):225–237

Stowe LL, Jacobowitz H, Ohring G, Knapp KR, Nalli NR (2002) The advanced very high resolution radiometer (AVHRR) pathfinder atmosphere (PATMOS) climate dataset: initial analyses and evaluations. J Clim 15(11):1243–1260

Vazquez JV, Tran AV, Sumagaysay R, Smith E, Hamilton M (1995) NOAA/NASA AVHRR oceans Pathfinder sea surface temperature data set user's guide version 1.2. See: http://pathfinder.nodc.noaa.gov/AVHRR_Pathfinder_Oceans_UsersGuidev1.2_27Sept1995.pdf

Walton CC, Pichel WG, Sapper JF (1998) The development and operational application of non-linear algorithms for the measurement of sea surface temperatures with the NOAA polar-orbiting environmental satellites. J Geophys Res 103:27999–28012

Chapter 17
Some Reflections on Thirty-Five Years of Ocean Color Remote Sensing

Howard R. Gordon

17.1 Some Historical Milestones

In the first 15 or so years of my 35-year involvement in ocean color remote sensing, I was fortunate to have witnessed and participated in much of the early development of this enterprise. In this paper I relate those that had a significant impact on the subject and on my own work, and try to describe the historical setting in which they took place. These were very exciting and sometimes trying times for the members of the CZCS Experiment Team. I hope I can convey some of that excitement and frustration, and along the way, a few details about the subject.

My ocean color initiation was in the early 1970s when I started investigating radiative transfer in natural waters with Otis Brown (then a graduate student), when I supervised the Ph. D dissertation of George Maul on the application of LANDSAT-1 imagery to optical oceanography, and when I tried (and failed) to make in-situ optical measurements in support of an aircraft experiment by Fabian Poulson (ERIM) to try to remotely measure bathymetry. The latter failure showed me the difficulty of validation exercises in support of remote sensing. Although ocean optics has a considerable history, I shall describe only the events that occurred during, or had an influence on, my own involvement with ocean color remote sensing. A more complete history of events leading up to the Coastal Zone Color Scanner's (CZCS) approval for flight by NASA is given by Austin (1992), and a comprehensive history of ocean color remote sensing is presently being prepared by Jim Acker.

My first contact with ocean optics was in late 1967, when I moved to the University of Miami. Most of what I learned about the subject in my first 2 years there came from Jerlov's book *Optical Oceanography* (1968) and from Jerzy Dera, who visited from the fall of 1967 to the summer of 1968. Reading *Optical Oceanography* now, one realizes that ocean color was barely on the horizon at that time.

H.R. Gordon (✉)
Department of Physics, University of Miami, Coral Gables, FL 33124, USA
e-mail: gordon@physics.miami.edu

V. Barale et al. (eds.), *Oceanography from Space*,

DOI 10.1007/978-90-481-8681-5_17,

Probably the most important event in establishing the potential of ocean color observations, and leading directly to CZCS, was the work of Clarke et al. (1970). They used a spectroradiometer mounted on an airplane to measure the radiance backscattered from the water, and the intervening atmosphere, off Georges Bank at an altitude of 305 m. Contemporaneous measurements of the chlorophyll *a* concentration were made in the surface waters along the airplane's flight track. The results provided a clear indication that the modification to the spectrum of upwelling radiance by chlorophyll variations, over the range 0.1–3.0 mg/m^3, could be observed at aircraft altitudes. The authors also pointed out that backscattering from the atmosphere between the sensor and the surface added light that seriously degraded the quality of the spectra. Referring to such "air light," they stated at the close of their paper: "If such interference can be eliminated or identified and allowed for, spectroscopic procedures from aircraft (and perhaps from satellites) will be of great value in the rapid investigation of oceanic conditions, including conditions important for biological productivity."

This work initiated several studies in which aircraft measurements of color were combined with surface radiometry, further demonstrating the potential of ocean color remote sensing, and resulting in the approval of the CZCS in 1973. Although the experiments did confirm the serious degrading effect on the apparent color of the water by the intervening atmosphere on aircraft spectra, insufficient data were collected for development of analysis algorithms. The development of such algorithms became the first task of the CZCS Experiment Team.

17.2 The CZCS Nimbus Experiment Team (NET)

The CZCS Nimbus Experiment Team (NET) was formed in mid-1975 based on competitive proposals for membership. Its primary responsibility in the prelaunch era was to develop algorithms for the processing of CZCS imagery. The team membership is provided in Table 17.1.

There were essentially three areas that needed to be addressed. The first was various questions concerning the sensor, which was actually being built at the time.

Table 17.1 CZCS nimbus experiment team (NET) membership

Member	Affiliation
W. Hovis (Leader)	NASA/GSFC
F. Anderson	NRIO, Capetown, South Africa
R.W. Austin	SIO, Visibility Laboratory
E.T. Baker	NOAA/PMEL
D.K. Clark	NOAA/NESS
S.Z. El-Sayed	Texas A&M University
H.R. Gordon	University of Miami
B. Sturm	JRC Ispra, Italy
R.C. Wrigley	NASA/Ames
C.S. Yentsch	Bigelow laboratory for ocean sciences

These issues were addressed by Warren Hovis, the NET leader (Sensor Scientist). The second was the development of an atmospheric correction algorithm (removal of the "air light" described by Clarke et al., 1970). The group charged with this consisted of Austin, Gordon, J.L. Mueller, Sturm, and W.H. Wilson. The third was the development of in-water algorithms. This group consisted of Anderson, Austin, Baker, Clark, El-Sayed, R.C. Smith, Wrigley, and Yentsch. Mueller, Smith, and Wilson were not NET members, but nevertheless directly participated in NET activities and made important contributions to the NET's algorithm development effort. It should be emphasized again that, at the time of the NET's formation, there were no algorithms available for processing the incoming data.

17.2.1 Bio-Optical Algorithms

The main thrust of the in-water algorithm group was to try to acquire as much data as possible, in a large variety of waters, relating the water-leaving spectral radiance, $L_W(\lambda)$, or the upwelling spectral radiance (propagating toward the zenith) just beneath the water surface, $L_u(\lambda)$, to the concentration of chlorophyll *a*, the total mass of suspended material, or Total Suspended Matter (TSM), the concentration of some detrital materials (e.g., phaeophytin *a*), etc. The NET's data collection began after its formation and continued through 1979.

The first time I ever saw anything resembling a bio-optical algorithm was at a NET meeting in Miami in December 1977. At that time Ray Smith presented preliminary data that convinced me that one might be able to discern perhaps 5–6 levels of chlorophyll *a* based on measurement of $L_W(\lambda)$. Prior to that, I had felt that the CZCS would be more useful in estimating sediment concentrations in coastal areas than in estimating chlorophyll *a*. From then on, I was a believer.

As part of the NET's efforts, there were two significant cruises prior to launch and three post-launch cruises. The data from both pre- and post-launch cruises were pooled for the final algorithm. Figure 17.1 provides the station locations of the NET cruises organized by Dennis Clark as chief scientist. Clearly the focus of algorithm development by the NET was the waters off the coast of the United States.

Most of these sites would be considered to be Case 1 waters (Morel and Prieur, 1977; Gordon and Morel, 1983); however, some Case 2 waters were observed (off the Mississippi Delta and off the Chesapeake Bay). The oligotrophic waters of the Sargasso Sea were included in the data set as well. The principal measurements on the pre- and post-launch cruises were the spectral water-leaving radiance (made with an in-water spectral radiometer designed at the SIO Visibility Laboratory) and the concentrations of chlorophyll *a* and phaeophytin *a* measured fluorometrically (Clark, 1981).

Figure 17.2 provides examples of the spectra obtained and the final "blue-green" algorithm for estimation of the pigment concentration (*C*, the sum of the concentrations of chlorophyll *a* and phaeophytin *a*). In the final analysis, the data from Clark's cruises suggested that, in Case 1 waters, given measurements of $L_W(\lambda)$, it

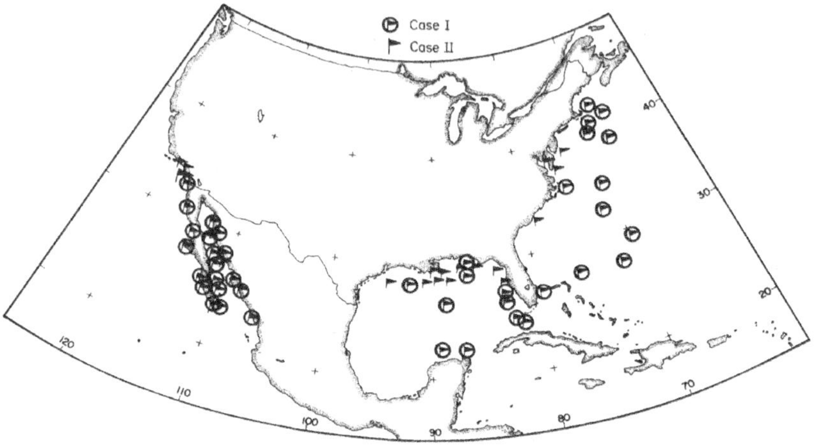

Fig. 17.1 Location of stations used in the development and the validation of CZCS imagery. From Gordon et al. (1983)

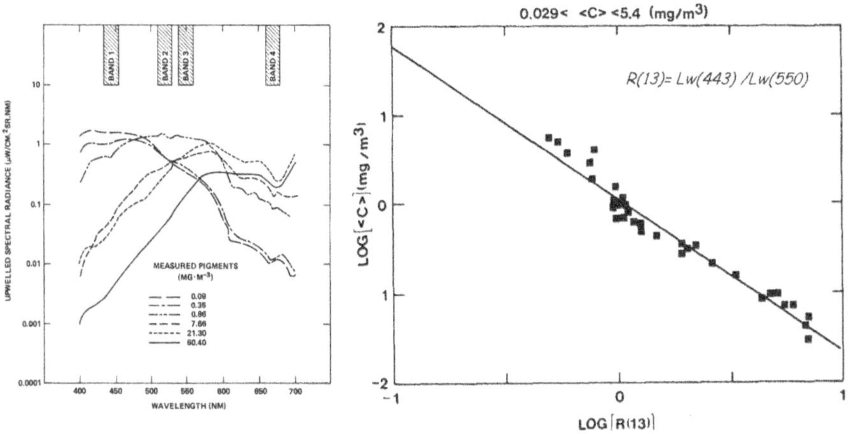

Fig. 17.2 *Left*: examples of water-leaving radiance spectra for several pigment concentrations. From Hovis et al. (1980). Reprinted with permission from the AAAS. *Right*: the *blue-green* algorithm used to retrieve the pigment concentration. From Gordon et al. (1983)

should be possible to retrieve C with an uncertainty of about 30% over the range $0.029 \leq C \leq 5.4$ mg/m^3. This was far better than I had thought possible.

17.2.2 Atmospheric Correction

My earliest thoughts about atmospheric correction (about 1976) were derived on the basis of solving the radiative transfer equation in the simplest of cases (single scattering). In this approximation the radiance (L_t) reaching the sensor can be decomposed into atmospheric components due to Rayleigh scattering (L_r), aerosol scattering (L_a) and a component due to the water-leaving radiance (L_w) transmitted (t) to the top of the atmosphere:

$$L_t(\lambda_i) = L_r(\lambda_i) + L_a(\lambda_i) + t(\lambda_i)L_w(\lambda_i), \tag{17.1}$$

where λ_i is the wavelength of the ith spectral band. I had discovered that the single scattering formula for the radiance L_r compared very favorably to the full multiple scattering result as long as the limit of small Rayleigh optical thickness (τ_r) was adopted, i.e., $\exp(-\tau_r) \approx (1 - \tau_r)$. The resulting formula is

$$L_r(\lambda_i) = \frac{\tau_r(\lambda_i)F_0(\lambda_i)p_r(\theta_v,\varphi_v;\theta_0,\varphi_0;\lambda_i)}{4\pi \cos\theta_v}, \tag{17.2}$$

where $F_0(\lambda_i)$ is the extraterrestrial solar irradiance at λ_i,

$$p_r(\theta_v,\varphi_v;\theta_0,\varphi_0;\lambda_i) = P_r(\Theta_-,\lambda_i) + [r(\theta_v) + r(\theta_0)]\,P_r(\Theta_+,\lambda_i),$$

$P_r(\Theta,\lambda_i)$ is the Rayleigh scattering phase function for scattering through an angle Θ, $r(\theta)$ is the Fresnel reflectance of the sea surface for light incident at an angle θ, and

$$\cos\Theta_\pm = \pm\cos\theta_v\cos\theta_0 - \sin\theta_v\sin\theta_0\cos(\phi_v - \phi_0),$$

with θ_0 and θ_v, respectively, the angle between a vector directed from the sea surface to the sun and the sensor, and ϕ_0 and ϕ_v, the corresponding azimuth angles of the two vectors. The phase function for Rayleigh scattering is

$$P_r(\Theta) = \frac{3}{4}\left(1 + \cos^2\Theta\right).$$

I felt this observation was very important, because it provided an analytical expression for L_r, significantly reducing the computation time required for its determination. More important, using the same approximation, it seemed clear that the aerosol contribution should be given by a similar formula with the terms having the subscript "r" being replaced by terms having the subscript "a" for aerosol (and p_r replaced by $\omega_a p_a$, where ω_a is the aerosol single scattering albedo – scattering coefficient ÷ extinction coefficient). Thus, the spectral variation of L_a should follow the spectral variation of $\omega_a\tau_a p_a$, i.e.,

$$S(\lambda_i,\lambda_j) \equiv \frac{L_a(\lambda_i)}{L_a(\lambda_j)} = \frac{F_0(\lambda_i)\,\omega_a(\lambda_i)\tau_a(\lambda_i)p_a(\lambda_i)}{F_0(\lambda_j)\,\omega_a(\lambda_j)\tau_a(\lambda_j)p_a(\lambda_j)} \equiv \frac{F_0(\lambda_i)}{F_0(\lambda_j)}\varepsilon(\lambda_i,\lambda_j). \tag{17.3}$$

This relationship gave me the idea for an atmospheric correction algorithm. Assuming that the aerosol size distribution could be described by a power law in particle diameter (in accordance with models at that time), the phase function should be almost independent of wavelength and $\tau_a(\lambda_i) \propto (\lambda_i)^{-\alpha}$, where α is called the Ångström exponent and typically, $0 \le \alpha \le 2$. Furthermore, when the aerosol is non-absorbing ($\omega_a = 1$),

$$\varepsilon(\lambda_i,\lambda_j) = \left(\frac{\lambda_j}{\lambda_i}\right)^{\alpha}. \tag{17.4}$$

Under these conditions, if we knew L_a at a single wavelength (and knew α) we could determine L_a at all other wavelengths. Knowing that $L_w(\lambda_{\text{Red}})$ in the red is often negligible compared to the blue and the green (Fig. 17.2),

$$L_a(\lambda_{\text{Red}}) = L_t(\lambda_{\text{Red}}) - L_r(\lambda_{\text{Red}}), \text{ and } L_a(\lambda_i) = \frac{F_0(\lambda_i)}{F_0(\lambda_{\text{Red}})}\left(\frac{\lambda_{\text{Red}}}{\lambda_i}\right)^{\alpha} L_a(\lambda_{\text{Red}}), \tag{17.5}$$

essentially reducing the atmospheric correction problem to determining a single parameter: α.

To test these ideas, in particular the formula for $\varepsilon(\lambda_i,\lambda_i)$, Equation (17.4), we attempted to use the Ocean Color Scanner (OCS, a CZCS simulator built by NASA/GSFC and designed to fly on a U-2 aircraft) based at NASA Lewis Research Center under the direction of Jack Saltzman. Data of $L_t(\lambda_i)$ were obtained by Jack on a flight of altitude $\sim$ 15 km over the Gulf of Mexico south of the Mississippi delta coincident with a CZCS NET prelaunch algorithm development cruise. I used the data and Equation (17.1) to form

$$L_a(\lambda_i) = L_t(\lambda_i) - L_r(\lambda_i) - t(\lambda_i)L_w(\lambda_i), \tag{17.6}$$

expecting the resulting spectral variation to be $L_a(\lambda_i) \sim F_0(\lambda_i) \times (\lambda_i)^{-\alpha}$, verifying that we were on the right track. Indeed, the spectral variation did follow the expected relationship, but with $\alpha = 8$: completely impossible! Recall that for the Rayleigh component $L_r(\lambda_i) \propto F_0(\lambda_i) \times (\lambda_i)^{-4}$, and the variation of the aerosol scattering must be a weaker function of wavelength than molecular scattering. Thus, the results made no sense whatsoever. I told Jack the results and he said he would get back to me in a few days. Later he called and said I should take the measured $L_t(443)$ and multiply it by 0.7, with corrections of a similar magnitude for the other spectral bands. Thus, Jack believed the OCS calibration was in error by as much as 30%. With errors of this magnitude, I saw no sense in trying to use the OCS to show that our formulas were reasonable approximations to reality. That was the last time I tried to use aircraft data to validate our atmospheric correction ideas, and there was *no* prelaunch test of the algorithms or even their underlying assumptions.

17.3 IUCRM Colloquium: "Passive Radiometry of the Ocean"

During this same time period (1976–1978), I had tested these ideas using simulated data derived from multiple scattering solutions to the radiative transfer problem in the ocean-atmosphere system (Gordon, 1978). They seemed to hold up well, so Dennis Clark and I decided to combine his proposed algorithm (Clark, 1981) for estimating the water's pigment concentration from radiance ratios, e.g.,

$L_w(443/L_w(550)$, with my atmospheric correction ideas to try to estimate the accuracy with which C could be estimated. This work led to three important conclusions, if the assumptions underlying the correction algorithm were correct. First, knowledge of the value of the parameter α is much more important than the actual aerosol concentration, i.e., given knowledge of α the error in C is only a weak function of the aerosol optical thickness. Second, for low values of C, e.g., ≤ 0.2 mg/m^3, accurate values can be retrieved as long as α is not overestimated. Third, as C increases, the accuracy with which α must be known also increases; however, accuracies in C considerably better than $\pm(1/4)\log_{10}(C)$, for $C \leq 1.0$ mg/m^3 are possible with only a coarse estimate of α. This work was reported at the IUCRM Colloquium on "Passive Radiometry of the Ocean" at the Institute of Ocean Sciences, Patricia Bay near Victoria, BC, Canada from June 14 to 21, 1978 (Gordon and Clark, 1980).

The IUCRM Colloquium was the precursor to "Oceanography from Space" Venice 1980. More important, it also provided one of the first opportunities for those interested in water color to exchange ideas. Most at the meeting understood that the CZCS was not optimally designed for ocean color. At the "Working Group on Water Color" session at the end of the meeting, several proposals were made concerning the specifications of a dedicated ocean color instrument (a CZCS follow-on sensor) with due respect for the requirements of atmospheric correction and of the absorption and scattering properties of water constituents. These are summarized in Table II in Morel and Gordon (1980).

With the exception of the "low priority" bands at 610 and 640 nm and the absence of a band near 670 nm, the specifications closely resemble the final configurations of SeaWIFS and MODIS. In addition, the Working Group also recommended that a program be established to assess the "spatial and temporal variation of the phytoplankton pigment concentration on a global scale. This Global Assessment of Phytoplankton Pigments (GAPP) would consist of utilizing the CZCS to prepare monthly or bi-monthly worldwide maps of the pigment concentration and hence provide an estimate of the total phytoplanktonic biomass of the world oceans as well as the spatial and temporal variation of the primary productivity" (Morel and Gordon, 1980). Thus, even before launch, when many considered CZCS to be a boondoggle being carried out by lunatics, those knowledgeable in oceanic optics could see important improvements that could be made in the system, and important applications of the ocean color data.

17.4 CZCS Launch, Initial Imagery and Validation Cruises

CZCS was launched in October 1978, and the first image I saw was from Orbit 116 (November 1, 1978). Because there was a history of space-borne instruments failing soon after launch, every effort was made to try to validate the CZCS data and the algorithms as soon after launch as possible. Thus, the NET organized two cruises in the Gulf of Mexico to commence as soon as imagery was available. R. Austin and C. Yentsch organized a cruise aboard the NOAA Vessel Researcher, with Team members R. Austin, C. Yentsch and S. El-Sayed participating. Dennis Clark chartered the Athena II, a 165 ft decommissioned patrol boat, from the Vietnam War, that was capable of speeds as high as 40 kt. The participating NET members on the Athena II were E. Baker, D. Clark and myself. Dennis' idea was to perform a hydrographic station at local noon, simultaneous with the satellite overpass. Measurements of upwelling spectral radiance, downwelling spectral irradiance, beam attenuation coefficient, phytoplankton pigments, TSM, particle size distribution, Secchi depth, and atmospheric transmittance were carried out at each station. I measured the backscattering coefficients of water samples in the blue (436 nm) and green (546 nm) with a light scattering photometer, and made hand-held atmospheric transmittance measurements with a Volz-like sun photometer.

The speed of the Athena II enabled us to proceed to the nadir point of the next day's satellite overpass with plenty of time to prepare for the station. With a ship proceeding at normal speeds ($\sim$10 kt) the best one could hope to accomplish would be to have the next day's station located near the edge of the scan, where atmospheric correction would be significantly more difficult because of the longer path through the atmosphere. During that cruise we made several stations in which the weather was sufficiently clear that excellent simultaneous CZCS imagery was obtained. In the next 8 months Dennis Clark organized two more cruises (Gulf of California and Middle Atlantic Bight) in support of the CZCS validation (and algorithm development). To underscore the difficulty of validating an ocean color sensor, one should note that of the 55 stations made underneath the CZCS, only 9 were usable for validation because of cloud contamination and the proximity of land.

At the time of launch, the proposed algorithms had yet to be implemented on the CZCS Processing System at GSFC. Processing a CZCS scene at that time (1979) was an enormous task, as the large mainframe computers were excruciatingly slow even by standards that would be set within the next 5 years. Under pressure to finish the validation, we had to scrounge computer time wherever we could find it. Dennis Clark, Jim Mueller, and I (assisted by Dave Ball of Computer Sciences Corporation, who worked with Jim) found an available computer coupled to an image display device at the AOIPS (Atmospheric and Oceanic Image Processing System) facility at GSFC. The computer was a PDP 1155 (minicomputer), and we were allowed to use it from time to time between the hours of about 6 PM–6 AM. The room was very cold and I always brought a hood from my parka to keep warm while processing the data.

I had developed a program that could take the ephemeris for the orbit and compute the Rayleigh scattering component for each pixel along a scan line. I did this computation using a UNIVAC 1106 mainframe computer for two overpasses coincident with our surface measurements, and stored the results on tape. I could also take the latitude and longitude along ship tracks and determine the line and pixel numbers along the track. We then took 512 $\times$ 512 pixel subscenes of CZCS images and applied the atmospheric correction algorithm (assuming $\varepsilon(\lambda_i,\lambda_i) = 1$) to the imagery to derive an estimate for $L_w(\lambda_i)$. This processing took several seconds per scan line. The monitor displayed the original $L_t(\lambda_i)$, which was then replaced by $L_w(\lambda_i)$ as each scan line was completed. The first image processed was from Orbit 130 (Fig. 17.3) near the Mississippi Delta.

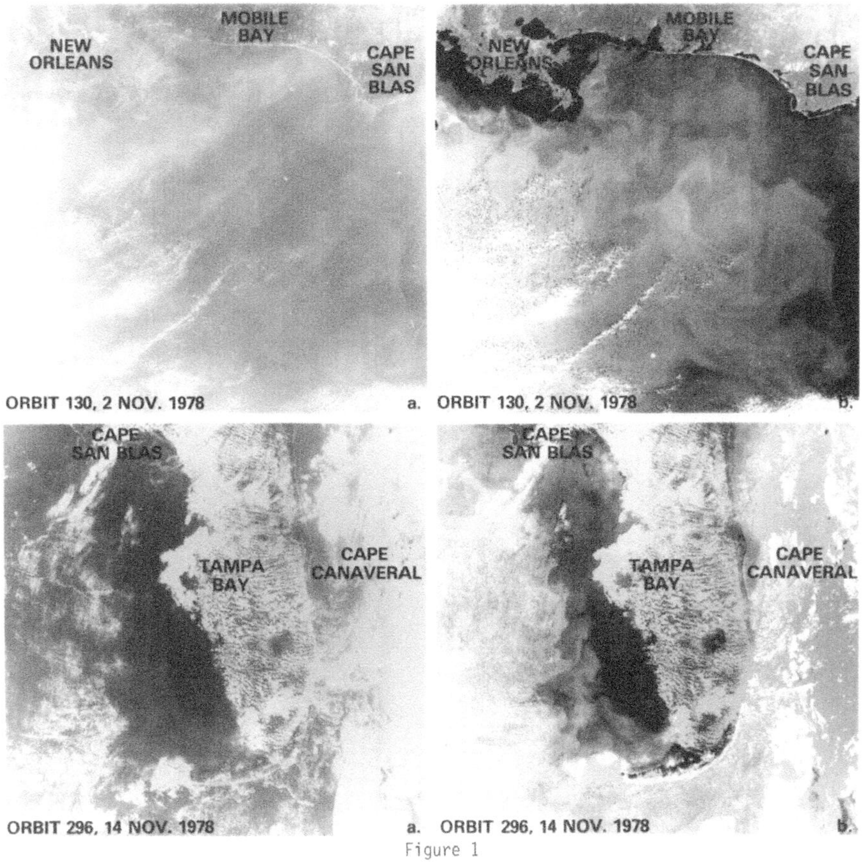

Fig. 17.3 Images of $L_t(443)$ (*left*) and $L_W(443)$ (*right*) for Orbits 130 (*top*) and 296 (*bottom*) over the Gulf of Mexico. From Gordon et al. (1980). Reprinted with permission of AAAS

This was a particularly hazy day, and as the computer was carrying out the correction, it appeared on the monitor that a veil was slowly being removed from the image. We were amazed at the clarity of the L_W image, which as can be seen, showed variability at a wide range of spatial scales, and strongly suggested that the biological activity was being driven by (or at least closely coupled to) the physical motion of the water. We applied the in-water algorithms to the retrieved $L_W(\lambda_i)$ to derive the pigment concentration. The resulting concentration for Orbit 296 off Tampa, FL is shown in Fig. 17.4, and the derived pigment concentration along the track is given in Fig. 17.5.

It was fortuitous that the ship track went through a particularly intense bloom (apparently caused by nutrient sources in the Everglades) off Fort Myers, FL. This in fact was a total accident. We did not know the bloom was there (no "real-time" color imagery to guide the ship), and expected the ship to traverse a track straight from the station near Key West to Tampa. However, the ship's crew wanted to watch

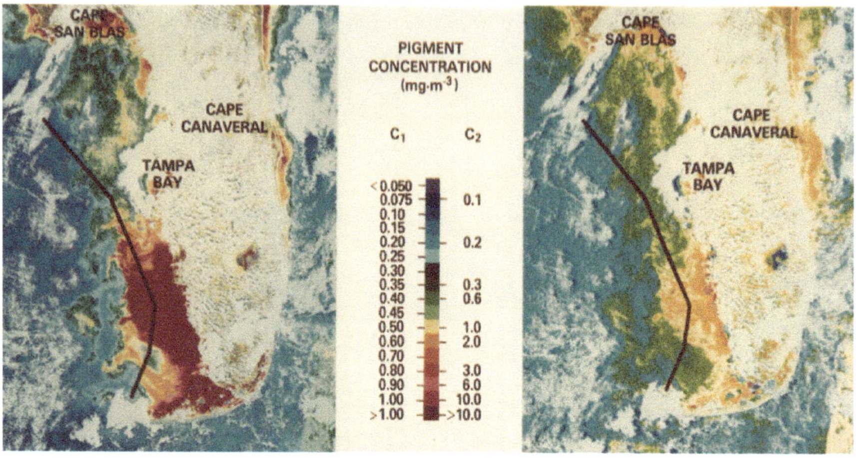

Fig. 17.4 Phytoplankton pigments for Orbit 296 processed with the two different algorithms: *Left* is using $R(13) = L_w(443)/L_w(550)$ and *right* is using $R(23) = L_w(520)/L_w(550)$. The *dark line* is the ship track of the Athena II. From Gordon et al. (1980). Reprinted with permission of AAAS

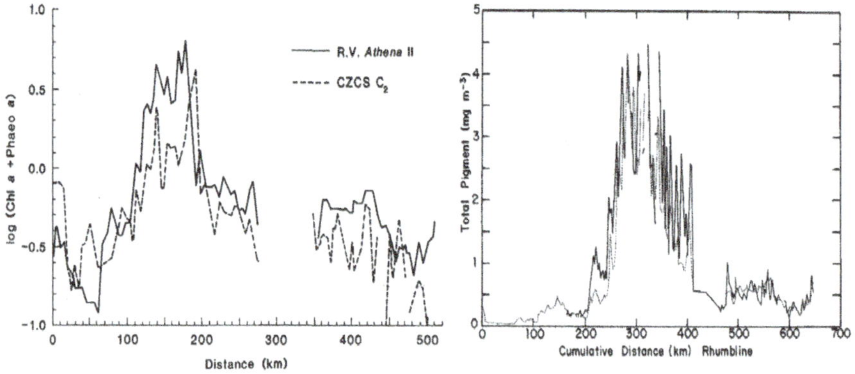

Fig. 17.5 *Left*: The retrieved pigment concentration using $R(23)$ along the track line in Fig. 17.4 (CZCS C_2) compared with that measured by the Athena II. From Gordon et al. (1980). Reprinted with permission of AAAS. *Right*: The same track line processed using the $R(13)$ algorithm 3 years later. From Gordon and Morel (1983)

Monday Night Football on television, and had to alter the course to take the ship close to Fort Myers in order to get good reception (Raiders 34, Bengals 21). This bloom provided an excellent test of the algorithm as is shown in Fig. 17.5 on the left comparing the surface-measured and CZCS-estimated pigment concentrations. This work constituted the initial validation of CZCS data and the processing algorithms. The rapid improvement of CZCS retrievals with time is underscored by the right panel in Fig. 17.5, which shows the same track line processed 3 years later with improved algorithms and improved calibration. Note that the scale on the *y*-axis is no longer logarithmic as it was in the first analysis of the data.

At that time it was clear that there were three problems that needed to be addressed before significant progress could be made in ocean color studies. First, we needed a method of estimating the value of $\varepsilon(\lambda_i,\lambda_j)$ for a given pixel. Second, we had no way to judge the quality of the sensor calibration, as it was determined prior to launch. Third, there were only enough NASA computational resources available to process a small number of CZCS scenes.

It seemed to me that it was reasonable to expect that $\varepsilon(\lambda_i,\lambda_j)$ would be almost independent of position within an image. My reasoning was based on the fact that for a given aerosol type (i.e., a given size frequency distribution and a given particle refractive index) $\varepsilon(\lambda_i,\lambda_j)$ can depend on the viewing direction only through differences in the shape (the variation of $P(\Theta)$ with Θ) of the aerosol scattering phase function with wavelength. Since these differences are assumed to be small, if the aerosol type is the same throughout an image, although the aerosol concentration may vary, $\varepsilon(\lambda_i,\lambda_j)$ should be constant.

I decided to test this hypothesis by looking at the variation of the apparent value of $\varepsilon(\lambda_i,\lambda_j)$ along a line on the image in Fig. 17.3 from Orbit 130. This is what I wanted to test earlier with the OCS. We had measured $L_w(\lambda_i)$ at many locations within the Gulf of Mexico and found that as long as extreme coastal areas were avoided, $L_w(\lambda_i)$ had relatively stable values of ~0.31 and 0.22 mW/cm^{2}μmSr, respectively, at 520 and 550 nm. In addition, $L_w(670)$ was found to be close to zero. Thus, given $L_t(\lambda_i)$ and computing $L_r(\lambda_i)$, we could use $L_w(\lambda_i)$ to compute L_a at 520, 550, and 670 nm as given in Equation (17.6).

The computation could not be done accurately at 443 nm because $L_w(443)$ depends strongly on the pigment concentration, which as the figures above show, is highly variable in the Gulf, especially near the coast. We selected a track starting from Choctawhatchee Bay, FL (the large bay approximately midway between Mobile Bay and Cape San Blas on the image) due south, running for approximately 400 km. The graph on the left in Fig. 17.6 provides the radiances $L_t(\lambda) - L_r(\lambda)$, along the track. As $L_t(670) - L_r(670) = L_a(670)$, and L_a is proportional to the

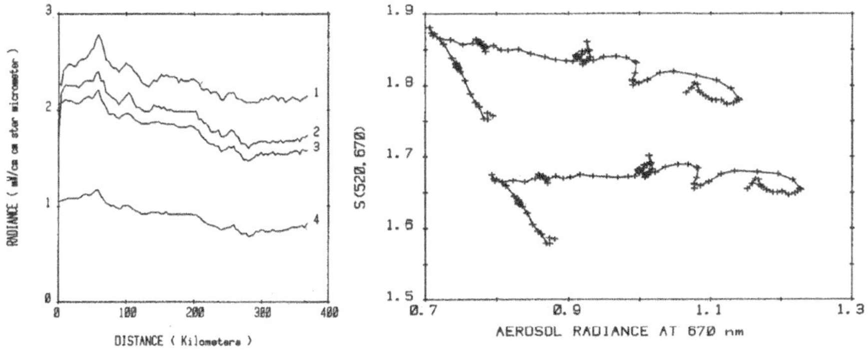

Fig. 17.6 *Left*: $L_t(\lambda) - L_r(\lambda)$, for the four CZCS bands, from Orbit 130 along a track from Choctawhatchee Bay, FL due south, running for approximately 400 km. *Right*: $S(520,670)$ along the same track before (*upper*), and after (*lower*), $F_0(670)$ adjustment as described in the text. From Gordon (1981)

aerosol optical thickness, the figure suggests that the aerosol concentration varies by a factor of 1.5 along the track. Using Equation (17.3) to form $S(\lambda_i,\lambda_j)$, the graph on the right in Fig. 17.6 (upper curve) shows the resulting variation of $S(520,670)$ along the track as a function of the aerosol radiance (concentration).

The result seems to show that $S(520,670)$ varies with aerosol concentration; however, there are two items in this estimate that carry significant uncertainties: the values of $F_0(\lambda_i)$ and the sensor calibration leading to the values of $L_t(\lambda_i)$. Error in either of these could lead to significant variations in $S(520,670)$ with aerosol concentration (Equations (17.2) and (17.6)). As an example, I decreased the value of $F_0(670)$ by about 6% and this simple change led to the lower curves in Fig. 17.6 (right), i.e., rendered $S(520,670)$ nearly constant along the track. This exercise, which was the subject of my paper at Oceans from Space, Venice 1980 (Gordon, 1981), convinced me (1) that it was probably reasonable to assume the S (and ε) is independent of aerosol concentration, and (2) that the sensor calibration (relative to whatever version of the extraterrestrial solar irradiance is being used) is of paramount importance, and we needed to address the CZCS calibration.

At approximately the same time, at the urging of Charlie Yentsch and Ros Austin, the NET agreed to devote a significant portion of the 2 h/day that CZCS operated to acquiring a global data set. Although the computer resources to analyze such a data set did not exist at the time, they would in a few years. This turned out to be an key decision and the results demonstrated the importance of ocean color to global marine ecology.

17.5 Calibration

Understanding the calibration (or at least the relative calibration) of CZCS was a difficult problem (and, without several assumptions, impossible). Calibration as used here refers to the conversion from the digital counts (DC) recorded by the sensor to top-of-atmosphere radiance L_t, i.e., $L_t(\lambda_i) = k(\lambda_i) \times DC(\lambda_i)$, where $k(\lambda_i)$ is the "calibration constant." The onboard calibration system did not work well and did not include the entire optical train, so its use was abandoned. The only calibration scheme that seemed possible was what is now referred to as *vicarious calibration*, estimating the sensor radiance based on theoretical considerations and measurements of $L_w(\lambda_i)$ made at the surface, or equivalently, ensuring that the application of the algorithms to (re)calibrated sensor radiances yielded the observed $L_w(\lambda_i)$'s within their expected uncertainties. For CZCS, we had only the measurements of $L_w(\lambda_i)$ carried out on the validation cruises. We made the leap-of-faith assumption that the atmospheric correction algorithm was valid (and $L_w(\text{Red}) \approx 0$). Thus, combining Equations (17.1), (17.2), (17.3), (17.4) and (17.5), the $L_w(\lambda_i)$'s are given by

$$t(\lambda_i)L_w(\lambda_i) = L_t(\lambda_i) - L_r(\lambda_i) - [L_t(\text{Red}) - L_r(\text{Red})] \times \left[\frac{F_0(\lambda_i)}{F_0(\text{Red})}\right] \varepsilon(\lambda_i,\text{Red}),$$

where $t(\lambda_i)$, the diffuse transmittance, can be computed with good accuracy by simply ignoring aerosols. However, we still needed an independent method of estimating $\varepsilon(\lambda_i,\text{Red})$. This was supplied by what Dennis Clark and I referred to as the "clear water radiance concept." This was based on our observation that when $C \leq 0.25$ mg/m^3, the normalized water leaving radiance, $[L_w(\lambda_i)]_N$, defined through $L_w = [L_w]_N \cos\theta_0 t_0/a_{\oplus}^2$, where t_0 is the diffuse transmittance of the solar beam to the sea surface, θ_0 is the solar zenith angle, and $a_{\oplus}$ is the Earth-Sun distance in astronomical units (the mean $a_{\oplus}$ over 1 year is unity), at 520 and 550 nm were constant and known, and that at 670 nm was essentially zero (Gordon and Clark, 1981). Thus, if clear water could be located in a scene containing the $L_w(\lambda_i)$ measurements, it would be possible to determine $\varepsilon(520,\text{Red})$ and $\varepsilon(550,\text{Red})$, and through extrapolation (using a power law) $\varepsilon(443,\text{Red})$. Then, assuming $\varepsilon(\lambda_i,\text{Red})$ is independent of position in the image under consideration (a much weaker assumption than assuming a constant aerosol concentration), and assuming the sensor calibration was correct at 670 nm, we could estimate the water-leaving radiance in the other bands. Note the assumptions required to perform this vicarious calibration assessment: (1) the atmospheric correction algorithm is correct; (2) the ε-values are independent of position and their variation with wavelength is given by a power (Ångström's) law; and (3) the calibration of the spectral band at 670 nm is correct. The procedure is then to fractionally change $L_t(\lambda_i)$, i.e., $k(\lambda_i)$ for the fixed $DC(\lambda_i)$, until the measured and retrieved $L_w(\lambda_i)$'s are brought into confluence. Note that the computation of $L_r(\lambda_i)$ requires the extraterrestrial solar irradiance (Equation (17.2)), so any error in this quantity will be interpreted as an error in sensor calibration, i.e., $k(\lambda_i)$, therefore the "calibration" will be dependent on the particular values used for the solar irradiance. With these assumptions, we determined adjustments to the sensor calibration that seemed to work well for imagery from *all* of the validation scenes obtained during June of 1979. Unfortunately, when we used this vicarious calibration and examined data from earlier cruises, we found that the agreement between the measured and retrieved $L_w(\lambda_i)$ values became increasingly poorer as we progressed backward in time. We interpreted this as a decrease in the sensitivity of the instrument with time. This decreasing sensitivity with time was a major problem for the analysis of CZCS data. It was apparently caused by residue accumulating on the scan mirror due to out gassing of the instrument. This made it clear that ensuring the stability of, or carefully monitoring the stability of, future ocean color sensors was paramount. It is interesting to note that most of the validation data that were obtained within a year of launch were also used to adjust the sensor calibration for its variation with time. This is likely the origin of the term "cal-val" in reference to such activities.

17.6 The NOSS Interlude

In the early 1980s, with the success of the CZCS, a proposal was made to include an expanded instrument on a new platform, the National Ocean Satellite System (NOSS). Armed with high quality CZCS imagery, and the IUCRM Water Color

Working Group's recommendations regarding optimal spectral bands, we felt that the proposed ocean color sensor would face little opposition for inclusion on the platform. Little did we know. There was fierce opposition to its inclusion. I recall a presentation to a NASA advisory committee regarding ocean color at which the most prominent member of the committee made the statement: "I know of no respectable biologist that thinks this [ocean color] is important." It is interesting to note that years later Dick Barber referred to CZCS as one of the seven most important developments in marine biology in the last 50 years! After much time was spent trying to get an ocean color instrument on NOSS, the entire NOSS program was canceled; the first of several such failures. However, the time and effort were not wasted.

17.7 Back to CZCS

One of the significant problems processing CZCS imagery was the intense amount of computational resources required. Basically, in 1980 the image processing systems were PDP mini-computers coupled to image display devices. I was fortunate to work with Otis Brown and Bob Evans at the University of Miami, who developed a system for SST image processing. However, one of the biggest breakthroughs for ocean color processing was the development of the VAX computer systems by DEC. These "super-min" computers enabled the processing of CZCS scenes with acceptable computation times, and made processing of the entire CZCS data set a possibility. The Brown-Evans image processing system was ported to the VAX along with all of the CZCS processing algorithms. This processing system was then duplicated on a larger scale at GSFC in a joint project with Wayne Esaias, Chuck McClain and Gene Feldman to process all of the CZCS imagery and demonstrate the full potential of ocean color remote sensing to marine ecology (Esaias et al., 1986).

17.8 SeaWiFS and MODIS

Between 1984 and 1988 there was much work devoted to flying an improved ocean color scanner. After several failures, this effort succeeded with the approval of SeaWiFS as a joint project between NASA and EOSAT, a private, for-profit company. The SeaWiFS sensor was designed solely for the purpose of ocean color, with a special emphasis on accurate radiometry. Special features included NIR spectral bands for atmospheric correction, a solar diffuser for on-board calibration, the facility for viewing the moon to monitor the long-term stability of the radiometry, and increased radiometric sensitivity for better resolution of the water-leaving signal. The sensor was designed to operate continuously, which greatly increased the data coverage over CZCS. In addition, in support of SeaWiFS (and later MODIS), a dedicated calibration facility was developed by Dennis Clark (Clark et al., 1997). This facilitated the vicarious calibration of these sensors over time. SeaWiFS, which is in its 11th year of operation, has been an enormous success.

From the end of CZCS to the approval of SeaWiFS, most of my effort was devoted to enhancements to the atmospheric correction algorithm, with some work on a semi-analytic model of water-leaving radiance (Gordon et al., 1988). We replaced the single scattering computation of $L_r(\lambda_i)$ with a full multiple scattering computation, including polarization (Gordon et al., 1988), and tried to understand the influence of sea surface roughness on atmospheric correction (Gordon and Wang, 1992a, b). We also examined calibration requirements and enhancements and signal-to-noise considerations for future sensors (Gordon, 1987, 1990). During this time André Morel and coworkers examined the influence of the variation of Ozone concentration and the variation of atmospheric pressure on atmospheric correction (André and Morel, 1989) and developed the first atmospheric correction algorithm that truly coupled a model of ocean color to first-order radiative transfer (Equations (17.1), (17.2), (17.3), (17.4), (17.5) and (17.6)) (Bricaud and Morel, 1987). However, even with all of the algorithm enhancements, it became clear that significant improvement in CZCS processing was unlikely simply because of instrument limitations.

With the improved radiometric sensitivity of SeaWiFS (and later MODIS) over that of CZCS, the atmospheric correction was still not up to the task because of the partial neglect of multiple scattering effects, particularly the interaction between aerosol and Rayleigh scattering. Menghua Wang and I set out to try to modify the correction algorithm to include multiple scattering effects. Our idea was (1) to use the basic structure of the algorithm, but to rewrite Equation (17.1) to explicitly include the Rayleigh-aerosol interaction ($L_{ra}(\lambda_i)$), i.e., $L_t(\lambda_i) = L_r(\lambda_i) + L_{ra}(\lambda_i) + L_a(\lambda_i) + t(\lambda_i)L_w(\lambda_i)$, (2) compute $L_t(\lambda_i)$ for various aerosol models and concentrations with $L_w(\lambda_i) = 0$ including all orders of multiple scattering, (3) compute $L_r(\lambda_i)$ as before using a full multiple scattering code, and (4) compute $L_t(\lambda_i) - L_r(\lambda_i) = L_{ra}(\lambda_i) + L_a(\lambda_i)$ as a function of the aerosol optical thickness and model and store the computations for later use in look-up-tables (LUTs). We tested such a scheme using (as before) an aerosol model for which the scattering phase function was independent of wavelength, and then re-evaluated it using more realistic aerosol models. Our final algorithm was still being used in SeaWiFS and MODIS processing into 2009 (Gordon and Wang, 1994).

CZCS, SeaWiFS, and MODIS are the only sensor programs with which I have had direct involvement. Although I have no first-hand knowledge of them, for completeness I mention the other sensors that were flown in the mid to late 1990s. These include OCTS (JAXA) and POLDER (CNES) on ADEOS (JAXA) and MOS (DLR) on IRS-P3 (ISRO).[1]

[1] For information on these sensors, see http://www.ioccg.org/sensors_ioccg.html

17.9 The Future

For classical ocean color remote sensing, i.e., utilizing sensors with a small number of spectral bands, there are still difficulties with atmospheric correction in the presence of absorbing aerosols (Gordon, 1997), which fortunately contaminate only a small fraction of the imagery. The difficulties associated with these aerosols are that (1) they cannot be identified using spectral bands in the NIR, (2) their effect in the visible depends strongly on their vertical distribution, and (3) their perturbation on L_t increases as λ decreases, so their effect is large where phytoplankton absorb light. I believe that dealing with absorbing aerosols for sensors of this type requires using coupled ocean-atmosphere algorithms such as those proposed by Moulin et al. (2001) or Chomko et al. (2003). They are also easily modified to operate in Case 2 waters (Kuchinke et al., 2009).

Beyond the classical instruments lies the promise of sensors with high spectral resolution. It has already been demonstrated that with such sensors separation of the total phytoplankton population into functional groups is possible (Bracher et al., 2008), even *without* what is traditionally though of as "atmospheric correction."

17.10 Some Closing Remarks

There are many people who have contributed to the success of ocean color remote sensing "behind the scenes" with little recognition. Early in the CZCS mission, Jack Sherman and Harold Yates at NOAA/NESS were strong supporters. Bob Kirk saw the SeaWiFS instrument to completion as project manager. Stan Wilson at NASA/HQ wisely instituted a temporary (2-year) rotating position at HQ to oversee ocean color activities and shepherd its development. To those who interrupted their own research to serve in this position: Ken Carder, Wayne Esaias, Curt Davis, Jim Yoder, Frank Muller-Karger, Marlon Lewis, Gregg Mitchell, Robert Frouin, Janet Campbell, John Marra, and Chuck Trees; we all owe a debt of gratitude. The position was finally made permanent and is now filled by Paula Bontempi.

Ackowledgments I am grateful for the research support received over the years from NASA, NOAA, and ONR. Also, I thank the organizers of this conference and of the three other "Oceanography from Space" conferences, in which I have participated, for providing us the opportunity to discuss ocean remote sensing in such a beautiful and historic city.

References

André JM, Morel A (1989) Simulated effects of barometric pressure and ozone content upon the estimate of marine phytoplankton from space. J Geophys Res 94:1029–1037

Austin RW (1992) Optical remote sensing of the oceans: BC (before CZCS) and AC (after CZCS). In: Barale V, Schlittenhardt PM (eds.) Ocean Colour: Theory and Applications in a Decade of CZCS Experience, Kluwer Academic, Dordrecht, pp. 1–15

Bracher A, Vountas M, Dinter T, Burrows JP, Rottgers R, Peeken I (2008) Quantitative observation of cyanobacteria and diatoms from space using PhytoDOAS on SCIAMACHY data. Biogeosci Discuss 5:4559–4590

Bricaud A, Morel A (1987) Atmospheric corrections and interpretation of marine radiances in CZCS imagery: use of a reflectance model. Oceanologica Acta No. SP:33–50

Chomko RM, Gordon HR, Maritorena S, Siegel DA (2003) Simultaneous retrieval of oceanic and atmospheric parameters for ocean color imagery by spectral optimization: a validation. Rem Sen Environ 84:208–220

Clark DK (1981) Phytoplankton pigment algorithms for the Nimbus-7 CZCS. In: Gower JFR (ed.) Oceanography from Space, Plenum, New York, pp. 227–237

Clark DK, Gordon HR, Voss KJ, Ge Y, Broenkow W, Trees C (1997) Validation of atmospheric correction over the oceans. Jour Geophys Res 102D:17209–17217

Clarke GL, Ewing GC, Lorenzen CJ (1970) Spectra of backscattered light from the sea obtained from aircraft as a measurement of chlorophyll concentration. Science 167:1119–1121

Esaias WE, Feldman GC, McClain CR, Elrod JA (1986) Monthly satellite-derived phytoplankton pigment distribution for the North Atlantic ocean basin. EOS Trans Am Geophys Union 67:835–837

Gordon HR (1978) Removal of atmospheric effects from satellite imagery of the oceans. Appl Opt 17:1631–1636

Gordon HR (1981) Preliminary assessment of the Nimbus-7 coastal zone color scanner atmospheric correction algorithm in a horizontally inhomogeneous atmosphere. In: Gower JFR (ed.) Oceanography from Space, Plenum, New York, pp. 257–265

Gordon HR (1987) Calibration requirements and methodology for remote sensors viewing the oceans in the visible. Rem Sens Environ 22:103–126

Gordon HR (1990) Radiometric considerations for ocean color remote sensors. Appl Opt 29: 3228–3236

Gordon HR (1997) Atmospheric correction of ocean color imagery in the earth observing system era. J Geophys Res 102D:17081–17106

Gordon HR, Brown JW, Evans RH (1988) Exact rayleigh scattering calculations for use with the Nimbus-7 coastal zone color scanner. Appl Opt 27:862–871

Gordon HR, Clark DK (1980) Atmospheric effects in the remote sensing of phytoplankton pigments. Bound Layer Meteor 18:299–313

Gordon HR, Clark DK (1981) Clear water radiances for atmospheric correction of coastal zone color scanner imagery. Appl Opt 20:4175–4180

Gordon HR, Clark DK, Brown JW, Brown OB, Evans RH, Broenkow WW (1983) Phytoplankton pigment concentrations in the middle Atlantic bight: comparison of ship determinations and coastal zone color scanner measurements. Appl Opt 22:20–36

Gordon HR, Clark DK, Mueller JL, Hovis WA (1980) Phytoplankton pigments derived from the Nimbus-7 CZCS: initial comparisons with surface measurements. Science 210:63–66

Gordon HR, Morel AY (1983) Remote Assessment of Ocean Color for Interpretation of Satellite Visible Imagery: A Review, Springer-Verlag, New York

Gordon HR, Wang M (1992a) Surface roughness considerations for atmospheric correction of ocean color sensors. 1: The rayleigh scattering component. Appl Opt 31:4247–4260

Gordon HR, Wang M (1992b) Surface roughness considerations for atmospheric correction of ocean color sensors. 2: Error in the retrieved water-leaving radiance. Appl Opt 31:4261–4267

Gordon HR, Wang M (1994) Retrieval of water-leaving radiance and aerosol optical thickness over the oceans with SeaWiFS: a preliminary algorithm. Appl Opt 33:443–452

Hovis WA, Clark DK, Anderson F, Austin RW, Wilson WH, Baker ET, Ball D, Gordon HR, Mueller JL, El Sayed SY, Sturm B, Wrigely RC, Yentsch CS (1980) Nimbus-7 coastal zone color scanner: system description and initial imagery. Science 210:60–63

Jerlov NG (1968) Optical Oceanography, Elsevier, New York

Kuchinke CP, Gordon HR, Franz BA (2009) Spectral optimization for constituent retrieval in Case 2 waters I: implementation and performance. Rem Sens Environ 113:571–587

Morel A, Gordon HR (1980) Report of the working group on water color. Bound Layer Meteor 18:343–355

Morel A, Prieur L (1977) Analysis of ocean color. Limnol Oceanogr 22:709–722

Moulin C, Gordon HR, Chomko RM, Banzon VF, Evans RH (2001) Atmospheric correction of ocean color imagery through thick layers of Saharan dust. Geophys Res Lett 28:5–8

Chapter 18
Field Radiometry and Ocean Color Remote Sensing

Giuseppe Zibordi and Kenneth J. Voss

18.1 Introduction

In the eighteenth century the experimental and theoretical work of Pierre Bauger and Johann Lambert in the area of light measurements led to the formulation of basic theories such as the law of addition, the inverse square law and the cosine law of illumination (Johnston, 2001). Despite these advances, the first successful measurements of marine light were only performed in the 1920s (Kundsen, 1922; Shoulejkin, 1924; Pettersson and Landberg, 1934; Jerlov and Liljequist, 1938) followed by a substantial progress in understanding marine optical processes, in producing theories to quantitatively describe the in-water light field and in defining fundamental laws and methods for underwater optics (Gershun, 1939; Le Grand, 1939).

After the first pioneering phase, advancements in marine light measurements were successively linked to progresses in quantitative optical radiometry culminating with the definition of basic designs for spectral radiometers (Jerlov, 1951; Steeman Nielsen, 1951) and in the realization of light detectors and of standards for absolute spectral calibration (Tyler and Smith, 1970; Slater, 1980).

In the 1980s, with the introduction of satellites for the remote observation of ocean color to map marine phytoplankton biomass at a global scale, in-situ optical radiometry became a basic component of the first ocean color mission (i.e., the Coastal Zone Color Scanner (CZCS)). While satellite radiometry provided global synoptic observations of the radiance emerging from the sea, in-situ radiometry was required to develop algorithms linking the satellite observations to the optically significant seawater components (Kirk, 1994; Mobley, 1994; Spinrad et al., 1994).

Major developments in in-situ marine optical radiometry were then driven by the need to reduce the uncertainties in field measurements for their application in advanced bio-optical modeling, vicarious calibration of satellite sensors and

G. Zibordi (✉)
Institute for Environment and Sustainability, Joint Research Centre, European Commission, Ispra 21027, Italy
e-mail: giuseppe.zibordi@jrc.ec.europa.eu

V. Barale et al. (eds.), *Oceanography from Space*,
DOI 10.1007/978-90-481-8681-5_18,

validation of space derived radiometric products. Impetus for these developments, which led to unprecedented accuracy in marine optical radiometry through the definition and assessment of new measurement methods and protocols, were the recent space missions for the global mapping of marine biomass (Robinson, 2004): the Sea-viewing Wide Field-of-view Sensor (SeaWiFS), the Moderate Resolution Imaging Spectroradiometer (MODIS), the Global Imager (GLI) and the Medium Resolution Imaging Spectrometer (MERIS).

This work is a brief introduction to in-situ marine optical radiometry with a view to its application to ocean color remote sensing.

18.2 Terminology

Optical radiometry is the science and technology of measuring radiant energy in the electromagnetic spectrum from the ultraviolet to the infrared wavelengths. Basic radiometric quantities are the solid angle, irradiance and radiance.

In the ideal case of a cone in a 3-dimensional space that resembles the typical geometry for radiometric measurements, the solid angle is the area intercepted by the cone on the surface of a unit sphere centered at the vertex of the cone itself. By indicating with ϕ and θ the azimuth and zenith angles in spherical coordinates, the solid angle Ω in units of sr for a right cone with vertex angle 2Θ and axis oriented in the direction $\theta = 0$, is obtained by integrating the element of solid angle $d\Omega = \sin\theta \, d\theta \, d\phi$ according to

$$\Omega = \int_0^{2\pi} d\phi \int_0^{\Theta} \sin\theta \, d\theta = 2\pi \, (1 - \cos\Theta). \qquad (18.1)$$

For a full hemisphere, $\Omega = 2\pi\,\mathrm{sr}$.

Irradiance and radiance are commonly expressed as spectral quantities. Spectral irradiance at the wavelength λ, denoted as $E(\lambda)$ and generally provided in units of $W/m^2/nm$, is the radiant flux incident on, passing through, or emerging from a surface per unit surface area and per unit wavelength interval:

$$E(\lambda) = \frac{d^2\Phi}{dS_0 d\lambda} \qquad (18.2)$$

where $d\Phi$ is the element of radiant flux (the time rate of flow of radiant energy), dS_0 is an element of area at the surface and $d\lambda$ is an element of wavelength centered at λ.

Spectral radiance, denoted by $L(\lambda)$ and generally provided in units of $W/m^2/sr/nm$, is the radiant flux at a given point and direction per unit solid angle, per unit projected area and per unit wavelength interval:

$$L(\theta,\phi,\lambda) = \frac{d^3\Phi}{d\Omega \, dS_0 \cos\theta \, d\lambda} \qquad (18.3)$$

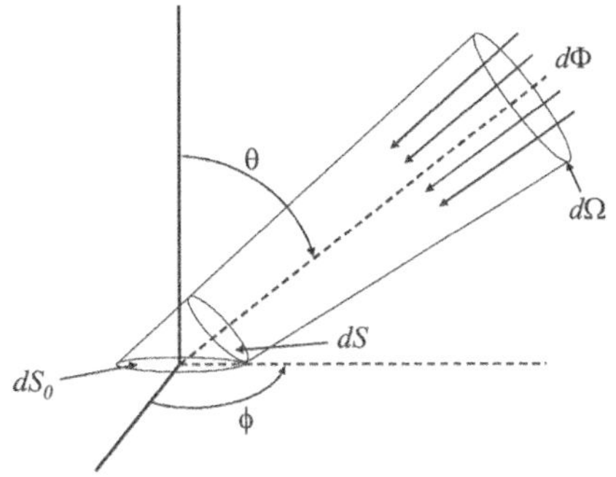

Fig. 18.1 Schematic of the radiance concept

where θ is the angle between the direction of the radiant flux and the normal to the surface at the specified point, and $dS = dS_0 \cos\theta$ is the projected area on the plane perpendicular to the direction of propagation (Fig. 18.1).

The integral of all radiance elements over the hemispherical solid angle gives the irradiance

$$E(\lambda) = \int_0^{2\pi} d\phi \int_0^{\pi/2} L(\theta,\phi,\lambda) \cos\,\theta\ \sin\,\theta\ d\theta. \tag{18.4}$$

If $L(\theta,\phi,\lambda)$ is isotropic, that is $L(\theta,\phi,\lambda)$ has constant value $L(\lambda)$ for any $\theta$and ϕ, then $E(\lambda) = \pi L(\lambda)$.

Particularly relevant to marine optics is the law of radiance invariance at a plane interface. This describes the change in radiance distribution across two media of refractive indices n_1 and n_2, assuming radiance is not absorbed at the interface between the two media. Explicitly, if ρ is the reflectance for the given angle of incidence at the interface for the radiance L_1 in the medium with refractive index n_1, the radiance that enters the medium with refractive index n_2 is

$$L_2 = (1 - \rho)\, L_1 \frac{n_2^2}{n_1^2}. \tag{18.5}$$

This relationship is usually called n^2 *law of radiance* and states that for a light beam crossing the interface between two media with different refractive indices, the ratio of radiance to the square of the refractive index of the medium remains invariant when ignoring reflection losses at the interface (i.e., $\rho = 0$).

18.3 In-situ Measurement Systems

A radiometer is composed of optics, detector(s) and associated electronics. The optics collect the input radiant flux and spectrally decompose it through spectral filters with specific wavelength bands (i.e., bandwidths) or alternatively disperse it through prisms or gratings. Detectors translate the radiant flux received from the optics into an electrical signal which is then converted into a digital number. Optics vary with the type of measurement. In the case of radiance measurements, the radiant flux is generally collected through field stops defining the full-angle field-of-view of the radiometer. This typically varies in the range of 1–20° depending on the application or detector features. In the case of irradiance measurements the input radiant flux is collected through a diffuser called collector. This is shaped to ideally gather the directional radiance contributions with a response varying as the cosine of the incident angle. Detectors include single- or multiple-detectors generally coupled with spectral filters in multispectral systems, or detector-arrays coupled with prisms and gratings in hyperspectral systems.

The elements commonly defining the performance of a radiometer are the spectral bandwidth and resolution, the responsivity (output counts per input of incident radiant flux), the detectivity (responsivity divided by the root mean square noise of the detector output) and the operational range (defined by the minimum radiant flux saturating the detector's output).

Systems for marine optical radiometry can be roughly separated into above- and in-water systems. Above-water systems provide the capability of determining the so called water-leaving radiance which carries information on the seawater optically significant constituents. In-water systems can provide comprehensive characterization of the in-water radiometric properties through radiance and irradiance measurements from a variety of configurations and deployment gears (e.g., profilers, buoys). Distinctive examples of in-water optical radiometer systems are provided by imaging devices utilized to map the radiance distribution.

18.3.1 Above-Water Systems

Above-water radiometry, when compared to the more consolidated and widely used in-water radiometry, had been almost unexploited up to the 1980s when specific measurement methods were proposed and applied (Morel, 1980; Carder and Steward, 1985). Relevant contributions to the refinement of the early methods came with the theoretical work of Mobley (1999) and Fougnie et al. (1999), followed by the experimental activities of Toole et al. (2000), Hooker et al. (2002a), Zibordi et al. (2002) and Deschamps et al. (2004).

Most of the published methods determine the water-leaving radiance, $L_W(\lambda)$, from measurements of the total radiance from above the sea, $L_T(\theta,\phi,\lambda)$, (which includes water-leaving, sky-glitter and sun-glint radiance contributions) and the diffuse radiance from the sky, $L_i(\theta',\phi,\lambda)$ (i.e., sky radiance), applying rigorous protocols (Deschamps et al., 2004; Hooker et al., 2004; Zibordi et al., 2004c). This implies the adoption of rigid measurement geometries (see the example in Fig. 18.2).

The accuracy of radiometric products determined from measurements performed with above-water systems heavily depends on the capability of minimizing glint perturbations in $L_T(\theta,\phi,\lambda)$. The methods currently applied utilize filtering schemes

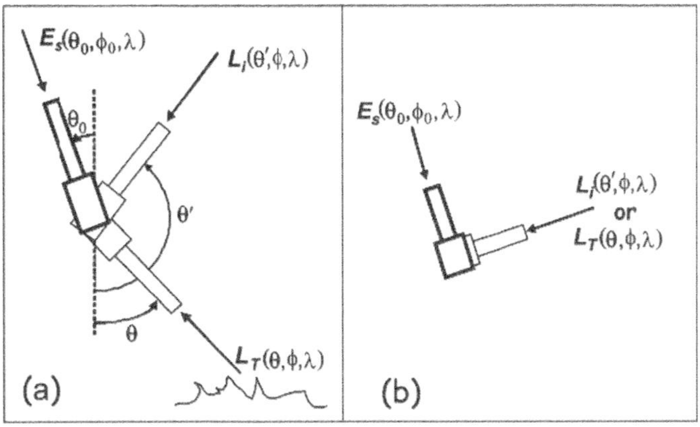

Fig. 18.2 Schematic of the measurement geometry for an above-water radiometer utilized for the direct sun irradiance E_s, sky radiance L_i and total radiance from above the sea L_T: (**a**) *side view* and (**b**) *top view* of the instrument (after Zibordi et al., 2009b). θ_0 and ϕ_0 indicate the sun zenith and azimuth angles, respectively; θ and ϕ are the viewing angle and azimuth for L_T; θ' the viewing angle for L_i

to remove data likely affected by sun-glint (Hooker et al., 2002a; Zibordi et al., 2002) or alternatively a polarizer to reduce the sky-glitter and sun-glint contributions (Fougnie et al., 1999).

18.3.2 In-Water Profiling Systems

In-water radiometry relies on the application of the method proposed by Smith and Baker (1984, 1986), which combines achievements of earlier experimental studies (Dera et al., 1972; Jerlov, 1976; Tyler, 1977). The method requires measurements in the water column at different depths as well as the above-water downward irradiance. The in-water radiometric measurements are used to extrapolate to 0^- depth (i.e., just below the water surface) radiometric quantities which cannot be directly measured because of the fluctuation of the sea surface due to waves. Above-water downward irradiance data are used to minimize the effects of illumination variations on in-water radiometric measurements during data collection.

In-water continuous profiles of radiometric quantities generally result from measurements performed with radiometers on winched and free-fall systems. Winched systems had extensive use in the past (Smith et al., 1984). However, since the late 1980s, the design of free-fall systems has provided the possibility of performing measurements at a greater distance from the deployment structure (Lewis et al., 1986; Waters et al., 1990), and has allowed the collection of continuous profiles ideally unaffected by perturbations due to ship-shading and ship-motion. Current free-fall systems (Hooker and Maritorena, 2000; Zibordi et al., 2004a) provide and the capability of measuring the upwelling radiance $L_u(z,\lambda)$, the downward irradiance

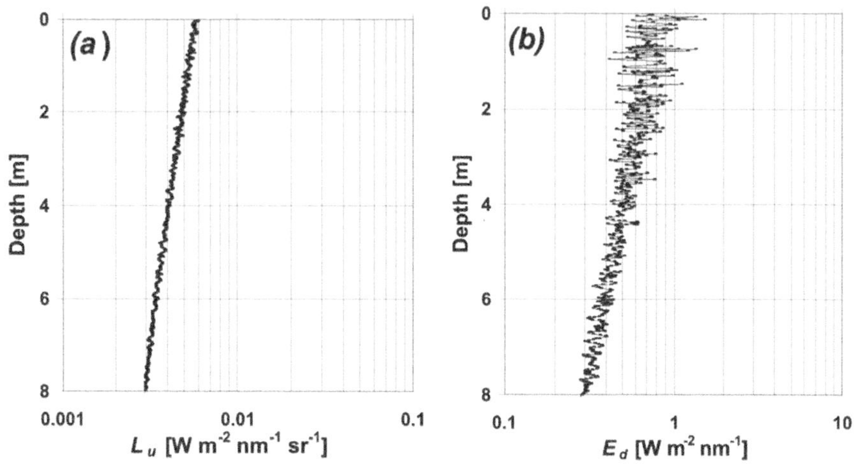

Fig. 18.3 Radiometric profiles of $L_u(z,\lambda)$ and $E_d(z,\lambda)$ at 555 nm performed with 40 cm wave height and diffuse attenuation coefficient $K_d(490)=0.09$/m (after Zibordi et al., 2004a)

$E_d(z,\lambda)$ and the upward irradiance $E_u(z,\lambda)$ as a function of depth z, in addition to the above-water downward irradiance $E_d(0^+,\lambda)$. An example of in-water radiometric profile of $L_u(z,\lambda)$ and $E_d(z,\lambda)$ is displayed in Fig. 18.3.

Due to perturbations caused by wave focusing and defocusing, the accuracy of the derived sub-surface radiometric products is a function of the sampling depth-interval and of the depth resolution as defined by the system acquisition rate and deployment speed. Thus the determination of highly accurate in-water radiometric products requires sampling near the surface (especially in coastal regions due to possible vertical inhomogeneities in the seawater optical properties), and the capability of producing a number of measurements per unit depth suitable to minimize the effects of wave perturbations and not significantly affected by tilt (Zibordi et al., 2004a). In oceanic waters, due to the near-surface homogeneity of the seawater bio-optical properties, the requirement of sampling near the surface can be relaxed. This allows for the use of buoy systems equipped with radiometers operated at fixed-depths at a few meters below the surface.

18.3.3 Buoy Systems

In recent times the concept of multiple radiometers deployed at different fixed depths was utilized by Dera et al. (1972) who envisaged its applicability to buoy systems. Since the late 1990s, the use of fixed-depth radiometers has become the basis for measurements performed through bio-optical buoys for satellite ocean color applications (Clark et al., 1997; Clark et al., 2002; Kishino et al., 1997; Pinkerton and Aiken, 1999; Antoine et al., 2008a; Kuwahara et al., 2008). These buoy-based systems generally provide the capability of measuring $L_u(z,\lambda)$ and $E_d(z,\lambda)$ at two or more depths (typically between 1 and 10 m) in addition to the above-water downward irradiance $E_d(0^+,\lambda)$. By neglecting the effects of system tilt, the accuracy of radiometric products determined with buoys is a function of the discrete depths selected for the radiometers, the acquisition rate and the logging interval (Zibordi et al., 2009a).

18.3.4 Imaging Systems

The radiance distribution (angular variation of the radiance field) is important to investigate the anisotropy of light in natural waters (Voss and Morel, 2005). These investigations were performed over several decades by designing and applying systems based on single field-of-view radiometers (Jerlov and Fukuda, 1960; Tyler, 1960; Aas and Hojerslev, 1999). With these radiometers, successive measurements over various directions were required to map the in-water light field distribution. A significant advance in measuring radiance distribution was marked by the development of an underwater camera equipped with a fisheye lens and a photopic filter (Smith et al., 1969). This measurement concept was later revisited and applied to electro-optics cameras (Voss, 1989; Voss and Chapin, 2005) allowing for mapping the spectral radiance distribution at several spectral bands to investigate the angular variation of the downwelling and upwelling radiance fields (see Fig. 18.4).

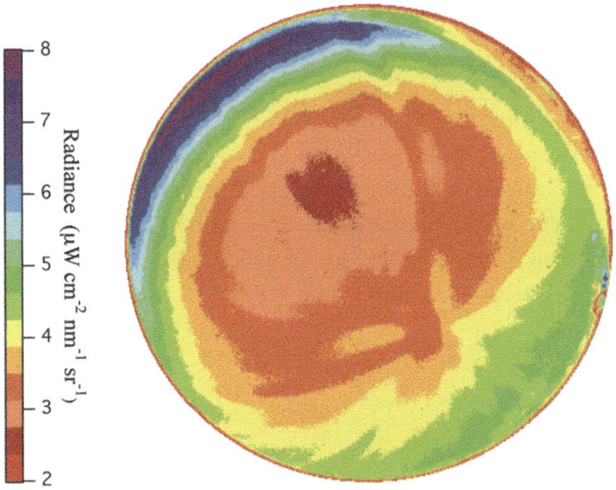

Fig. 18.4 Map of the upwelling radiance distribution at 490 nm produced with an electro-optics camera system (Voss and Chapin, 2005) on March 12, 2007 at 19:33 GMT off of Honolulu, Hawaii. Center of image is nadir direction while edge of *circle* is horizon (90° nadir angle). Anti-solar point is towards the *lower right* from the *center* of the map. Some contours follow the solar refracted rays which cause high and low radiance regions in the image and extend radially from the anti-solar point

18.4 Calibration of Optical Radiometers

Absolute calibration and careful characterization of measuring instruments are crucial for determining physical quantities which are independent of the particular instrument used in the data collection.

18.4.1 Radiometric Calibration

Absolute calibration of radiometers requires defining the mathematical transformation that relates the sensor output to the appropriate radiometric quantity. By applying the concept of measurement equation (Wyatt, 1978) to yield the sensor output for a specific source configuration and by assuming that the radiometer has ideal spectral performance and linear response in the operational range, the conversion from relative to physical units (called calibration) of the radiometric quantity $\Im(\lambda)$ (i.e., $E(\lambda)$ or $L(\lambda)$) at wavelength λ is given by

$$\Im(\lambda) = C_{\Im}(\lambda) I_f(\lambda)[\mathrm{DN}(\lambda) - \mathrm{D0}(\lambda)] \tag{18.6}$$

where $\mathrm{C}_{\Im}(\lambda)$ is the in-air absolute calibration coefficient, $I_f(\lambda)$ is the so-called immersion factor accounting for the change in response of the sensor when immersed in water with respect to air, $\mathrm{DN}(\lambda)$ is the digital output for a given input signal and $\mathrm{D0}(\lambda)$ is the dark value measured by obstructing the entrance optics. Assuming $I_f(\lambda) = 1$, $\mathrm{C}_{\Im}(\lambda)$ is determined by applying Equation (18.6) to in-air measurements of a known source whose radiant flux falls in the operational range of the sensor.

In-air absolute calibration coefficients for irradiance sensors, $\mathrm{C_E}(\lambda)$, are generally achieved using an irradiance standard, $E_L(\lambda)$, for instance obtained with a FEL 1,000 W calibrated lamp (Grum and Becherer, 1979). Assuming a sensor with narrow bandwidth centered at λ, a point-source and a point-detector, $\mathrm{C_E}(\lambda)$ is determined from the reading of $\mathrm{DN}(\lambda)$ related to the input irradiance $E(\lambda)$. For a source positioned on axis and normal to the collector of the irradiance sensor

$$E(\lambda) = E_L(\lambda)\frac{d_0^2}{d^2} \tag{18.7}$$

where d is the distance between source and sensor, and d_0 the distance at which the value $E_L(\lambda)$ is defined.

Similar to $\mathrm{C_E}(\lambda)$, the in-air absolute calibration coefficient of radiance sensors, $\mathrm{C_L}(\lambda)$, is determined using a known radiance source, $L(\lambda)$. This can be obtained with integrating spheres or systems composed of an irradiance standard (i.e., a FEL 1,000 W) on axis and normal to the faceplate of a reflectance standard (i.e., a plaque with calibrated directional-directional reflectance). The adoption of the lamp-plaque system, instead of an integrating sphere, helps to reduce the relative uncertainties between radiance and irradiance calibrations when using the same irradiance standard for both types of calibration. With the lamp-plaque system, $C_L(\lambda)$ is determined from DN(λ) related to the input radiance $L(\lambda)$. Under the assumption of a radiance sensor with narrow bandwidth centered at λ and a narrow field-of-view viewing the plaque at an angle θ with respect to the normal to the plaque,

$$L(\lambda) = E(\lambda)\rho_d(\lambda,\theta)\pi^{-1} \tag{18.8}$$

where $\rho_d(\lambda,\theta)$ is the directional-directional reflectance of the plaque for the specific viewing configuration (generally $\theta = 45°$) and $E(\lambda)$ is given by Equation (18.7) with distance d between lamp and plaque. It is recalled that while increasing d augments the homogeneity of the radiance field within the field-of-view of the sensor, the intensity decreases with $1/d^2$. Thus, d needs to be chosen to satisfy both intensity and homogeneity requirements for the sensor under calibration. Finally, if the directional-hemispherical reflectance $\rho_h(\lambda)$ is provided for the plaque as opposed to the directional-directional reflectance $\rho_d(\lambda,\theta)$, a suitable correction coefficient is required to relate these two factors.

Extended analysis of calibration uncertainties were made within the framework of the SeaWiFS Intercalibration Round Robin Experiments (SIRREX) which addressed absolute calibrations of field radiometers (see Hooker et al., 2002b and references therein). In particular, during SIRREX-7 comprehensive efforts were focused on the evaluation of uncertainties of lamp radiant fluxes (based on multiple measurements performed on a set of lamps), calibration repeatability (as affected by power supply, lamp stability, and radiometer alignment) and plaque reflectance (due to spatial inhomogeneity and uncertainty in directional-directional reflectance). Results suggested ranking calibration uncertainties as primary (minimum), secondary (average) and tertiary (high) based on the difficulty of reducing the size of uncertainties from different individual sources. These values vary from 1.1 to 3.4% for irradiance and from 1.5 to 6.3% for radiance.

18.4.2 Cosine Response of Irradiance Sensors

Irradiance sensors should ideally collect the directional radiance contributions with a response varying as the cosine of the incident angle. Real collectors, nevertheless, have an angular response which deviates from this ideal cosine. Consequently, the error in the cosine response is a source of uncertainty in irradiance measurements. This generally increases with the angle of incidence on the collector and depends on wavelength, sun zenith (i.e., geographic position, season and time), atmospheric optical conditions (i.e., cloudiness, aerosol type and load), and additionally the seawater optical properties and depth for in-water measurements.

The cosine error, denoted as $f_c(\theta,\phi,\lambda)$ and expressed in percent, is conveniently described through the normalized angular response – the response divided by the cosine of the angle of incidence and by the response at normal incidence – at the center-wavelength λ of each spectral band

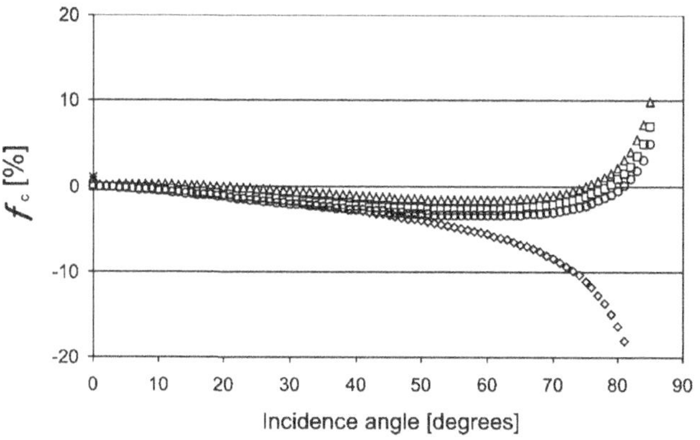

Fig. 18.5 Cosine errors f_c at 412 nm (*diamonds*), 490 (*triangles*), 555 (*squares*) and 665 nm (*circles*), determined for a class of above-water irradiance sensors (after Zibordi and Bulgarelli, 2008)

$$f_c(\theta,\phi,\lambda) = 100\left[\frac{E(\theta,\phi,\lambda)}{E(0,\lambda)\cos\theta} - 1\right] \quad (18.9)$$

where $E(\theta,\phi,\lambda)$ is the measurement taken at incidence angle θ and azimuth ϕ, and $E(0,\lambda)$ is the measurement taken at $\theta = 0$, with $E(0,\lambda)\cos\theta$ indicating measurements for an ideal cosine response. The cosine error is indicated as $f_c(\theta,\lambda)$ when assumed independent of the azimuth. An example of cosine error functions is presented in Fig. 18.5 for a class of above-water irradiance sensors.

Operational correction schemes for quantifying the error $\varepsilon_c(\theta_0,\lambda)$ as a function of $f_c(\theta,\lambda)$ in above-water downward irradiance measurements (Zibordi and Bulgarelli, 2008) include the use of empirical relationships relying on the assumption of an isotropic sky radiance distribution, the knowledge of sun zenith θ_0 and diffuse to direct irradiance ratio $I_r(\theta_0,\lambda)$. Specifically,

$$\varepsilon_c(\theta_0,\lambda) = \langle f_c(\lambda)\rangle \frac{I_r(\theta_0,\lambda)}{I_r(\theta_0,\lambda)+1} + f_c(\theta_0,\lambda)\frac{1}{I_r(\theta_0,\lambda)+1} \quad (18.10)$$

where the two terms on the right side of Equation (18.10) account for the effects of cosine error on diffuse and direct irradiance, respectively, with

$$\langle f_c(\lambda)\rangle = \int_0^{\pi/2} f_c(\theta,\lambda)\sin(2\theta)d\theta. \quad (18.11)$$

An analysis of the uncertainties associated with the application of such a scheme to a series of radiometers (Zibordi and Bulgarelli, 2008) has shown the capability of reducing measurement errors from 10–15% down to 1.5% for $\theta_0 > 60°$. It is likely that a similar approach is applicable to in-water data using modeled or measured radiance distributions.

18.4.3 Immersion Factor of In-Water Radiometers

The refractive index of the diffuser material, $n_d(\lambda)$, used for manufacturing irradiance collectors is always larger than the refractive index of water, $n_w(\lambda)$, and of air, n_a. Since $n_w(\lambda) > n_a$, the Fresnel reflectance of the external water-diffuser interface is smaller than that of the air-diffuser interface. Consequently, the transmission of light through the external interface of the diffuser is larger in water than in air. However, the internal diffuser-water interface also reflects less of the internal diffuse light back into the diffuser, when compared to the corresponding diffuser-air interface. Thus because the increase in light transmitted back into the water is greater than the increased amount of light transmitted from the water into the diffuser, there is a decrease in the transmittance of the diffuser when the instrument is in water with respect to in air.

Early studies on immersion effects for irradiance sensors were carried out by Atkins and Poole (1933) who made an attempt to describe the internal and external reflection factors for an opal glass diffuser. Successive studies were performed by Berger (1958, 1961) attempting a rigorous description of the physical processes involved with the optics of immersed radiometers. These results were later used by Westlake (1965) to extensively illustrate the reflection-refraction processes occurring at the air-diffuser and at the water-diffuser interfaces.

A comprehensive description of protocols for the experimental characterization of the immersion factor of in-water irradiance collectors, was given by Tyler and Smith (1970) based on the use of a collimated source, and by Aas (1969) based on a point source. This latter method was implemented by Petzold and Austin (1988) and applied by Mueller (1995) and Zibordi et al. (2004b), and later subject to refinements (Hooker and Zibordi, 2005a). These studies highlighted (Mueller, 1995) and afterward confirmed (Zibordi et al., 2004b) the need for the experimental characterization of each individual irradiance collector as opposed to applying class-based immersion factors.

In agreement with the latter studies the immersion factor, $I_f(\lambda)$, of an irradiance sensor is determined from in-air and in-water measurements performed with the sensor vertically illuminated by a point source at fixed distance. Specifically,

$$I_f(\lambda) = \frac{E(0^+,\lambda)}{E(0^-,\lambda)} t_{wa}(\lambda) \tag{18.12}$$

where $E(0^+,\lambda)$ is the irradiance measured with the instrument dry (i.e., in-air), $t_{wa}(\lambda)$ the transmittance of the air-water interface, and $E(0^-,\lambda)$ the in-water subsurface irradiance determined from the least square fit – as a function of the water level above the sensor – of log transformed in-water measurements. The latter are computed with $I_f(\lambda) = 1$ and corrected for the geometric perturbation induced by the finite distance between point source (i.e., a lamp) and irradiance sensor. This correction accounts for the radiant flux change at the collector surface as a function of the water depth and source-collector distance (Aas, 1969). Zibordi et al. (2004b) showed the possibility of experimentally determining $I_f(\lambda)$ for irradiance sensors with an uncertainty of approximately 0.5% for values falling in a typical range of 1.35–1.40.

The responsivity of radiance sensors also changes when operated in water as opposed to air due to the relative change in the solid angle field–of–view (depending on radiometer design) and the variation of reflectance and transmittance of the optical window (because the refractive index $n_g(\lambda)$ of the window is different from that of the air or water). $I_f(\lambda)$ for radiance sensors is commonly derived theoretically unlike that of irradiance sensors which can only be determined experimentally. A general model proposed by Austin (1976) for determining $I_f(\lambda)$ for radiance sensors simply relies on the knowledge of the refractive indices of seawater and that of the radiometer's window.

The alternative possibility of experimentally characterizing $I_f(\lambda)$ for radiance sensors was recently documented by Zibordi (2006). The application of the proposed methodology did not show any appreciable instrument to instrument dispersion of $I_f(\lambda)$ for radiometers with identical optical design (e.g., see Fig. 18.6 for a class of radiometers successively manufactured with optical windows made of different materials and thus having different refractive indices (after Zibordi, 2006)). However, differences varying from less than 1% up to several percent were observed between the theoretical determinations and experimental characterisations of $I_f(\lambda)$ for radiometers with different optical designs (Zibordi and Darecki, 2006). This suggests that the experimental characterization of $I_f(\lambda)$ for sample radiance sensors of each class should become part of their quality assurance process to quantify the deviation of actual immersion factors from their theoretical determinations.

The experimental characterization of the immersion factor is also a requirement for each fisheye radiance distribution camera because of the hemispherical dome window and its interactions with the various components of the optical system (Voss

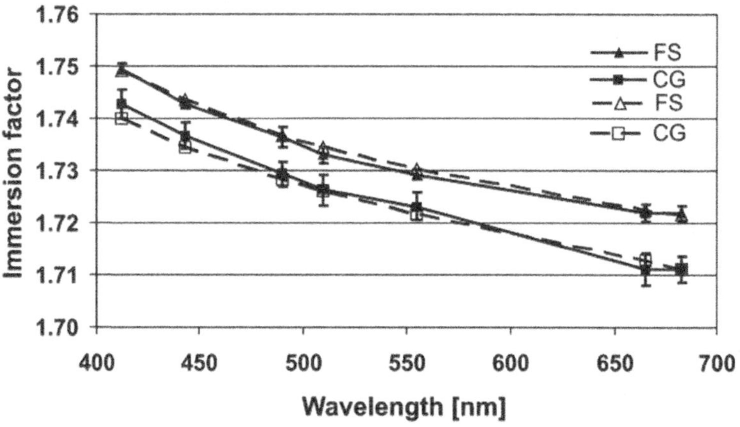

Fig. 18.6 Comparison of theoretical (*dashed lines*) and experimental (*continuous lines*) immersion factors I_f for a class of in-water radiance sensors having Fused Silica (FS) and Optical Crown Glass (CG) windows. Vertical error bars for the experimental values indicate the standard deviation of I_f from different sensors

and Chapin, 2005). This entails the characterization of each camera system and its verification after any system change.

The method proposed by Zibordi (2006) for the experimental characterization of $I_f(\lambda)$ for radiance sensors, relies on in-water and in-air measurements of a homogeneous and Lambertian source virtually immersed in the water. Measurements need to be performed keeping the sensor-source distance constant with the sensor looking vertically at the source. Specifically, $I_f(\lambda)$ is determined from

$$I_f(\lambda) = \frac{L(0^+,\lambda)}{L(0^-,\lambda)} \frac{\Omega_a}{\Omega_w(\lambda)} \frac{1}{t_{\mathrm{wa}}(\Omega_w,\lambda)} \tag{18.13}$$

where $L(0^+,\lambda)$ is the above-water value, computed as the intercept of the least squares regression – as a function of the distance of the sensor from the water surface – of in-air measurements made with different water levels and corrected for the different air-water optical paths. The term $L(0^-,\lambda)$ is the spectral in-water radiance determined from measurements taken with the instrument immersed in the water and computed with $I_f(\lambda) = 1$. The terms Ω_a and $\Omega_w(\lambda)$ indicate the solid angle field-of-view in air and in water, respectively. The term $t_{\mathrm{wa}}(\Omega_w,\lambda)$ indicates the water-air transmittance averaged over the solid angle $\Omega_w(\lambda)$.

The values of $I_f(\lambda)$, for both irradiance and radiance sensors, should be experimentally characterized using pure water to ensure best reproducibility of measurements. Clearly correction factors need to be applied to experimental values of $I_f(\lambda)$ obtained with pure water to account for the different refractive index of seawater.

18.5 Radiometric Products of In-situ Optical Radiometers

Data products from in-water radiometric measurements generally include spectral values of: irradiance reflectance, remote sensing reflectance, normalized water-leaving radiance, diffuse attenuation coefficient and the so called Q-factor. Data products from above-water radiometric measurements are generally restricted to the normalized water-leaving radiance and the remote sensing reflectance.

Elements on the reduction of former in-situ radiometric products are hereafter presented in agreement with consolidated protocols (e.g., see Mueller and Austin, 1995 and successive revisions).

18.5.1 Products from In-Water Measurements

The following data reduction process equally applies to fixed-depth and continuous profile radiometric data $\Im(z,\lambda,t)$ (i.e., $L_u(z,\lambda,t)$, $E_u(z,\lambda,t)$ and $E_d(z,\lambda,t)$) at wavelength λ, with z expressing dependence on depth and t on time. The first required step is the minimization of the effects of light change during data collection. This is performed by applying above-water downward irradiance $E_d(0^+,\lambda,t)$ data according to

$$\Im_0(z,\lambda,t_0) = \frac{\Im(z,\lambda,t)}{E_d(0^+,\lambda,t)} E_d(0^+,\lambda,t_0) \quad (18.14)$$

where $\Im_0(z,\lambda,t_0)$ indicates radiometric quantities as if they were taken at each depth z at the same time t_0, and $E_d(0^+,\lambda,t_0)$ specifies the above-water downward irradiance at time t_0 (with t_0 generally chosen to coincide with the beginning of the acquisition sequence).

Omitting the variable t, the sub-surface quantities $\Im_0(0^-,\lambda)$ (i.e., $L_u(0^-,\lambda)$, $E_u(0^-,\lambda)$ and $E_d(0^-,\lambda)$ are then determined as the exponentials of the intercepts resulting from the least-squares linear regressions of $\ln \Im_0(z,\lambda)$ versus z within the extrapolation interval identified by $z_0 < z < z_1$ and chosen to satisfy the requirement of linear decay of $\ln\Im_0(z,\lambda)$ with depth. The negative values of the slopes of the regression fits are the so-called diffuse attenuation coefficients $K_{\Im}(\lambda)$ (i.e. $K_l(\lambda)$, $K_u(\lambda)$ and $K_d(\lambda)$) for the selected extrapolation interval.

Derived radiometric data products are then the dimensionless irradiance reflectance at depth 0^-, $R(0^-,\lambda)$, defined as the ratio of $E_u(0^-,\lambda)$ to $E_d(0^-,\lambda)$, and the Q-factor at nadir, $Q_n(0^-,\lambda)$ in units of sr, defined as the ratio of $E_u(0^-,\lambda)$ to $L_u(0^-,\lambda)$.

Additional data products are the remote sensing reflectance, $R_{\mathrm{rs}}(\lambda)$, in units of sr^{-1}

$$R_{\mathrm{rs}}(\lambda) = \frac{L_w(\lambda)}{E_d(0^+,\lambda)} \quad (18.15)$$

and the normalized water-leaving radiance, $L_{\mathrm{wn}}(\lambda)$, in units of $\mathrm{W/m^2/nm/sr}$

$$L_{\mathrm{wn}}(\lambda) = R_{\mathrm{rs}}(\lambda) E_0(\lambda) \quad (18.16)$$

where $E_0(\lambda)$ is the average extra-atmospheric solar irradiance (Thuillier et al., 2003) and $L_{\mathrm{w}}(\lambda)$ the so called water-leaving radiance, i.e., the radiance leaving the sea and quantified just above the surface as

$$L_w(\lambda) = 0.543 L_u(0^-,\lambda). \quad (18.17)$$

The factor 0.543 has been computed assuming n_w is independent of wavelength (Austin, 1974), and accounts for the reduction in radiance from below to above the water surface due to the change in the refractive index at the air-water interface.

Both $R_{\mathrm{rs}}(\lambda)$ and $L_{\mathrm{wn}}(\lambda)$ are then quantities corrected for the illumination effects dependent on the sun zenith angle, Sun-Earth distance and the atmospheric transmittance (Mueller and Austin, 1995).

18.5.2 Products from Above-Water Measurements

Data products from above-water radiometry are $L_{wn}(\lambda)$ and $R_{rs}(\lambda)$ derived from $L_T(\theta,\phi,\lambda)$ and $L_i(\theta',\phi,\lambda)$ whose measurement geometry defined by θ, θ' and ϕ (see Fig. 18.2) is chosen to minimize the wave perturbation effects: generally $\theta = 40°$, $\theta' = 140°$ and $\phi = \phi_0 \pm 90°$ with sun azimuth ϕ_0. Specifically, the water-leaving radiance $L_w(\lambda,\theta,\phi)$ for the chosen viewing geometry is computed as

$$L_w(\theta,\phi,\lambda) = L_T(\theta,\phi,\lambda) - \rho(\theta,\phi,\theta_0,W)L_i(\theta',\phi,\lambda), \tag{18.18}$$

where $\rho(\theta,\phi,\theta_0,W)$ is the sea surface reflectance which can be theoretically determined as a function of the geometry identified by θ, ϕ, θ_0, and of the sea state expressed through the wind speed, W. The need to minimize the effects of wave perturbations in $L_T(\theta,\phi,\lambda)$ and eventually the effects of cloud perturbations in $L_i(\theta',\phi,\lambda)$, has suggested determining these values from the average of n-independent measurements satisfying strict filtering criteria (Zibordi et al., 2009b). Ideally, similar to the scheme applied for in-water data, the individual values of $L_T(\theta,\phi,\lambda)$ and $L_i(\theta',\phi,\lambda)$, should be corrected for illumination changes as a function of time using $E_d(0^+,\lambda)$ measurements.

The normalized water-leaving radiance $L_{wn}(\lambda)$ is determined as

$$L_{wn}(\lambda) = L_w(\theta,\phi,\lambda)\frac{E_0(\lambda)}{E_d(0^+,\lambda)}C_{\Re Q}(\theta,\phi,\theta_0,\lambda,\tau_a,IOP,W) \tag{18.19}$$

where the term

$$C_{\Re Q}(\theta,\phi,\theta_0,\lambda,\tau_a,IOP,W) = \frac{\Re_0}{\Re(\theta,W)}\frac{Q(\theta,\phi,\theta_0,\lambda,\tau_a,IOP)}{Q_n(\theta_0,\lambda,\tau_a,IOP)} \tag{18.20}$$

is introduced to remove the viewing angle dependence on $L_w(\theta,\phi,\lambda)$. The quantities $\Re(\theta,W)$ and $\Re_0$ (i.e., $\Re(\theta,W)$ at $\theta = 0$) account for the sea surface reflectance and refraction, and mostly depend on θ and W. The quantities $Q(\theta,\phi,\theta_0,\lambda,\tau_a,IOP)$ and $Q_n(\theta_0,\lambda,\tau_a,IOP)$ are the Q-factors at viewing angle θ and at nadir (i.e., $\theta = 0$), respectively, describing the anisotropic distribution of the in-water radiance field and depending on θ, ϕ, θ_0 and, τ_a and the seawater inherent optical properties (*IOPs*) as a function of λ.

In case measurements of $E_d(0^+,\lambda)$ are not available, the ratio $E_0(\lambda)/E_d(0^+,\lambda)$ can be replaced by $\left[D^2 t_d(\lambda)\cos\theta_0\right]^{-1}$(see Zibordi et al., 2009b), where D_2 accounts for the variations in the Sun-Earth distance as a function of the day of the year, and $t_d(\lambda)$ is the atmospheric diffuse transmittance computed from measured or estimated values of the aerosol optical thickness $\tau_a(\lambda)$ (Gordon and Clark, 1981).

18.5.3 Exact Normalized Water-Leaving Radiance

Morel and Gentili (1996) introduced an additional term, the exact normalized water-leaving radiance, $L_{WN}(\lambda)$, defined as

$$L_{WN}(\lambda) = L_{wn}(\theta)\ C_{f/Q}(\theta_0,\lambda,\tau_a,IOP) \tag{18.21}$$

where

$$C_{f/Q}(\theta_0,\lambda,\tau_a,IOP) = \frac{f_0(\lambda,\tau_a,IOP)}{Q_0(\lambda,\tau_a,IOP)}\left[\frac{f(\theta_0,\lambda,\tau_a,IOP)}{Q_n(\theta_0,\lambda,\tau_a,IOP)}\right]^{-1} \tag{18.22}$$

and, $Q_0(\lambda,\tau_a,IOP)$ and $f_0(\lambda,\tau_a,IOP)$ are the values of $Q_n(\theta_0,\lambda,\tau_a,IOP)$ and $f(\theta_0,\lambda,\tau_a,IOP)$ at $\theta_0 = 0$, respectively. $f(\theta_0,\lambda,\tau_a,IOP)$ relates the irradiance reflectance to the seawater backscattering to absorption ratio, $b_b(\lambda)/a(\lambda)$.

The dependence of bi-directional effects on *IOP*s can be conveniently expressed through chlorophyll *a* concentration, *Chla*, in Case-1 waters only, i.e., chlorophyll dominated waters (Morel et al., 2002).

Sample values of $L_{WN}(\lambda)$ from above-water radiometry are presented in Fig. 18.7 for two measurement sites representing different water types: the Acqua Alta Oceanographic Tower (AAOT) site in the Adriatic Sea characterized by moderately sediment dominated waters; and the Helsinki Lighthouse Tower (HLT) site in the Baltic Sea characterized by waters dominated by colored dissolved organic matter (see Zibordi et al., 2009b).

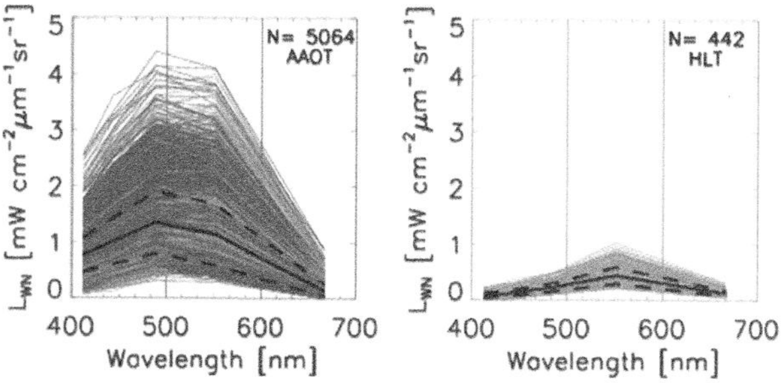

Fig. 18.7 Exact normalized water leaving radiance $L_{WN}(\lambda)$ spectra from above water radiometric measurements performed at the AAOT and HLT sites. *Thick continuous lines* indicate averages while *thick dashed lines* indicate ±1 standard deviation. *N* indicates the number of measurements

18.6 Measurement Perturbations

The accuracy of radiometric measurements carried out at sea is likely to be affected by various perturbing effects. Above-water measurements may be perturbed by the shading and reflection of deployment superstructures (i.e., ships, oceanographic towers), changes in the illumination conditions and wave effects. In addition to these perturbations, in-water measurements are also affected by instrument self-shading.

The superstructure perturbations largely vary with the illumination conditions, the seawater inherent optical properties and the deployment geometry (Gordon, 1985; Doyle and Zibordi, 2002; Hooker and Morel, 2003; Piskozub, 2004; Hooker and Zibordi, 2005b). Wave effects can produce quite large uncertainties as a function of sea state and seawater optical properties (Mobley, 1999; Hooker et al., 2002a; Zibordi et al., 2004a; Zibordi et al., 2009a). Self-shading produces a decrease in the in-water measurements of the upward light field (Gordon and Ding, 1992; Zibordi and Ferrari, 1995; Aas and Korsbø, 1997). This effect scales with the seawater absorption coefficient and the size of the instrument case.

18.6.1 Super-Structure Perturbations

In the early 1970s data produced with a photographic system measuring the in-water radiance distribution visually documented the effects of ship perturbations in underwater light measurements (Smith, 1974). Approximately 10 years later the ship-shading effects were quantitatively investigated through Monte-Carlo simulations (Gordon, 1985). This theoretical study led to the proposal of collecting in-water radiometric data at distances larger than 10 m from the ship, to minimize the superstructure perturbations. Later work (Voss et al., 1986; Helliwell et al., 1990; Saruya et al., 1996; Weir et al., 1994; Piskozub, 2004) confirmed that in-water radiance and irradiance measurement uncertainties increase substantially when reducing the deployment distance of the instrument from the ship.

In developing operational protocols for optical radiometric measurements to support satellite ocean color calibration and validation activities, Mueller and Austin (1995) suggested the minimum ship distance from the instrument deployment point as a function of the seawater diffuse attenuation coefficient. More recently the need to accurately quantify uncertainties in measurements taken at the AAOT led to extended investigations of tower perturbations in optical radiometric data collected near its superstructure with in- and above-water radiometers (Zibordi et al., 1999; Hooker and Zibordi, 2005b). Results from these analyses became the rationale for the development and implementation of operational methods for the minimization of superstructure perturbations. In the case of above-water radiometry, the minimization of perturbing effects due to deployment superstructures can be obtained through the adoption of rigid measurement geometries (Hooker and Zibordi, 2005b). For in-water radiometers the first choice is to make measurements sufficiently far from superstructures. When this is not feasible, it may be alternatively possible to remove the perturbation effects by simulating the radiance fields for each specific measurement condition accounting for the deployment geometry, and the marine and atmospheric optical properties (Doyle and Zibordi, 2002).

18.6.2 Wave Effects

The focusing and defocusing of sunrays refracted by surface waves produce large light fluctuations in the upper sea layer. The origin, amplitude, frequency and depth

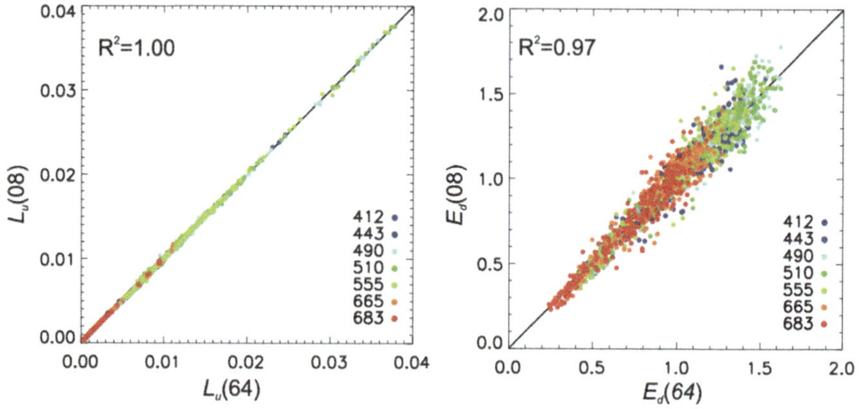

Fig. 18.8 Scatter plot of $L_u(0^-,\lambda)$ and $E_d(0^-,\lambda)$ obtained with decreased resolution profiles (i.e., 8 measurements per meter) versus reference values from full resolution profiles (i.e., 64 measurements per meter). Radiances $L_u(0^-,\lambda)$ are in units of W/m^2/nm/sr while irradiances $E_d(0^-,\lambda)$ are in units of W/m^2/nm. R^2 indicates the determination coefficient (after Zibordi et al., 2004a)

extension of these fluctuations were addressed both theoretically (e.g., Schenck, 1957; Snyder and Dera, 1970; Stramski and Dera, 1988; Walker, 1994; Zaneveld et al., 2001) and experimentally (e.g., Dera and Stramski, 1986; Weidemann et al., 1990; Dera et al., 1993).

Intuitively, for a given in-water continuous profiling system, any increase in the acquisition rate and decrease in the deployment speed is expected to produce an increase in the accuracy of the extrapolated subsurface optical quantities due to a more extended averaging of the wave effects over time as a function of depth (see Fig. 18.8).

In the case of fixed-depth in-water systems, minimization of focusing and shading effects can be achieved by averaging data over time (Zibordi et al., 2009a). In the case of above-water radiometry, wave effects can be minimized by filtering techniques simply based on the removal of the individual measurements highly affected by glint (Hooker et al., 2002a; Zibordi et al., 2002). It is important though that the individual measurements entering into the average do not exceed the instruments saturation signal.

18.6.3 Self-Shading

The finite size of in-water radiometers affects the radiance field and induces errors in the measured upwelling radiance and upward irradiance. Gordon and Ding (1992) evaluated the self-shading effects through numerical simulations. They estimated errors ranging from a few up to several tens percent as a function of the size of the radiometer, the absorption coefficient of the medium, and the type of illumination. For a given radiometer, the error is much higher in the near-infrared than in the

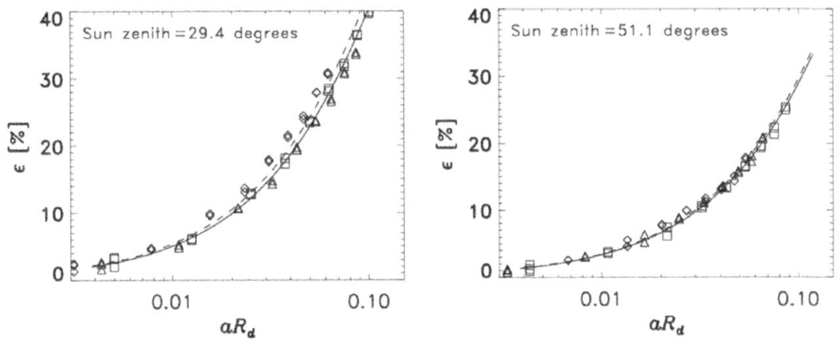

Fig. 18.9 Radiance self-shading error ϵ (in percent) as a function of seawater absorption, a, times the radius of the radiometer case, R_d, at different sun zeniths: 29.4 (*left panel*) and 51.1 (*right panel*) degrees (after Zibordi and Ferrari, 1995). Symbols indicate experimental data at 550 (*diamond*), 600 (*triangle*) and 640 nm (*square*). *Continuous lines* indicate the best fit of experimental data while *dashed lines* indicate the theoretical values computed as in Gordon and Ding (1992)

visible because of the stronger pure water absorption. In the blue and green spectral regions, the error increases with the concentration of absorbing particles and colored dissolved organic matter.

While a specific experimental investigation (see Fig. 18.9) confirmed the theoretical results by Gordon and Ding (1992), additional studies addressed the effects of asymmetries in radiometers shape, deployment methods and concurring bottom perturbations (Doyle and Voss, 2000; Piskozub et al., 2000; Leathers et al., 2001; Leathers et al., 2004).

Aside from these investigations, self-shading perturbations also triggered the need for smaller and smaller in-water radiometer systems (e.g., see Voss and Chapin, 2005; McClain et al., 2004).

18.7 Uncertainty Budgets

Optical radiometric data have direct application in the development and assessment of theoretical models describing the seawater light extinction processes (e.g., Bulgarelli et al., 2003; Chang et al., 2003) and of empirical algorithms linking the seawater apparent optical properties to the optically significant constituents expressed through their inherent optical properties or concentrations (e.g., O'Reilly et al., 1998; Maritorena et al., 2002; D'Alimonte and Zibordi, 2003; Darecki and Stramski, 2004; D'Alimonte et al., 2004; Lee et al., 2005). In addition, radiometric data are essential for the vicarious calibration of space sensors and the validation of remote sensing products (e.g., Mélin et al., 2005; Bailey and Werdell, 2006; Zibordi et al., 2006; Franz et al., 2007; Mélin et al., 2007; Bailey et al., 2008; Antoine et al., 2008b). The most accurate input data is always the most desirable for any bio-optical modeling and calibration or validation activity. However, accuracy requirements impact methodological and instrumental investment which should be weighed against the specific need for each application. Because of this, on the basis of target accuracies in derived products (i.e., 35% for *Chla* in open ocean), a maximum uncertainty of 5% has been specified for $L_{WN}(\lambda)$ determined from space in the blue spectral region over oligotrophic waters (Gordon, 1997). This requirement has been streamlined through the so called 1% radiometry concept (McClain et al., 2004): each major and independent uncertainty affecting the determination of in-situ $L_{WN}(\lambda)$ should be lower than 1–2% to ensure the overall uncertainty budget of $L_{WN}(\lambda)$ does not exceed the 5% threshold.

The major uncertainties affecting in-water subsurface optical data can be summarized as: (i) uncertainty from the absolute in-air calibration and uncertainty in the determination of the immersion factor; (ii) uncertainty in the correction factors applied for removing self-shading perturbations in the absence of any superstructure perturbation; (iii) uncertainty in the determination of corrections factors for removing the effects of the anisotropy of the seawater light field; (iv) uncertainty in the determination of $E_d(0^+,\lambda)$; (v) uncertainty in the determination of $E_0(\lambda)$ at the center-wavelength λ when ignoring the actual bandwidths; (vi) uncertainty in the extrapolation of sub-surface data; (vii) environmental variability resulting from the combination of wave induced perturbations with seawater variability and illumination changes. Values of these uncertainties are presented in Table 18.1 as estimated for $L_{WN}(\lambda)$ at 443, 555 and 665 nm from measurements performed in moderately sediment dominated waters with radiometers having 9 cm diameter and operated on an optical profiler with 6 Hz acquisition rate and 0.1 m/s deployment speed (after, but not exclusively, Zibordi et al., 2004a).

The quadrature sum of the major sources of uncertainty shows, with the exception of 665 nm, values close to the 5% target established for the absolute radiometric uncertainty of $L_{WN}(\lambda)$ for satellite applications.

Uncertainties affecting above-water radiometric measurements can be summarized as: (i) uncertainty resulting from the in-air absolute calibration; (ii) uncertainty in the correction factors applied to remove the effects of off-nadir viewing angle and anisotropy of the seawater light field; (iii) uncertainty in the determination of the diffuse atmospheric transmittance $t_d(\lambda)$ utilized to compute $E_0(\lambda)/E_d(0^+,\lambda)$; (iv)

Table 18.1 Uncertainties (in percent) for L_{WN} determined from in-water optical profiler data collected in costal waters

Uncertainty source	443	555	665
Absolute calibration	2.8	2.8	2.8
Self-shading corrections	0.9	0.6	2.5
Anisotropy corrections	0.4	0.9	0.5
$E_d(0^+,\lambda)$	2.8	2.8	2.8
$E_0(\lambda)$	1.9	0.1	0.2
Extrapolation	1.0	0.9	2.4
Environmental variability	1.1	1.3	2.8
Quadrature sum	4.7	4.4	6.0

Table 18.2 Uncertainties (in percent) for L_{WN} determined from above-water radiometric data collected in coastal waters

Source of uncertainty	443	555	670
Absolute calibration	2.7	2.7	2.7
Viewing angle & f/Q corrections	2.0	2.9	1.9
$t_d(\lambda)$	1.5	1.5	1.5
$\rho(\theta,\phi,\theta_0,W)$	1.5	0.7	2.5
Environmental variability	2.1	2.1	6.4
Quadrature sum	4.5	4.8	7.8

uncertainty in the determination of the sea surface reflectance $\rho(\theta,\phi,\theta_0,W)$ as a result of uncertainties in the wind speed and of the filtering applied to $L_T(\theta,\phi,\lambda)$ to minimize the wave effects; (v) environmental variability resulting from the combination of wave induced perturbations with changes in seawater optical properties and illumination conditions during measurements (this latter perturbation source is implicitly assumed to include uncertainties in the determination of $L_i(\theta',\phi,\lambda)$ and polarization effects of the sea surface). Estimates for the identified uncertainties are presented in Table 18.2 at 443, 555 and 670 nm (as determined, but not exclusively, by Zibordi et al., 2009b).

Similar to uncertainties determined for $L_{WN}(\lambda)$ from in-water data, the resulting uncertainties for $L_{WN}(\lambda)$ from above-water data are close to the 5% target except for the values at 670 nm which reach 8% mostly because of the effects of wave perturbations.

Tables 18.1 and 18.2 do not include uncertainties associated with sensitivity changes over time of sensors. These should be traced through pre- and post-deployment calibrations or, more comprehensively through the use of portable sources during field activities (Hooker and Aiken, 1998).

18.8 Application: The Validation of Primary Remote Sensing Products

The primary quantity of interest for ocean color remote sensing is $L_{WN}(\lambda)$ from which higher level products are derived. As a result of this, the assessment of satellite derived $L_{WN}(\lambda)$ using in-situ data is a fundamental requirement for any ocean color mission.

Most ocean color validation programs rely on the combination of field observations from many different and fully independent sources (Werdell et al., 2003). This solution, clearly driven by the need to produce large data sets of in-situ measurements representative of the various marine bio-optical regimes, is however likely affected by differences between the various field instruments utilized to perform measurements, the use of diverse sampling methods, the adoption of a variety of calibration sources and protocols, and the application of assorted processing schemes.

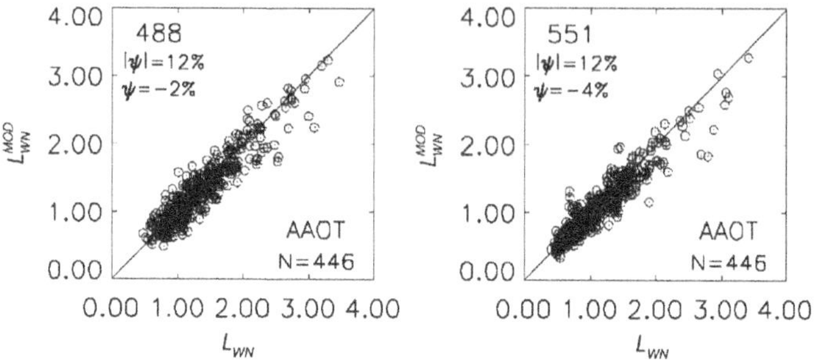

Fig. 18.10 Scatter plots of normalized water leaving radiances obtained at the AAOT site from MODIS top-of-atmosphere radiances (L_{WN}^{MOD}) and in-situ radiometric measurements (L_{WN}) at 488 and 551 nm. Symbols $|\psi|$ and ψ indicate uncertainties and biases expressed through the average of absolute differences and the average of differences, respectively. *N* indicates the number of match-ups

It is then expected that the effectiveness of ocean color applications increases using measurements from networks of standardized instruments operating at different sites representative of distinct water types, for which uncertainties have been thoroughly assessed and a major effort has been made to reduce systematic errors (see Zibordi et al., 2009b).

An example of validation exercise is shown in Fig. 18.10 for satellite ocean color products using in-situ above-water radiometric data from a coastal site. The analysis is presented through scatter plots of normalized water-leaving radiance determined from MODIS (L_{WN}^{MOD}) and in-situ (L_{WN}) measurements at 488 and 551 nm (the selected center-wavelengths are those frequently applied with empirical bio-optical algorithms). Comparison matchups were produced according to Zibordi et al. (2006) using pairs of in-situ and satellite data collected within 1 h of each other to minimize perturbations induced by the temporal variability of the sea and atmosphere. Spectral differences between remote sensing and in-situ data were minimized by applying band-shift corrections to in-situ $L_{WN}(\lambda)$ products (Zibordi et al., 2009b).

The assumption of equivalence of radiometric products determined at the very different in-situ and satellite spatial resolutions commonly applies to open sea regions. It can be however extended to moderately spatially inhomogeneous coastal waters when the validation exercise is supported by a number of match-ups capable of capturing the effects of random changes in seawater optical properties.

The results given in Fig. 18.10, indicating uncertainties slightly above 10% and negative biases of a few percent for both center-wavelengths, are in full agreement with independent investigations made for open ocean regions (Bailey and Werdell, 2006). Analysis like this, performed for various geographic regions characterized by different atmospheric and marine optical properties, are a required step to assess the accuracy of remote sensing products and eventually suggest revisions of the algorithms and methods applied to satellite data.

18.9 Closing Note

Since the 1920s marine optical radiometry progressed in conjunction with advances in measurement technology and absolute calibration methods. Fundamental steps in quantitative in-situ marine optical radiometry were the definition of the basic design of spectral radiometers in the 1950s and the availability of accurate spectral irradiance standards in the 1960s and 1970s. From the 1980s, the previous incremental progress was followed by exceptional developments driven by the major accuracy requirements set for in-situ optical radiometry supporting satellite missions for the global mapping of marine biomass.

This work has briefly summarized the status of in-situ marine optical radiometry through a synopsis of measurement means, calibration techniques, analysis methods, uncertainty estimates and products applications, as consolidated during 3 decades of satellite ocean color investigations. It is recognized that the space available for this overview has made impossible to further address such a complex and extended matter. It is however expected that the comprehensive list of references accompanying the work can help any interested reader to identify additional and specific sources of information.

Acknowledgments Acknowledgments are due to the European Commission (EC), the European Space Agency (ESA), the Office of Naval Research (ONR) and the National Aeronautics and Space Administration (NASA), which through EC Framework Programs, the ESA Envisat Program, the ONR Ocean Optics Program and the NASA Biogeochemistry Program, supported the recent research and developments in marine radiometry presented in this chapter. Finally, a sincere appreciation is expressed to all those scientists who put their efforts in progressing in marine optical radiometry.

References

Aas E (1969) On submarine irradiance measurements. Technical Report 6, Institute of Physical Oceanography, University of Copenhagen, Copenhagen, Denmark

Aas E, Højerslev NK (1999) Analysis of underwater radiance observations: apparent optical properties and analytic functions describing the angular radiance distribution. J Geophys Res 104:8015–8024

Aas E, Korsbø B (1997) Self-shading effect by radiance meters on upward radiance observed in coastal waters. Limnol Oceanogr 42:968–974

Antoine D, Guevel P, Desté JF, Bécu G, Louis F, Scott AJ, Bardey P (2008a) The "BOUSSOLE" Buoy – A new transparent-to-swell taut mooring dedicated to marine optics: design, tests, and performance at sea. J Atmos Oceanic Technol 25:968–989

Antoine D, D'Ortenzio F, Hooker SB, Bécu G, Gentili B, Tailliez D, Scott AJ (2008b) Assessment of uncertainty in the ocean reflectance determined by three satellite ocean color sensors (MERIS, SeaWiFS and MODIS-A) at an offshore site in the Mediterranean Sea (BOUSSOLE project). J Geophys Res 113:C07013, doi:10.1029/2007JC004472

Atkins WRG, Poole HH (1933) The photo-electric measurement of the penetration of light of various wavelengths into the sea and the physiological bearing of results. Phil Trans Roy Soc Lon (B) 222:129–164

Austin RW (1974) The remote sensing of spectral radiance from below the ocean surface. In: Jerlov NG, Nielsen ES (eds.) Optical aspects of oceanography, Academic Press, New York

Austin RW (1976) Air-water radiance calibration factor. Technical Memorandum ML – 76 – 004T, Scripps Institution of Oceanography, La Jolla, CA

Bailey SW, Hooker SB, Antoine D, Franz B A, Werdell PJ (2008) Sources and assumptions for the vicarious calibration of ocean color satellite observations. Appl Opt 47:2035–2045

Bailey SW, Werdell PJ (2006) A multi-sensor approach for the on-orbit validation of ocean color satellite data products. Remote Sen Environ 102:12–23

Berger F (1958) Uber die ursache des "oberfl¨acheneffekts" bei lichtmessungen unter wasser. Wetter u Leben 10:164–170

Berger F (1961) Uber den "taucheffekt" bei der lichtmessung ber and unter wasser. Arch Meteorol Wien (B) 11:224–240

Bulgarelli B, Zibordi G, Berthon JF (2003) Measured and modeled radiometric quantities in coastal waters: towards a closure. Appl Opt 42:5365–5381

Carder KL, Steward RG (1985) A remote sensing reflectance model of a red tide dinoflagellate off west florida. Limnol Oceanogr 30:286–298

Chang GC, Dickey TD, Mobley CD, Boss E, Pegau S (2003) Toward closure of upwelling radiance in coastal waters. Appl Opt 42:1574–1582

Clark DK, Feinholz ME, Yarbrough MA, Johnson BC, Brown SW, Kim YS, Barnes RA (2002) Overview of the radiometric calibration of MOBY. In: Barnes WL (ed.) Proceedings of SPIE, Earth Observing Systems VI, vol 4483, pp. 64–76

Clark DK, Gordon HR, Voss KJ, Ge Y, Broenkow W, Trees C (1997) Validation of atmospheric correction over the oceans. J Geophys Res 102:17209–17217

D'Alimonte D, Zibordi G (2003) Phytoplankton determination in an optically complex coastal region using a multi layer perceptron neural network. IEEE Trans Geosc Rem Sens 41: 2861–2868

D'Alimonte D, Zibordi G, Berthon JF (2004) Determination of CDOM and NPPM absorption coefficient spectra from coastal water remote sensing reflectance. IEEE Trans Geosc Rem Sens 42:1770–1777

Darecki M, Stramski D (2004) An evaluation of MODIS and SeaWiFS bio-optical algorithms in the Baltic Sea. Remote Sens Environ 89:326–350

Dera J, Sagan S, Stramski D (1993) Focusing of sunlight by the sea surface waves: new results from the Black Sea. Oceanologia 34:13–25

Dera J, Stramski D (1986) Maximum effects of sunlight focusing under a wind-disturbed sea surface. Oceanologia 23:15–42

Dera J, Wensierski W, Olszewski J (1972) A two-detector integrating system for optical measurements in the sea. Acta Gephysica Polonica 20:3–159

Deschamps PY, Fougnie B, Frouin R, Lecoomte P, Verwaerde C (2004) SIMBAD: a field radiometer for satellite ocean-color validation. Appl Opt 43:4055–4069

Doyle JP, Voss KJ (2000) 3D instrument self-shading effects on in-water multi-directional radiance measurements. Proceedings of the Ocean Optics XV, Monte Carlo, available from the Office of Naval Research, Arlington, VA

Doyle JP, Zibordi G (2002) Optical propagation within a 3-dimensional shadowed atmosphere-ocean field: application to large deployment structures. Appl Opt 41:4283–4306

Fougnie B, Frouin R, Lecomte P, Deschamps PY (1999) Reduction of skylight reflection effects in the above-water measurement of diffuse marine reflectance. Appl Opt 38: 3844–3856

Franz BA, Bailey SW, Werdell PJ, McClain CR (2007) Sensor-independent approach to the vicarious calibration of satellite ocean color radiometry. Appl Opt 46:5068–5082

Gershun A (1939) The light field. J Math Psychol 18:51–151 (translated by Moon P, Timoshenko G)

Gordon HR (1985) Ship perturbation of irradiance measurements at sea. Part 1: Monte Carlo simulations. Appl Opt 24:4172–4182

Gordon HR (1997) Atmospheric correction of ocean color imagery in the Earth observing system era. J Geophys Res 102:17081–17106

Gordon HR, Clark DK (1981) Clear water radiances for atmospheric correction of coastal zone color scanner imagery. Appl Opt 20:4175–4180

Gordon HR, Ding K (1992) Self-shading of in-water optical instruments. Limnol Oceanogr 37:491–500

Grum F, Becherer RJ (1979) Optical radiation measurements, Academic Press, New York

Helliwell WS, Sullivan GN, Macdonald B, Voss KJ (1990) Ship shadowing: model and data comparison. Proceedings of Ocean Optics X, SPIE vol 1302, pp. 55–71

Hooker SB, Aiken J (1998) Calibration evaluation and radiometric testing of field radiometers with the SeaWiFS quality monitor (SQM). J Atmos Oceanic Technol 15:995–1007

Hooker SB, Lazin G, Zibordi G, McClean S (2002a) An evaluation of above and in-water methods for determining water leaving radiances. J Atmos Oceanic Technol 19:486–515

Hooker SB, Maritorena S (2000) An evaluation of oceanographic radiometers and deployment methodologies. J Atmos Oceanic Technol 17:811–830

Hooker SB, McLean S, Sherman J, Small M, Lazin G, Zibordi G, Brown JW (2002b) The Seventh SeaWiFS Intercalibration Round-Robin Experiment (SIRREX-7), TM-2003-206892, vol 17, p. 69. NASA Goddard Space Flight Center, Greenbelt, MD

Hooker SB, Morel A (2003) Platform and environmental effects on above-water determinations of water-leaving radiances. J Atmos Oceanic Technol 20:187–205

Hooker SB, Zibordi G (2005a) Advanced methods for characterizing the immersion factor of irradiance sensors. J Atmos Oceanic Technol 22:757–770

Hooker SB, Zibordi G (2005b) Platform perturbation in above-water radiometry. Appl Opt 44: 553–567

Hooker SB, Zibordi G, Berthon JF, Brown JW (2004) Above-water radiometry in shallow coastal waters. Appl Opt 21:4254–4268

Jerlov NG (1951) Optical studies of ocean water. Report of the Swedish Deep-Sea Expedition, 3

Jerlov NG (1976) Marine Optics, vol 14 of Oceanography, Elsevier, Amsterdam

Jerlov NG, Fukuda M (1960) Radiance distribution in the upper layers of the sea. Tellus 12: 348–355

Jerlov NG, Liljequist G (1938) On the angular distribution of submarine deylight and the total sumbarine illumination. Sven Hydrogr Biol Komm Skr, Ny Ser Hydrogr 14:1–15

Johnston SF (2001) A history of light and color measurement: science in the shadow. Institute of Physics Publishing, Bristol

Kirk JTO (1994) Light & photosynthesis in aquatic ecosystems, Cambridge University Press, Cambridge

Kishino M, Ishizaka J, Saitoh S, Senga J, Utashima M (1997) Verification plan for ocean color and temperature scanner atmospheric correction and phytoplankton pigment by moored optical buoy system. J Geophys Res 102:1719–1727

Kundsen M (1922) On measurements of the penetration of light into the sea. Pub de Circ 76:1–15 (Cons Perm Internat Explor Mer)

Kuwahara VS, Chang G, Zheng X, Dickey T, Jiang S (2008) Optical moorings of opportunity for validation of ocean color satellites. J Oceanogr 64:691–703

Leathers RA, Downes TV, Mobley CD (2001) Self-shading correction for upwelling sea-surface radiance measurements made with buoyed instruments. Opt Express 8:561–570

Leathers RA, Downes TV, Mobley CD (2004) Self-shading correction for oceanographic upwelling radiometers. Opt Express 12:4709–4718

Lee ZP, Darecki M, Carder KL, Davis CO, Stramski D, Rhea WJ (2005) Diffuse attenuation coefficient of downwelling irradiance: evaluation of remote sensing methods. J Geophys Res 110:C02017, doi:10.1029/2004JC002573

Le Grand Y (1939) La pénétration de la lumière dans la mer. Ann Inst Ocèanogr 19:393–436

Lewis MR, Harrison WG, Oakey NS, Herbert D, Platt T (1986) Vertical nitrate fluxes in the oligotrophic ocean. Science 234:870–873

Maritorena S, Siegel DA, Peterson AR (2002) Optimization of a semianalytical ocean color model for global-scale applications. Appl Opt 41:2705–2714

McClain CR, Feldman GC, Hooker SB (2004) An overview of the SeaWiFS project and strategies for producing a climate research quality global ocean biooptical time-series. Deep Sea Res 51:5–42
Mélin F, Berthon J-F, Zibordi G (2005) Assessment of apparent and inherent optical properties derived from SeaWiFS with field data. Remote Sens Environ 97:540–553
Mélin F, Zibordi G, Berthon J-F (2007) Assessment of satellite ocean color products at a coastal site. Remote Sens Environ 110:192–215
Mobley CD (1994) Light and water. Radiative transfer in natural waters, Academic Press, San Diego, CA
Mobley CD (1999) Estimation of the remote sensing reflectance from above-water methods. Appl Opt 38:7442–7455
Morel A (1980) In-water and remote measurements of ocean color. Bound Layer Meteorol 18: 177–201
Morel A, Antoine D, Gentili B (2002) Bidirectional reflectance of oceanic waters: accounting for raman emission and varying particle scattering phase function. Appl Opt 41:6289–6306
Morel A, Gentili B (1996) Diffuse reflectance of ocean waters III: implication of bidirectionality for the remote-sensing problem. Appl Opt 35:4850–4862
Mueller JL (1995) Comparison of irradiance immersion coefficients for several marine environmental radiometers (MERs). In: Hooker S, Firestone E, Acker J (eds.) Case studies for SeaWiFS calibration and validation, TM-1995-104566, Part 3, vol 27, p. 46, NASA Goddard Space Flight Center, Greenbelt, MD
Mueller JL, Austin RW (1995) Ocean optics protocols for SeaWiFS validation, rev 1, TM-1995-104566, vol 25 of SeaWiFS Technical Report Series, p. 66, NASA Goddard Space Flight Center, Greenbelt, MD
O'Reilly JE, Maritorena S, Mitchell BG, Siegel DA, Carder KL, Garver SA, Kahru M, McClain CR (1998) Ocean color chlorophyll algorithms for SeaWiFS. J Geophys Res 103:24937–24953
Pettersson H, Landberg S (1934) Submarine daylight. Medd Oceanogr Inst Göteborg 6:1–13
Petzold TJ, Austin RW (1988) Chracterization of MER-1032. Tech. Memo. EN-001-88T, Visibility Laboratory of the Scripps Institution of Oceanography, University of California, San Diego, CA
Pinkerton MH, Aiken J (1999) Calibration and validation of remotely-sensed observations of ocean colour from a moored data buoy. J Atmos Oceanic Technol 16:915–923
Piskozub J (2004) Effect of ship shadow on in-water irradiance measurements. Oceanologia 46:103–112
Piskozub J, Weeks AR, Schwarz JN, Robinson IS (2000) Self-shading of upwelling irradiance for an instrument with sensors on a sidearm. Appl Opt 39:1872–1878
Robinson IS (2004) Measuring the oceans from space: the principles and methods of satellite oceanography, Springer, Berlin
Saruya Y, Oishi T, Kishino KKM, Jodai Y, Tanaka A (1996) Influence of ship shadow on underwater irradiance fields. Proceedings of the Ocean Optics XIII, SPIE, vol 2963
Schenck H (1957) On the focusing of sunlight by ocean waves. J Opt Soc Am 47:653–657
Shoulejkin W (1924) On the color of the sea. Phys Rev 23:744–751
Slater PN (1980) Remote sensing: optics and optical systems, Addison-Wesley Publishing Company, Reading, MA
Smith RC (1974) Structure of solar radiation in the upper layers of the sea. In: Jerlov NG, Steeman Nielson E (eds.) Optical aspects of oceanography, Academic Press, New York
Smith RC, Austin RW, Tyler JE (1969) An oceanographic radiance distribution camera system. Appl Opt 27:341–351
Smith RC, Baker KS (1984) The analysis of ocean optical data. Proc Ocean Opt VII, SPIE 478:119–126
Smith RC, Baker KS (1986) Analysis of ocean optical data II. Proc Ocean Opt VIII, SPIE 637: 5–107
Smith RC, Booth CR, Star JL (1984) Oceanographic biooptical profiling system. Appl Opt 23:2791–2797

Snyder RL, Dera J (1970) Wave-induced light-field fluctuations in the sea. J Opt Soc Am 60: 1072–1079

Spinrad RW, Carder KL, Perry MJ (1994) Ocean optics, Oxford University Press, Oxford

Steeman Nielsen E (1951) Conditions of light in the fjord. Medd Denmarks Fiskeri-og Havunders 5:21–27

Stramski D, Dera J (1988) On the mechanism for producing flashing light under a wind disturbed water surface. Oceanologia 25:5–21

Thuillier G, Herse M, Labs D, Foujols T, Peetermans W, Gillotay D, Simon PC, Mandel H (2003) The solar spectral irradiance from 200 to 2400 nm as measured by the SOLSPEC spectrometer from the ATLAS and EURECA missions. Solar Phys 214:1–22

Toole DA, Siegel DA, Menzies DW, Neumann MJ, Smith RC (2000) Remote-sensing reflectance determinations in the coastal ocean environment: impact of instrumental characteristics and environmental variability. Appl Opt 39:456–469

Tyler JE (1960) Radiance distribution as a function of depth in an underwater environment. Bull Scripps Inst Oceanogr 7:363–412

Tyler JE (1977) Light in the sea, Dowden, Hutchinson and Ross, Inc, Stroudsburg, PA

Tyler JE, Smith RC (1970) Measurements of spectral irradiance underwater, Gordon and Breach Science Publishers, New York

Voss KJ (1989) Use of the radiance distribution to measure the optical absorption coefficient in the ocean. Limn Oceanogr 34:1614–1622

Voss KJ, Chapin AL (2005) Upwelling radiance distribution camera system NURADS. Opt Express 13:4250–4262

Voss KJ, Morel A (2005) Bidirectional reflectance function for oceanic waters with varying chlorophyll concentrations: measurements versus predictions. Limn Oceanogr 50:698–705

Voss KJ, Nolten JW, Edwards GD (1986) Ship shadow effects on apparent optical properties. Proc Ocean Opt VIII, SPIE 637:186–190

Walker RE (1994) Marine light field statistics, John Wiley & Sons, Inc, New York

Waters KJ, Smith RC, Lewis MR (1990, November) Avoiding ship-induced light-field perturbation in the determination of oceanic optical properties. Oceanography 3:18–21

Weidemann A, Hollman R, Wilcox M, Linzell B (1990) Calculation of near surface attenuation coefficients: the influence of wave focusing. Proc Ocean Opt X, SPIE 1302:492–504

Weir CT, Siegel DA, Michaels AF, Menzies DW (1994) In situ evaluation of a ships shadow. Proc Ocean Opt XII, SPIE 2258:815–821

Werdell PJ, Bailey S, Fargion G, Pietras C, Knobelspiesse K, Feldman G, McClain CR (2003) Unique data repository facilitates ocean color satellite validation. Eos Tran 84:377–387

Westlake DF (1965) Some problems in the measurement of radiation under water: a review. Photochem Photobiol 4:849–868

Wyatt CL (1978) Radiometric calibration: theory and methods, Academic Press, New York

Zaneveld JRV, Boss E, Barnard A (2001) Influence of surface waves on measured and modeled irradiance profiles. Appl Opt 40:442–449

Zibordi G (2006) Immersion factor of in-water radiance sensors: assessment for a class of radiometers. J Atmos Oceanic Technol 23:302–313

Zibordi G, Berthon J-F, D'Alimonte D (2009a) An evaluation of radiometric products from fixed-depth and continuous in-water profile data from moderately complex coastal waters. J Atmos Oceanic Technol 26:91–186

Zibordi G, Bulgarelli B (2008) Effects of cosine error in irradiance measurements from field ocean color radiometers. Appl Opt 46:5529–5538

Zibordi G, D'Alimonte D, Berthon JF (2004a) An evaluation of depth resolution requirements for optical profiling in coastal waters. J Atmos Oceanic Technol 21:1059–1073

Zibordi G, Darecki M (2006) Immersion factor for the RAMSES series of hyper-spectral underwater radiometers. J Opt A also Appl Opt 8:252–258

Zibordi G, Doyle GP, Hooker SB (1999) Offshore tower shading effects on in-water optical measurements. J Atmos Oceanic Technol 16:1767–1779

Zibordi G, Ferrari G (1995) Instrument self-shading in underwater optical measurements: experimental data. Appl Opt 34:2750–2754

Zibordi G, Holben B, Slutsker I, Giles D, D'Alimonte D, Mélin F, Berthon J-F, Vandemark D, Feng H, Schuster G, Fabbri B, Kaitala S, Seppälä J (2009b) AERONET-OC: a network for the validation of ocean color primary radiometric products. J Atmos Oceanic Technol 26: 1634–1651

Zibordi G, Hooker SB, Berthon J-F, D'Alimonte D (2002) Autonomous above-water radiance measurement from an offshore platform: a field assessment experiment. J Atmos Oceanic Technol 19:808–819

Zibordi G, Hooker SB, Mueller JL, McLean S, Lazin G (2004b) Characterization of the immersion factor for a series of in water optical radiometers. J Atmos Oceanic Technol 21:501–514

Zibordi G, Mélin F, Berthon J-F (2006) Comparison of SeaWiFS, MODIS, and MERIS radiometric products at a coastal site. Geophys Res Lett 33:L06617, doi:10.1029/2006GL025778

Zibordi G, Mélin F, Hooker SB, D'Alimonte D, Holben B (2004c) An autonomous above-water system for the validation of ocean color radiance data. IEEE Trans Geosci Remote Sens 42:401–415

Chapter 19
Forecasting the Coastal Optical Properties Using Satellite Ocean Color

Robert Arnone, Brandon Casey, Sherwin Ladner, and Dong-Shang Ko

19.1 Introduction

Algorithms for ocean color bio-optical properties from space have advanced significantly in the last 20 years. Improved alogorithms have advanced beyond chlorophyll to characterize coastal optical properties such as absorption from phytoplankton, colored dissolved organic matter and detritus, in addition to backscattering from the particle distribution. These properties have provided new insights into the changing conditions along our coast. However, for many coastal management applications these satellite derived properties are insufficient to make real-time decisons (Arnone and Parsons, 2004). Daily satellite ocean color imagery represents a nowcast of bio-optical conditions. Although these near real-time bio-optical products can be made available within hours of a satellite over pass, they may be inadequate for operations. In the coastal waters, changes are occuring within hours as a result of the tidal fluxuations, river discharges, precipation, and local wind events so that the nowcast of the bio-optical properties may not be representative of local conditions within the 24 h period. Real-time coastal decisons on processes such as:

- dissipation of a coastal plume,
- movement of a Harmful Algal Bloom,
- river plume dispersion,
- turbidity frontal movement,
- chlorophyll bloom dispersion,
- larval fish migration,

all may require hourly forecast of bio-optical properties on a daily basis.

A sensible forecast of bio-optical properties along coastal waters requires an initialization field. This field can be best represented by the nowcast from ocean color bio-optical properties. Forecast and prediction of these properties can be defined

R. Arnone (✉)
Oceanography Division, Naval Research Laboratory, Stennis Space Center, MS 39529, USA
e-mail: arnone@nrlssc.navy.mil

V. Barale et al. (eds.), *Oceanography from Space*,
DOI 10.1007/978-90-481-8681-5_19,

by coupling these properties with a forecast circulation model. Circulation models have advanced significantly in the last 10 years and are approaching the confidence of weather forecasting. Circulation models are driven by winds and boundary conditions and are data assimilative of sea surface temperatures and sea surface height.

A major forcing in coastal enviroments is the result of the physical processes such as tides and winds and river discharge (Arnone et al., 2007). These physical processes change on scales of hours and results in advection of water masses. Although bio-optical processes are different than physical processes, we argue that bio-optical time scales occur on longer (on order of several days) compared to hourly time scales of the physical forcing (Stramska et al., 1995). The coupling of the bio-optical properties with the physical circulation models should provide a capability to forecast bio-optical properties on short times scales (days), where as at longer times scales (several days to weeks) the bio-optical processes may be decoupled from physical processes. For these longer time scale forecasting, other more complex bio-optical models may be required (Jolliff et al., 2008).

Satellite ocean color products are available from several satellites such that daily imagery is available along most coastlines. The initialization field of the coastal environment can be reinitialized daily using updated satellite bio-optical products such that a 24 h forecast should be possible for coastal decisons.

Our objective is to demonstrate the ability to derive bio-optical products along coastal waters on short time scales based on coupling ocean color bio-optical products with forecast ocean circulation models. We evaluate the bio-optical forecast using satellite imagery to determine an effective forecast probability.

19.2 Satellite Ocean Color and Circulation Models

Ocean color imagery from MODIS-Aqua was used to determine the bio-optical properties along the coast of northern Gulf of Mexico at the Mississippi River delta, USA for a month period in October, 2009. The Quasi Analytical Algorithm (QAA) (Lee et al., 2002; Martinolich, 2006) was used to process the imagery into backscattering (551) and absorption (443) and chlorophyll products. We used the 1 km products in our example, although new algorithms have been developed to determine these properties at 250 m (Ladner et al., 2007) which show improved capability for coastal management requiring the high resolution. The backscattering coefficient has been used to estimate an effective particle concentration, if we assume a specific size and composition and this property can be treated as a water mass tracer only affected by particle settling. The total absorption properties are influenced by phytoplankton, detritus and colored dissolved organic matter. The absorption properties are less of a conservative tracer since they are influenced by biological processes such as growth, decay photo-oxidation, etc. However, on these short time scales, we argue they also are minimal since diurnal changes in phytoplankton concentration can be small. This is especially appropriate in the coast

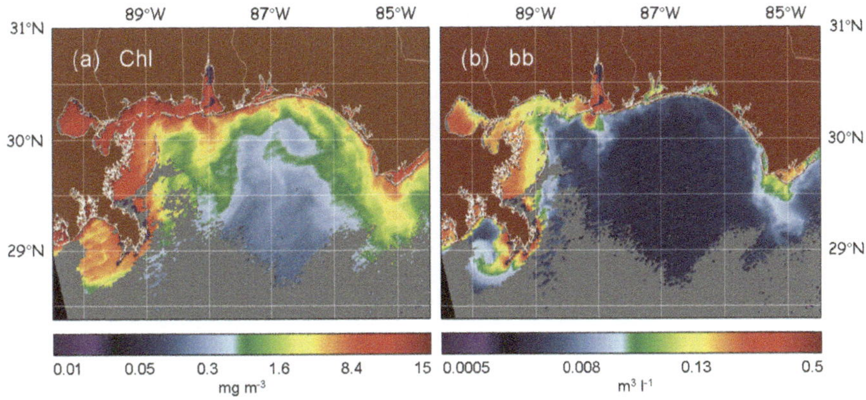

Fig. 19.1 MODIS derived coastal bio-optical properties of (**a**) chlorophyll [mg/m^3] and (**b**) backscattering (551) [m^3/l] from QAA, for 19 October 2009, along the Mississippi Delta USA. The clouds in the *lower part* of the images appear in *gray*

such as the Mississippi delta region which represents different bio-optical processes which include high sediment discharge from the river mouth, sediment resuspension along the coastal waters and offshore biological activity. We can examine the role of conservative and non-conservative processes within a 24 h forecast. Examples of the chlorophyll and backscattering a (551) are shown in Fig. 19.1 (panels a and b) for October 19, 2009.

The Navy Coastal Ocean Model (NCOM) was used to define the forecast circulation of the region. NCOM is a 41 layer model used for Navy operations. The model is forced with COAMPS winds and obtains boundary conditions from the "Global NCOM". This model assimilates both sea surface height fields from satellite altimeters and sea surface temperature from AVHRR (Martin, 2000). The model is run daily and reinitialized with updated winds fields and updated Sea Surface Height fields from satellite altimeters and Sea Surface Temperature. NCOM includes river discharge from 40 rivers entering the Gulf of Mexico. The nowcast and forecast out to 48 h is available from NCOM on a 3 h basis. Regional NCOM for the Intra-Americas Sea is the relocatable version of the NCOM. The horizontal current field (u, v) is required as an input to the bio-optical forecast. The expected format for these data files is NetCDF with one time step per file and u and v in separate files. The modeled current field for each time step is required to perform the advection.

If model fields are provided at a lower temporal resolution than the forecast, then an interpolation will be performed between each model time step, to produce the desired time steps. In the example shown in Fig. 19.2 (panel a), NCOM is at a resolution of 4 km and interpolated to 1 km, in order to match the satellite grid. NCOM has been shown to accurately represent the tides and circulation along the coast and will not be described in detail here (Ko et al., 2003).

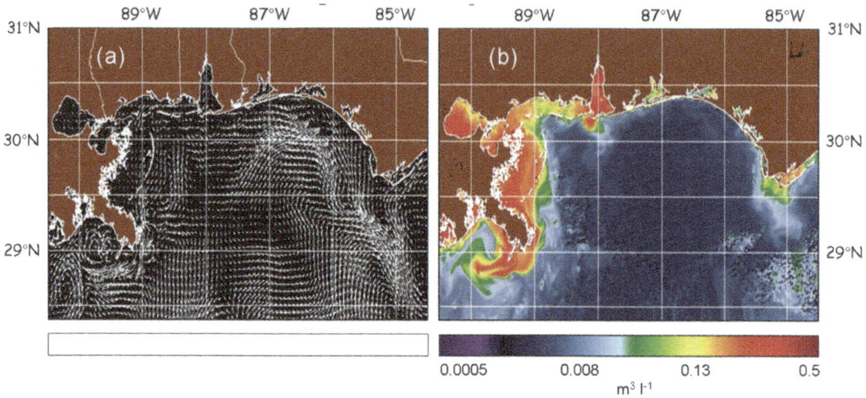

Fig. 19.2 *Panel* (**a**): NCOM surface currents of 19 October 2009. Currents are interpolated to the resolution of the imagery. *Panel* (**b**): initialization field of 19 October 2009 for backscattering coefficient. Notice the removal of the clouds using the steps outlined in the text

19.2.1 Bio-optical Initialization Field: "Nowcast"

The initialization bio-optical field is a critical step in the forecast. A bio-optical forecast requires complete spatial coverage in the initialization daily field. This is needed for several reasons: (1) to prevent gaps or No Data (ND) from creating artificial local boundary conditions in the advective forecast and (2) to prevent artificial bio-optical boundaries between ND and known data. The initialization field is constructed daily using all known data and the previous forecast field to provide a sensible bio-optical forecast. A realistic initialization field prevents abrupt data recreation and boundaries from artistically propagating into the forecast.

As seen in Fig. 19.1, a major limitation of the ocean color imagery is that complete coverage of bio-optical properties is not available daily. There are many times when limited coverage results from either cloud cover, algorithms failure (atmospheric, high turbidity, saturated radiance, etc.), no satellite coverage. These limitations require a procedure to "fill in the gaps". The initialization field is based on a series of additive steps to establish the "most recent" bio-optical coverage used to set the forecast.

The additive pixel procedure is used to determine the complete spatial distribution of coastal bio-optical properties. The hierarchy and details of the procedure are discussed in detail in Casey and Arnone (2007). The "gap filler" procedure is summarized in Steps 1–4:

1. today's bio-optical properties (best and most accurate);
2. spatial convolution of bio optical data in step 1 out several pixels (fills small holes, speckle, etc.);
3. triangular interpolation of data in Step 2 to fill in gaps;
4. yesterday's 24 h bio-optical forecast.

Depending on the area and cloud coverage, the bio-optical coverage takes about 1–2 weeks before a coherent bio-optical field initialization and forecast field can be established. Additionally, as increased satellite bio-optical data enters the initialization process, the bio-optical forecast improves, and it is then used in the next day's initialization field. This forecast "spin-up" time has been used in a variety of "cloudy" coastal regions with partial ocean coverage. A similar procedure to generate the initialization field is used in weather forecasting. The procedure has the advantage that for each day the "best" and most recent data enters the forecast. However, the entire bio-optical properties along a coast line can change from 1 day to the next, for example, if a "cloud free" scene enters the initialization field when the previous days initialization was based only on a forecast. Because the initialization is performed daily, reinitialization to observation conditions is rapid and the forecast improves. An example of the initialization field of backscattering is shown in Fig. 19.2 (panel b).

19.3 Forecasting Using Eulerian Advection

The forecasting of bio-optical properties is performed by applying a simple advection approach to satellite derived bio-optical products using the NCOM forecast circulation model in order to forecast the surface optical properties. As was described earlier, the theoretical basis for this, assumes bio-optical properties are controlled solely by the physical circulation within a 24 h cycle. The bio-optical processes such as phytoplankton growth and decay and CDOM production and oxidation are not considered and are remissive. Previous efforts used a Lagrangian advection approach; however, this process was shown to produce significant errors along coastal boundaries in addition to the high computing requirements. For these reasons we switched to an eulerian approach (Arnone et al., 2006).

The NCOM model and the satellite derived bio-optical initialization field pixel grid are established based on the grid resolution of the image (in this case 1 km). The vertical fluxes into and out of the grid cells of the bio-optical properties are estimated from the horizontal fields. We apply a "thin-layer approximation" to extend surface 2D advection to 3 dimensions. The advection is performed on a surface layer, for convenience, at 1 m thickness, which goes up and down with free sea surface such that the vertical velocity is 0 at the surface. The vertical velocity at the base of the layer can be determined from divergence/convergence of horizontal currents following the volume conservation. The vertical flux is estimated assuming a uniform concentration of field at vertical. Forward time stepping with first-order upwind advection is applied to the vertical advection. For the horizontal advection of the satellite field, a third-order upwind advection scheme with flow adjustment is applied to reduce diffusion and to prevent "a numerical overshoot".

So, e.g., as the surface layer bio-optical properties move offshore and diverge from coastal boundaries, the vertical flux replenishes the bio-optical concentration from a vertical upwelling (flux) of the subsurface bio-optical property. Similarly, the vertical flux of bio-optical concentration can account for downwelling flux into

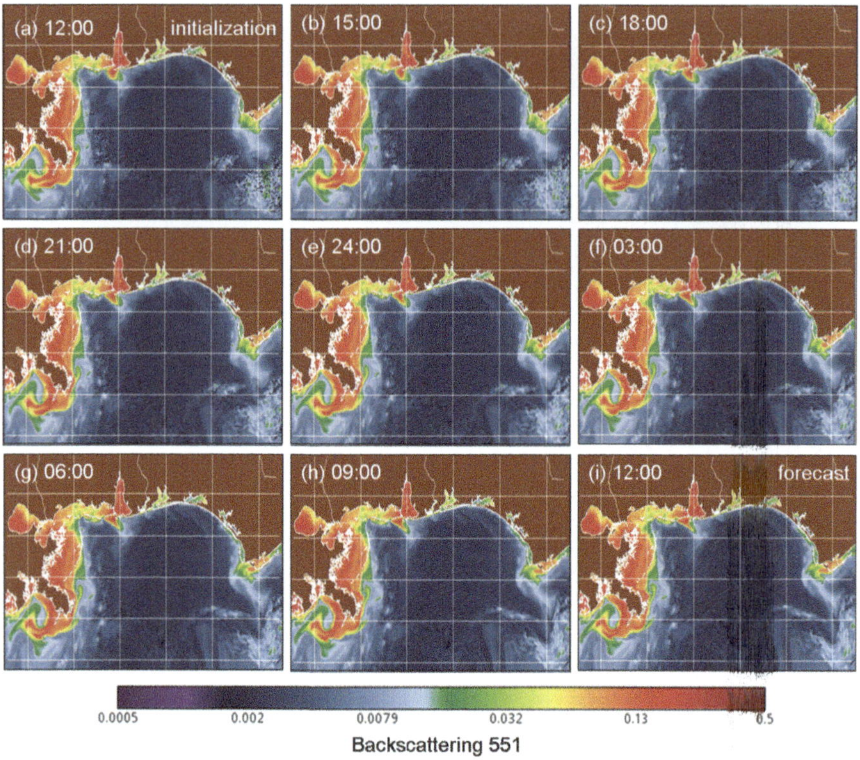

Fig. 19.3 Forecast of the surface backscattering coefficient for 19 October 2009, at 3 h intervals for 24 h. Time of the MODIS overpass was ~12:00 local. The various fileds show (**a**) 12:00 local (initialization), (**b**) 15:00 local, (**c**) 18:00 local, (**d**) 21:00 local; next day, (**e**) 24:00 local, (**f**) 03:00 local, (**g**) 06:00 local, (**h**) 09:00 local; (**i**) 12:00 local (24 h forecast)

the subsurface layers. However, these processes are not associated with bio-optical processes but are simply a conservative tracer of the bio-optical concentration physical flux.

The initialization field of backscattering (551) for October 19, 2009, was advected on an hourly basis and the surface bio-optical forecast is shown in Fig. 19.3 (panels a–j). This sequence represents the forecast of the surface bio-optical properties for every 3 h out to 24 h; i.e. next day. The initial field is created for approximately 11 AM local at the time of MODIS Aqua overpass and forecast out to 11 AM the next day.

19.4 Validation of the Bio-optical Forecast

The validation and estimate of the uncertainty in the bio-optical forecast can be quickly assessed by a pixel-by-pixel comparison of the 24 h forecast with the next day satellite bio-optical product. This is assuming that the next day bio-optical

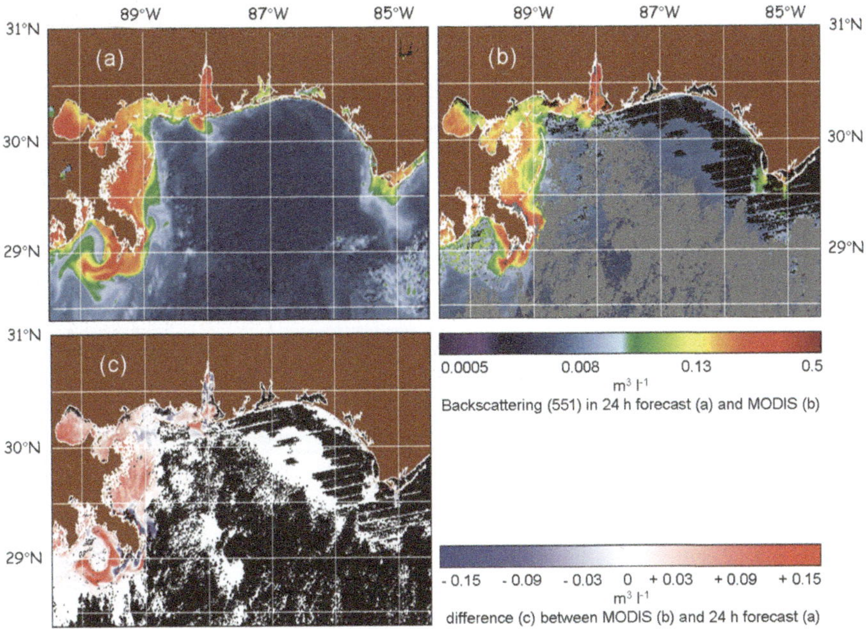

Fig. 19.4 Comparison and validation of forecast backscattering (551) for Oct 20, 2009: (**a**) 24 h forecast; (**b**) MODIS 20 October 2009, including clouds and atmospheric failures; (**c**) difference between MODIS and 24 h forecast (**b–a**), for *valid* retrievals only. Note the color coding in (**c**): *white* represents zero difference, *red* overestimated values, *blue* underestimated values. *Black* pixels are ND or clouds

property of interest can be retrieved (i.e. cloud free field). Satellite retrieved bio-optical properties are assumed valid. We assess the forecast accuracy based on a match up with these values. A difference comparison is computed only at locations where satellite retrieved properties occur (i.e. we do not use "gap filled"-derived bio-optical properties, described earlier in the validation procedure).

The difference between the backscattering forecast and next day's image indicates both over-estimating and under-estimating of the backscattering coefficient. We call this difference the "forecast error". In the example shown in Fig. 19.4, we notice that the Mississipi river plume was forecast to be advected to the west more so than what was observed in the next day's image. The difference image (Fig. 19.4c) shows higher backscattering (color coded red) than observed to the west of the river plume.

The satellite images of bio-optical properties provide an input into both the "nowcast" or initialization field and the validation field. Notice that the forecast error, (i.e. over or under estimated values) are spatially distributed and that certain regions have greater differences than others. These differences, however, are for an individual day's forecast and evaluation of a forecast is typically based on the forecast performance over a longer time sequence. As in weather predictions, a statistical approach to forecasting is used for an evaluation.

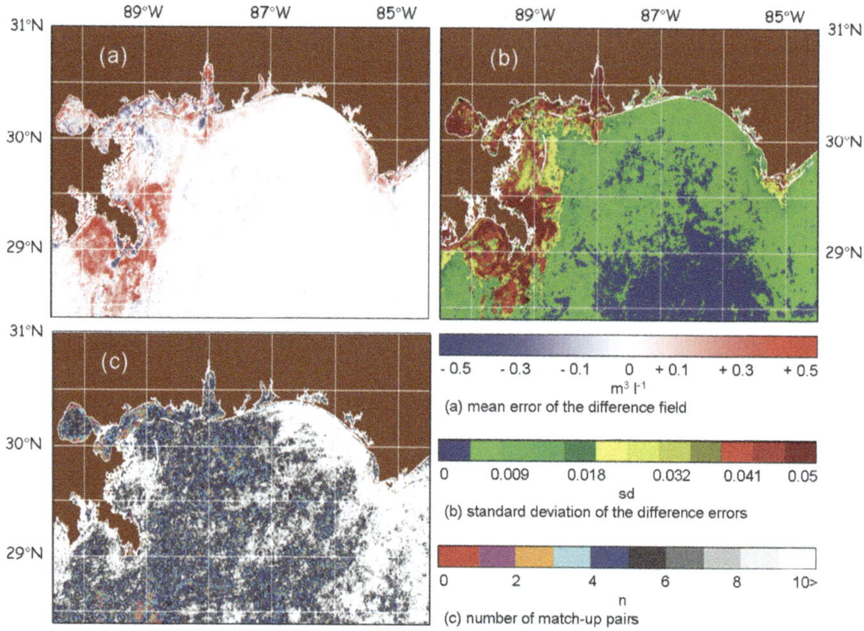

Fig. 19.5 Backscattering forecast error for month of October 2009 (~30 day): (**a**) mean error of the difference field; (**b**) standard deviation of the difference errors, (**c**) number of match-up pairs where the differences were computed (1–10+ days) in the month

The validation procedure described for computing the "forecast error" – which includes: (a) developing a daily initialization field, (b) 24 h forecast of the backscattering coefficient and (c) the comparison with the next day's image – was run for a 31 day sequence from October 1 to October 31, 2009. As expected, the daily forecast error was seen to change spatially. Certain regions showed more skill in bio-optical forecast than others. To assess this regional forecast skill, we determined the mean error of the difference field for the backscattering coefficient between the forecast and the next day's image (Fig. 19.5). The mean error field for this monthly period identifies the locations of higher and lower uncertainty of the forecast (Fig. 19.5a). White areas represent zero difference, red overesimated error and blue underestimated error. Note the overestimates in the location of the Mississipi river plume. There are regions where both over- and under-estimates of the forecast error occur on a monthly basis. However these errors are not clearly represented just by the monthly mean difference – i.e. positive and negative means can result in zero mean (white area).

To address these forecast errors, one must consider the standard deviation (sd) of the difference field (Fig. 19.5b). The standard deviation of this field illustrates the error about the mean and is perhaps a better way to represent forecast error. Along dynamic frontal locations, where both over- and under-estimates of the backscattering occur during the month, in response to the errors due to frontal movements, e.g., there will be a high standard deviation compared to the low mean error. Significant standard deviation errors (darker red) are observed in coastal areas associated with strong tidal regimes and the dissipation of the Mississippi River plume.

However, the total number of points used to determine the mean and standard deviation is critical to determine statistical validity. The number of matchup's pairs used to compute the mean and standard deviation for the month is represented in Fig. 19.5c at each location. This number represents the number of times during the month that a difference between the forecast field and a satellite retrieved backscattering product was computed. The greater number of matchup pairs, i.e. >10+ (color coded in white), represents valid statistical relationships,compared to the low numbers such as 4 and 5 (color coded in light and dark blue).

Notice that in areas where there are high number of data pairs (white in Fig. 19.5c), the monthly mean forecast errors are low (white in Fig. 19.5a) especially in the western region. The Mississippi River plume statistics are mixed. Representative errors occur where there are high numbers (white in Fig. 19.5c) in addition to a high mean error (red or blue in Fig. 19.5a). In the areas where there are lower number of match ups (light or dark blue in Fig. 19.5c) and the forecast error is high (red in Fig. 19.5a) the forecast errors are not representative. At these locations where the numbers are low, the statistical forecast error is unreliable.

In order to determine how the forecast of the backscattering compares with monthly "climatology", we examined the monthly mean and standard deviation of the backscattering coefficient for October 2009, which was computed based only on satellite derived backscattering (Fig. 19.6). As expected, the mean backscattering distributions do not show the small scale plumes and eddies along the coast as observed in October 19, and in the optical forecast. The monthly mean distribution is much different from the individual day's imagery and the forecast. The monthly standard deviation of backscattering (Fig. 19.6b) represents substantial changes in coastal backscattering which we believe is primarily resulting from the monthly

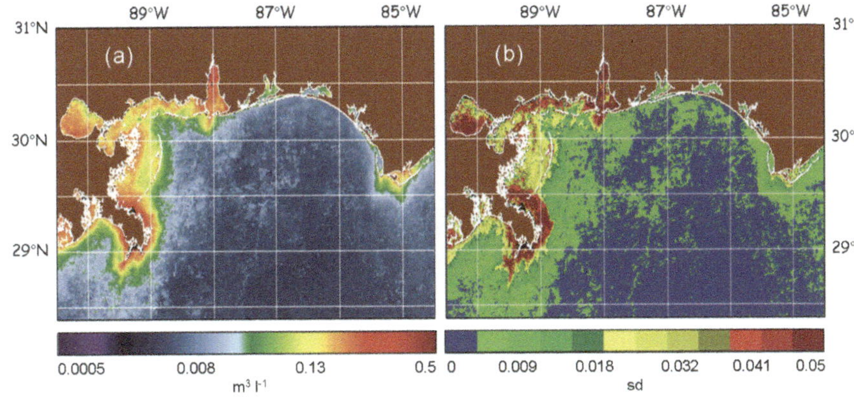

Fig 19.6 Monthly October 2009 backscattering (551) coefficient derived from MODIS –Aqua: (**a**) monthly mean and (**b**) standard deviation from the mean

variability in the physical processes and secondarily to the monthly bio-optical processes. Notice that regions with the highest standard deviation of backscattering (Fig. 19.6b) are locations associated with physical forcing in strong tidal regimes and coastal plumes.

We define the 24 h "persistence error" of the backscattering product as the difference within a 24 h change in backscattering. High changes indicate high persistence error, or lack of persistence. A lower error suggests that small changes occur and a similar backscattering coefficient can be used from 1 day to the next to represent the forecast. The persistence error is computed by differencing locations using two sequential (within 24 h) satellite backscattering fields.

The difference is computed daily and then averaged over the month. The persistence error represents change resulting from both physical and bio-optical processes, since it is based only on observed satellite retrievals of backscattering products and not on the physical circulation model. We computed this monthly persistence error field for October 2009, as mean and standard deviation (Fig. 19.7a, b).

To evaluate the representativeness of the monthly "persistence errors" of the mean and standard deviation, the number of match up samples (or satellite observed pairs) for each grid point is shown in Fig. 19.7c. This is the number of data points

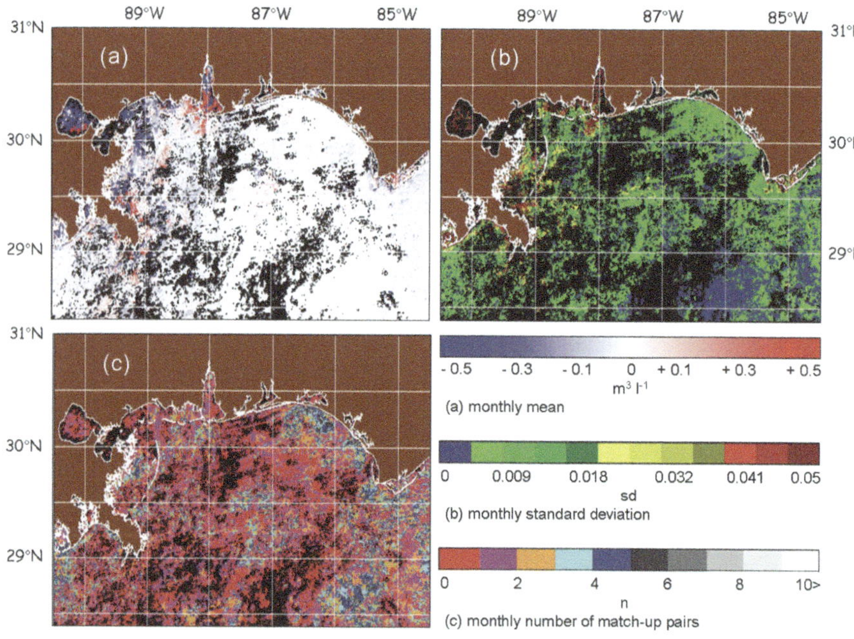

Fig. 19.7 The backscattering coefficient "persistence error" was computed as difference between two satellite backscattering retrievals within a 24 h period and statistics assembled for October 2009: (**a**) monthly mean (*black areas* indicate no pair of 24 h difference were observed); (**b**) monthly standard deviation; and (**c**) number of observational pairs used to generate the monthly statistics

used to compute the mean and requires two satellite products within 24 h. The large amount of black, red and pick (0,1,2) areas in Fig. 19.7c, indicate that there were few samples to compute the mean for the 31 day period and the mean and standard deviation persistence errors not representative. Therefore, the low persistence error observed (Fig. 19.7a) is misleading since there were very few points. A longer time period or more sequential observations is required to define the daily "persistence error"; although these statistics are difficult to obtain.

19.4.1 Persistence and the Forecast Error

The comparison between the persistence (Fig. 19.7) and forecast (Fig. 19.5) errors suggests persistence is better than forecast. However, several issues should be considered in these results. The number of samples used to compute persistence error (Fig. 19.7c) is small compared to those used for forecast error (Fig. 19.5c) because the calculation of the persistence error is dependent on cloud free "observations" from "two" sequential days, whereas forecast error is dependent on cloud free "observation" in one image. Although the persistence error has a lower mean and lower standard deviation than the forecast error, the small sample size in the persistence errors is not statistically valid and requires more data points.

Additionally, the forecast error was computed based on the initialization field which can include gap filled observations that were based on a previous forecast. For multiple cloudy or no observation days, the initialization field and the forecast would be based on old data that is greater than 24 h. This essentially assumes an older than 24 h observation used in the initialization field and the forecast error represents the error in many cases much greater than 24 h. Therefore, the forecast errors would be based on the age of the initialization field. This accounts for the higher error for the 24 h forecast (Fig. 19.5a). The advantage using the forecast error is that this statistics error is computed based on a much larger region and has a greater statistical number compared to the persistence error.

19.5 Uncertainty in the Forecast

We argued that for short time scales, the physical processes are responsible for controlling the distribution of surface bio-optical properties. Based on this, can we assess the uncertainty in the bio-optical forecast and determine where the error can occur? The first and perhaps largest uncertainty is from the physical circulation model. Although this model has been shown to represent the surface ocean conditions accurately, there is some temporal and spatial uncertainty of these processes which we did not represent. As discussed previously, the forecast physical models have several methods to assess their uncertainty which were not addressed here. A physical model can be set up to run with different initialization conditions which consider (1) grid resolution of the wind forcing fields (2) grid resolution of the bathymetry (3) model grid spacing (4) vertical layer spacing (5) boundary conditions etc. The physical model can run a set of ensemble (typically ~40) to determine the uncertainty in the currents forecast field (Rixen et al., 2009). Using the current uncertainty, we can estimate the optical forecast for each of the ensemble currents and define the mean and spread of the optical forecast. We did not do this in these examples and assume the model currents were valid and correct. This is one source of optical forecast uncertainty that was not taken into account.

Satellite retrievals of the bio-optical properties (backscattering coefficient or chlorophyll) have an uncertainty based on the uncertainty of the algorithms. We implemented a set of uniform optical relationships that are used in the QAA algorithms which are considered standard. However, these relationships can change with location and bio-optical processes. The QAA algorithm has been shown to have some degree of uncertainty, (Lee et al., 2010) however these relationships should work well in the Gulf of Mexico. The uncertainty from these algorithms can influence the initialization and forecast. We assume the uncertainty of the algorithms is similar from 1 day's satellite image to another. Lastly, satellite retrievals require processing for atmospheric correction in addition to the in-water algorithms. The uncertainty in the aerosol models used for atmospheric correction is spatially and temporally changing, especially in coastal areas where aerosol optical depths are variable. We did not include this source of uncertainty of the atmospheric correction in the MODIS backscattering retrievals that were used in the forecast.

19.6 Conclusion

The retrievals of bio-optical properties from ocean color satellite have made significant advances in defining a "nowcast" of coastal conditions. New capability is required for coastal operations, coastal managers and researchers to forecast these properties on time scales of hours to weeks. The coastal environment changes on scales of hours, mostly as a result of the physical forcing associated with tidal, discharge and currents. We coupled daily surface ocean color properties of the backscattering coefficient at 551 nm and chlorophyll concentration to a physical circulation model and advected the field to determine an hourly forecast of the satellite derived properties.

The forecast of bio-optical properties is reinitialized daily as new satellite observations enter the forecast. The methods to construct a gap filled initialization field and integrate it into a coastal bio-optical forecast system required approximately 1–2 weeks spin-up time, and is based on availability of cloud free observations. We illustrate a 1 month daily forecast of the surface particle backscattering properties for October 2009.

A daily 24 h forecast of the backscattering coefficient was evaluated based on comparison with the "next day's" derived product. We conducted the validation and uncertainty of daily bio-optical forecast for a 1 month period to estimate spatial statistical relationships of the forecast uncertainty. The October 2009 statistics (mean and standard deviation) of the backscattering coefficient forecast were characterized by the "forecast error", and "persistence error".

The forecast error showed higher errors in locations where there are strong currents and significant changes in the dynamics. We attributed these to uncertainty in the physical processes of the model forecast. The statistics of the persistence error (daily changes in satellite bio-optical properties), although appearing low, are not representative of the changing bio-optical conditions since the number of satellite observations used to generate the persistence statistics is not representative. We presented the forecast errors and persistence errors as absolute values. Perhaps a better way to describe the error would be using a percent difference in the forecast error. This may be more useful for managers using the forecast to look for maximum changes.

Coastal managers and researchers require new methods similar to weather forecasting to assess the coastal ecological conditions on time scale of hours. We demonstrated a new capability for using ocean color remote sensing to provide both the initialization and validation for bio-optical forecasting. On short time scales the physical processes are shown to be representative of the distributions of bio-optical properties. Future capabilities in ecological forecasting will rely on some degree of physical forcing in addition to bio-optical processes. We have shown an initial capability for coastal optical forecasting. However, improved methods to characterize the forecast uncertainty which are used in weather forecasting can be applied to bio-optical forecasting in the immediate future.

Acknowledgements This research was supported by the Naval Research Laboratory 6.1 "BIOSPACE "program and NASA REASoN program "Integrated analyses of ocean products in the Gulf of Mexico". Thanks to reviewers for useful suggestions and comments.

References

Arnone RA, Casey B, Ko DS, Flynn P, Carollo L, Ladner S (2007) Forecasting coastal optical properties using Ocean Color and coastal circulation models. SPIE Optics and Photonics, Conference 6680-43, San Diego, CA

Arnone RA, Flynn P, Ko DS, Martinolich P, Gould R, Haltrin V (2006) Forecasting optical properties from satellite derived optical properties. Proceedings of the Ocean Optics XVIII, Montreal, Canada

Arnone RA, Parsons AR (2004) Real-time use of ocean color remote sensing for coastal monitoring. In: Miller RL, Del Castillo CE, McKee BA (eds.) Remote sensing of the coastal environments, Chapter 14, Springer Publishing, Kluwer Academic, Dordrecht, pp. 317–337

Casey B, Arnone RA (2007) Simple and efficient technique for spatial and temporal composite imagery. SPIE Optics and Photonics, Conference 6680-43, San Diego, CA

Jolliff JK, Kindle J, Shulman I, Penta B, Friedrichs M, Helber R, Arnone RA (2008) Summary diagrams for coupled hydrodynamic – ecosystem model skill assessment. J Mar Sys 76(1–2):64–82

Ko DS, Preller RH, Martin PJ (2003) An Experimental Real time Intra Americas Sea Ocean Nowcast/Forecast System for Coastal Prediction. AMS 5th Conference on Coastal Atmospheric & Oceanic Prediction & Processes, New York

Ladner SD, Sandidge JC, Lyon PE, Arnone RA, Gould RW, Lee ZP, Martinolich PM (2007) Development of finer spatial resolution optical properties from MODIS. Proceedings, SPIE Optics and Photonics Meeting, San Diego, CA

Lee Z, Arnone R, Hu C,Werdell PJ, Lubac B (2010) Uncertainty of optical parameters and their propagations in an analytical ocean color inversion algorithm. Appl Opt 49(3):369–381

Lee Z, Carder KL, Arnone R (2002) Deriving inherent optical properties from water color: a multiband quasi-analytical algorithm for optically deep waters. Appl Opt 41(27):5755–5772

Martin PJ (2000) A description of the Navy Coastal Ocean Model, Version 1.0. NRL Report: NRL/FR/7322-009962. p. 39

Martinolich PM (2006) The Automated Satellite Data Processing System. NRL website, http://www7333.nrlssc.navy.mil/docs/aps_v4.0/user/aps/ version 4.0

Rixen M, Book JW, Carta A, Grandi V, Gualdesi L, Stoner R, Ranelli P, Cavanna A, Zanasca P, Baldasserini G, Trangeled C, Lewis C, Trees C, Grasso R, Giannechini S, Fabiani A, Merani D, Berni A, Leonard M, Paul Martin P, Rowley C, Hulbert M, Quaid A, Goode W, Preller R, Pinardi N, Oddo P, Guarnieri A, Chiggiato J, Carniel S, Russo A, Tudor M, Lenartz F, Vandenbulcke L (2009) Improved ocean prediction skill and reduced uncertainty in the coastal region from multi-model super-ensembles. J Mar Syst Coast Processes Chall Monit Prediction 78(S1):S282–S289

Stramska M, Dickey TD, Plueddemann A, Weller R, Langdon C, Marra J (1995) Bio-optical variability associated with phytoplankton dynamics in the North Atlantic ocean during spring and summer of 1991. J Geophys Res 100(C4):6621–6632

Chapter 20
Ocean-Colour Radiometry: Achievements and Future Perspectives

Shubha Sathyendranath and Trevor Platt

20.1 Introduction

Our understanding and interpretation of variations in ocean colour have matured and coalesced during the three decades of satellite-based remote sensing of spectral visible radiometry to create a separate discipline. Ocean-colour products from various satellite-sensors that reveal, in intricate detail, a wealth of information about phytoplankton distribution in the world oceans have by now become standard tools for oceanographers, and it is difficult to remember the under-sampled world with chlorophyll maps produced using sparse ship data that was the lot of biological oceanographers in the pre-ocean-colour era. Spectrally-resolved water-leaving radiance in the visible domain, and chlorophyll concentration derived from it, are now recognised as essential climate variables for studies of climate change (GCOS, 2004). Many national space agencies have their own ocean-colour programmes. The International Ocean Colour Co-ordinating Group (IOCCG) is working to coordinate and stimulate activities in the ocean-colour arena, and is advocating the various ways in which ocean-colour data can be used for the advancement of science as well as for the benefit of society at large.

These signs of maturity contrast with other signs of a field in its youth, such as continued rapid growth: the technology is still advancing, the interpretation is being taken to new heights, the product line is diversifying, and the applications are expanding (IOCCG, 2008). All of these are signs of a field that has not yet fully realised or exploited its true potential, or circumscribed its limitations. So we might ask: What are the promising lines for growth and development for ocean colour during, say, the next decade? In this chapter, we explore lines of research that provide scope for exciting developments in the future. These may be categorised under technical developments, improved interpretation, new products and novel applications. We also examine areas of challenge that are likely to set the limits on growth.

S. Sathyendranath (✉)
Plymouth Marine Laboratory, Plymouth PL1 3DH, UK; Dalhousie University, Halifax Nova Scotia, Canada
e-mail: shubha@dal.ca

V. Barale et al. (eds.), *Oceanography from Space*,
DOI 10.1007/978-90-481-8681-5_20, © Springer Science+Business Media B.V. 2010

Science advances in fits and starts, and often takes off in directions that were totally un-anticipated. Therefore, the authors have some latitude in thinking about the future, and can proceed knowing that the anticipated developments might differ from the eventual reality. The views presented in this paper are not intended to be exhaustive; instead it is very much a personal view, moulded by personal experience.

20.2 Technological Developments

The first ocean-colour satellite, the Coastal Zone Colour Scanner (CZCS), launched in 1978, has been succeeded since 1996 (after a gap of 10 years) by a series of satellite sensors for observing ocean colour, which have all been marked by significant improvements to their technical specifications, compared with the CZCS (IOCCG, 1999). These include better radiometric calibration, enhanced spectral resolution, and improved signal-to-noise ratio. All these contribute to higher radiometric precision of retrieved quantities such as water-leaving radiances, and they are all essential to improve the precision of the products derived from the observed radiometric properties. But of all these technological developments, it is the increased spectral resolution with high radiometric precision that offers the potential for development of novel products from ocean-colour data.

As is well known, ocean-colour radiometers were designed initially for the retrieval of a single variable: chlorophyll concentration. However, it was soon recognised that visible spectral radiometers with limited spectral wavebands, such as the CZCS, worked reasonably well only in the so-called Case 1 waters, where phytoplankton could be considered the single, independent variable responsible for changes in optical properties of sea water, and hence, in ocean colour. Applications to complex coastal and inland waters required additional wavebands. The time and space scales associated with coastal processes are typically smaller than those of open-ocean processes, and interrogating the coastal ocean at the appropriate scales requires higher spatial and temporal resolution than for the open ocean (IOCCG, 2000).

Recognition of these limitations led to improved spectral resolution in subsequent satellite sensors, which, along with improved radiometric precision to allow better atmospheric correction, were designed for application of ocean-colour data in the so-called Case-2 waters – optically-complex waters often encountered in coastal systems and fresh-water bodies – in which substances other than phytoplankton, such as suspended sediments and yellow substances, have an important and independent influence on optical properties. Pixel resolution at ground level has also improved over time, also contributing to improved applications in coastal waters.

Technological developments have also taken satellite sensors beyond multispectral sensors into the realm of hyper-spectral sensors: a term reserved for sensors with spectral resolution approaching a few nanometres. Geostationary ocean-colour sensors that allow high-frequency observations, once just a pipe dream, are now close to becoming a reality, with the anticipated launch of the Korean satellite GOCI in November 2009.

Given all these technological breakthroughs, especially in the last decade, we might ask: What are the additional developments that might be realised in the near future? If we compare what might be on the wish list of scientists against the impressive list of what has been already realised, it might be speculated that the next technological breakthroughs would come from sensors designed to yield nanometre, or sub-nanometre spectral resolution at least in selected wavebands. Such sensors would allow development and use of algorithms that exploit either small variations in the spectral form of reflectance, or that examine filling of Fraunhofer lines through trans-spectral processes, both to improve atmospheric correction and interpretation of water-leaving radiances (note: Fraunnhofer lines, the dark features in the solar spectrum, may get filled to smaller or greater extent in the water-leaving radiance, depending on the trans-spectral processes in the ocean, such as fluorescence or Raman scattering, which is commonly referred to as Fraunhofer line filling). Other technological innovations might come from success in combining features in satellite sensors that are at present mutually exclusive: such as bringing together high spatial resolution with high repeat frequency and global coverage. Perhaps such goals would be realised through constellations of satellites, which would also facilitate minimising loss of coverage due to clouds.

The CZCS was launched as a proof-of-concept mission. All the subsequent ocean-colour satellites have been improvements over the CZCS, with technical specifications that differed from those of the CZCS, and from each other, such that each of them has been pioneers in their own right. Whereas these innovations have promoted new interpretations and applications of ocean-colour data, it has also made it difficult to merge of data from different satellites (IOCCG, 2007). In the climate context, it is essential to create the longest possible time series of ocean-colour data in a seamless and continuous fashion. It is to be hoped that, in the future ocean-colour missions, the need for continuity would not be forgotten in the urge to innovation and creativity.

20.3 New Products

The brief overview of the advances in recent years clearly demonstrates that the technology has been able to respond in a very effective manner to our improved appreciation of the potential of ocean-colour sensors, over and above what had been conceived of in the initial years of development of the field. But arguably, the interpretation of data from the improved sensors has not yet reached its full potential.

The atmospheric signal, and not the ocean signal, remains always the dominant signal at the level of the satellite, such that it is atmospheric correction, and not radiometric accuracy at the level of the satellite, that will always determine the inherent precision of derived radiometric quantities at the sea level. In spite of the tremendous progress already made in atmospheric correction of ocean-colour data (Gordon, Chapter 17, this volume), it remains one of the major hurdles in the use of ocean-colour data in the sense that the accuracy and precision of retrieved water-leaving radiances is often significantly lower than what could be expected from the sensor specifications which determine the precision and accuracy at the top of the atmosphere. It is therefore essential to continue to improve atmospheric correction procedures, if we are to derive full benefits from the advanced ocean-colour sensors now available.

But have we already reached the peak in this area of research? Perhaps not. Scientists are still finding new ways to improve atmospheric correction: multi-spectral approaches, use of longer wavebands in the infrared, Fraunhofer line filling (Vountas et al., 2007), and the use of UV wavebands in addition to those in the infrared. All these are promising routes, as are inversion techniques that treat the ocean and atmosphere as a coupled system. But is it likely that when the limit of atmospheric correction is reached, it would set the limit to in-water properties or quantities that might be retrieved from ocean-colour data?

Not necessarily. The spectral characteristics of atmospheric constituents are quite distinct from those of many oceanic constituents, such that errors resulting from atmospheric correction are likely to have a spectral form different from those of in-water constituents. There is, therefore, scope for development of algorithms that exploit these differences: the need is for algorithms that would be insensitive to potential systematic errors in the water-leaving radiances arising from atmospheric correction. When atmospheric correction errors are relatively high, one anticipates that algorithms for retrieval of concentrations of substances such as phytoplankton pigments that have distinctive peaks and troughs in their inherent optical properties would work better than those designed to quantify substances such as yellow substances or suspended material whose inherent optical properties are monotonic functions of wavelength in the visible domain. The fluorescence line height might be considered one such algorithm that exploits a distinctive peak associated with chlorophyll fluorescence. In principle, the fluorescence line height would be insensitive to systematic offsets in the background signal, due, for example, to errors in atmospheric correction, or to the presence of non-pigmented scattering particles in the water. Thus, when the limits of atmospheric correction are known, it would be essential to understand the nature of the residual errors and incorporate them into the in-water algorithms.

The algorithms for Case-2 waters have been improving over the years (IOCCG, 2000), facilitated by the higher spectral resolution in the visible, improvements in atmospheric-correction procedures, and application of novel mathematical and statistical tools such as neural networks (Doerffer and Schiller, 1998). The need to address Case-2 problems provided impetus for development of in-water algorithms that are based on inversion of theoretical models of ocean colour that rely on our understanding of the inherent optical properties of oceanic constituents and of radiative transfer in the ocean (Sathyendranath et al., 1989; IOCCG, 2006), whereas, initially, models developed for Case-1 waters were mostly empirical in nature. These theoretical developments helped extend the line of products from ocean-colour data to include concentrations of suspended sediments and yellow substances. Improvements in Case-2 algorithms have also helped extend applications of ocean-colour data to many issues related to coastal zone management.

But the development of algorithms for Case-2 waters cannot yet be considered as finished: unresolved issues include development of algorithms that work across many geographic regions (IOCCG, 2000). In Case-2 waters, the water-leaving radiance is influenced by a number of optically-active substances varying independently of each other. The nature of these substances and the range of their concentrations vary across location. Differences in their inherent optical properties, and insufficient information on the variability in these properties, hinder further progress. It may well be that a global algorithm that would perform equally well in all regions is not feasible. If such be the case, then branching algorithms would be needed, each optimised for various conditions, such that regional algorithms could be stitched together in a seamless manner. In Case-2 waters, one or more sea-water constituents may mask the signals from other substances to such a level that the retrieval of the minor constituents may become impossible.

Such limits on retrieval algorithms have to be established clearly, along with the precision of the retrieved variables under all realistic conditions. Such information would facilitate further applications of Case-2 algorithms: users need to understand exactly what the satellite products are revealing, and the level of confidence that can be placed in the products. Case-2 waters are influenced by phytoplankton, yellow substances and suspended particulate material other than phytoplankton, all varying independently of each other, whereas Case-1 waters are defined as those waters in which optical variability is determined primarily by phytoplankton and substances co-varying with them. As such, Case-1 waters can be considered a sub-set of Case-2 waters. Therefore, in principle, Case-1 algorithms can be subsumed within Case-2 algorithms, provided Case-2 algorithms are tested as rigorously as has been the case for Case-1 algorithms.

Though the initial drive to improve spectral resolution in ocean-colour satellites came from the desire to improve applications to coastal waters, high spectral resolution has also allowed us to move beyond the detection of just the concentration of phytoplankton as indexed by chlorophyll concentration, towards identification of various phytoplankton types that have distinct optical signatures (Nair et al., 2008). Algorithms for identifying a number of phytoplankton types, such as diatoms, coccolithophores and certain types of blue-green algae have emerged in the last few years. Algorithms that discriminate between phytoplankton on the basis of their size have also been proposed. Algorithms for discriminating between types of phytoplankton may be classified as those that are based on abundance and those that are based on spectral signatures. The abundance-based methods relate certain ranges in the concentrations in chlorophyll (trophic status) with a particular phytoplankton community. Methods based on spectral signatures use discriminatory traits in the optical properties of certain phytoplankton to distinguish them from other phytoplankton.

Both approaches have their limitations and their advantages. The spectral-signature methods have the potential to improve in accuracy and expand in range of phytoplankton types that may be so-identified, as instruments with higher spectral resolution become available. The non-linear nature of the algorithms, the plasticity in the optical properties of phytoplankton functional types and the small signals on the basis of which the discriminations have to be made, will set limits on what is possible in this new and emerging direction. But we have yet to reach those limits. Abundance-based methods rely on correlations, and it is essential to establish whether such correlations hold for different regions of the world ocean, and whether they hold over time in a given location.

Measuring phytoplankton physiological rate processes from space remains a goal. It has long been suggested that the solar-induced fluorescence signal of chlorophyll can be used to infer some physiological rate processes of phytoplankton. New methods have also emerged that relate environmental conditions revealed for example through sea-surface temperature, chlorophyll concentration and day length, to infer photosynthetic rate parameters from satellites. One anticipates further development in these directions in the near future, but they would have to go hand in hand with improved understanding of factors responsible for variations in algal physiological processes in the natural environment.

20.4 Secondary Products and Novel Applications

The growth in the field may be judged not only from the new products that are emerging, but also from the innovative ways in which the products are used to generate new information. For example, ocean colour data are now being used to characterise the phenology of phytoplankton dynamics in a systematic manner at the global scale. Primary production (Platt et al., 2008) and new production (Sathyendranath et al., 1991) were among the earliest and most valued secondary products of ocean-colour remote sensing. But now the product range is being expanded to include various phytoplankton loss terms, and to improve our understanding of the cycle of carbon in the ocean. These products are indictors of the state of the marine ecosystem (Platt and Sathyendranath, 2008), extremely valuable in ecosystem-based management of marine resources.

Ecological indicators are objective metrics, each quantifying some relevant characteristic of the pelagic ecosystem. They are intended for operational (routine) application on serial data with a view to detecting ecosystem change in response to perturbations such as global warming or over fishing. Remote sensing is particularly suitable for construction of ecological indicators, with superior attributes such as rapidity of coverage, resolution, repeat frequency, cost-effectiveness and ability to produce information on fundamental ecosystem properties quantified in standard units (Table 20.1). At any instant, the state of the ecosystem can be represented as a vector whose elements are selected from the list of possibilities, according to the question under consideration. Such a list of indicators that can be produced using ocean colour and sea-surface temperature is given in the Table 20.2 (adapted from Platt and Sathyendranath, 2008). The vector of ecosystem status is time-dependent, and the elements will evolve in an informative way that can be revealed through analysis of time-series data. The indicators, as well as being of intrinsic scientific interest, are useful in the implementation of ecosystem-based management of marine resources, a principle

Table 20.1 Pelagic indicators and remote sensing potential

(a) **Ideal characteristics of pelagic indicators**
- Represent a well-understood and widely-accepted ecosystem property
- Quantifiable unambiguously in standard units
- Measurable rapidly at low incremental cost
- Repeat frequency compatible with intrinsic time scale of properties
- Measurable at a variety of scales
- Possibility to create long (multi-decadal) time series

(b) **Remote sensing for operational metrics**
- Requisites of speed, resolution, repeat frequency, cost-effectiveness are easily met
- Autotrophic biomass important ecosystem property
- Primary production fields can also be generated
- SST and chlorophyll obtainable at same resolution
- Time series possible: seasonal dynamics can be quantified objectively
- Allows interannual comparisons

Table 20.2 List of some useful indicators amenable to remote sensing

Indicator	Label	Dimensions
Initiation of spring bloom	b_i	[T]
Timing of spring maximum	b_t	[T]
Amplitude of spring bloom	b_a	[ML^{-3}]
Duration of spring bloom	b_d	[T]
Total production in spring bloom	b_p	[ML^{-2}]
Annual phytoplankton production	P_Y	[ML^{-2}]
Annual new production	P_N	[ML^{-2}]
Initial slope of light-saturation curve	α^{B}	[L^2]
Assimilation number	$P^B{}_{\mathrm{m}}$	[T^{-1}]
Particulate organic carbon	C_T	[ML^{-3}]
Phytoplankton carbon	C_p	[ML^{-3}]
Carbon-to-chlorophyll ratio	χ	Dimensionless
Phytoplankton growth rate	μ	[T^{-1}]
Generalised phytoplankton loss rate	L	[ML^{-3}/T]
Integrated phytoplankton loss	L_T	[ML^{-3}]
Spatial variance in biomass field	$\sigma_B{}^2$	[M^2/L^6]
Spatial variance in production field	$\sigma_P{}^2$	[M^2/L^4]
Phytoplankton functional types	NA	NA
Phytoplankton size	d	[L]
Delineation of biogeochemical provinces	NA	NA

generally accepted (subject of a Declaration of the United Nations), but difficult to put in practice.

An important entry in the table indicators is the partition of the ocean into ecological (biogeochemical) provinces. Discussions on ecosystem-based management usually call for definition of the geographic extent of ecosystems. In the case of marine ecosystems, the problem is further complicated by their fluid and mobile status, such that it becomes important to recognise the dynamic nature of the boundaries of ecosystems (Platt and Sathyendranath, 1999). The concept of ecological provinces of the ocean, developed initially as an aid to remote sensing of primary production (Longhurst, 2007), has proved extremely useful in this context, and methods have now emerged that use remote sensing to map the boundaries of ecological provinces of the ocean at scales that are relevant to the problem at hand (Devred et al., 2007).

Remote sensing of ocean colour probes the phytoplankton population in the aquatic environment. But recent studies have shown how the phenology in the phytoplankton dynamics can be linked to processes at higher trophic levels, such that ocean colour is becoming an important tool in various fisheries applications, ranging from understanding fish recruitment to streamlining fishing operations (IOCCG, 2009). With the advent of methods to study phytoplankton functional types, and with better understanding of how to link what is observed at the level of phytoplankton to what is happening at the higher trophic levels, ocean-colour is beginning to contribute to some discussions of marine biodiversity.

Secondary products and applications based on them are likely to continue to grow in the future. But it is worth considering whether the most interesting and novel application of remote sensing of ocean colour in the near future might not be in the arena of high-seas governance and implementation of marine policy and law. The concept of straddling stocks of commercial fish is well established, and the implications for setting fisheries policies and settling fishing disputes have been the

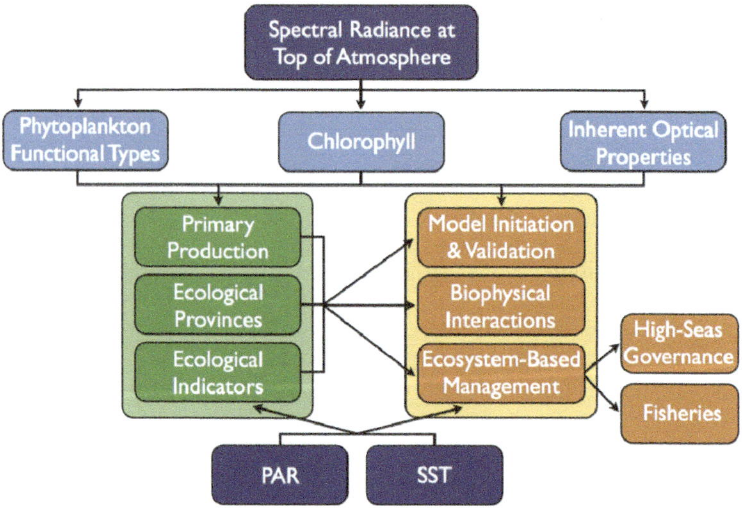

Fig. 20.1 Schematic diagram showing some of the spectral radiance-derived oceancolour products (*top row*), secondary products (*left box*) and some applications (*right box* and derived items). Note that the product line and applications are enhanced with incorporation of additional satellite-derived inputs such as photosynthetically available radiation (PAR) and sea-surface temperature (SST)

subject of much debate. But discussions of straddling ecosystems may only be just beginning. If integrity of ecosystems is to be maintained when marine resources are exploited, and if an ecosystem is spread over more than one national jurisdiction, then international agreements would have to be in place to implement ecosystem-based management. In implementation of such agreements, remote sensing would have a key role to play in mapping ecosystems, and their movements over time.

It is worth recalling that most of the ocean lies outside the jurisdiction of any nation. The high seas are a commons whose welfare is in jeopardy unless some form of high-seas governance is brought to bear. Information, continuously updated, will be the key to success in this regard. Because remote sensing of ocean colour provides our only window into the ocean ecosystem on synoptic scales, we can expect that, through the ecological indicators it provides, it will become of increasing importance in future years for the rational management of the high seas.

Applications are enhanced when ocean-colour data are complemented with other satellite-derived data, for example sea-surface temperature and photosynthetically-available radiation, PAR. Some of the ocean-colour products and their applications are presented schematically in Fig. 20.1.

20.5 Conclusion

In addition to advances in the technology, the interpretation and the applications, what other developments may we see in the future? It is interesting to speculate that there will be new developments and improvements in validation of the satellite-derived products using in-situ data. In validation exercises, the in-situ observations are considered the "truth". But it may be argued that, at the scale of the satellite observations determined by the pixel resolution at ground level, what the satellite seas is the "truth", in the sense that only the satellite is able to sample the entire area in an instant. Furthermore, the satellite observations yield information without the need for collecting sea-water samples, handling them or extracting pigments, all of which add errors to the in-situ observations. Though the satellite observations have also their own sources of error associated with radiometric data collection and application of algorithms, the satellite observations may be more representative of what is there at the scale of the pixel, than information derived from small volume analyses from in-situ observing platforms. Will we get to the stage where the radiometric quantities derived from satellites will be treated as the primary standard at that scale observation for some of the retrieved properties? Perhaps, but before we attain that level of confidence in satellite observations, the issue of incompatibilities in scales that continues to plague validation exercises has to be addressed fully. It may be that some new thinking in our approaches to validation is needed, before the issue is laid to rest.

The main thrust of the paper is on the potential for further development in the field of ocean-colour research. But what factors are likely to set limits on further growth? These would include the plasticity in inherent optical properties of phytoplankton, which makes it difficult to assign unique optical properties to different phytoplankton types. The influence of the atmosphere will ensure that ocean-colour studies would always have to deal with a small signal in a high noise environment. The non-linearity in the relationships between the radiometric quantities derived from satellites and the in-water constituents of interest will always be a problem to contend with.

The technological developments go hand in hand with innovative products. On the other hand, many of the secondary products depend on the availability of long time series of standardised products. So there will always be demands for sustained, long-term, consistent observations as well as for technological advancement and development of novel products. In a world with limited resources, some of these requirements may at times be in conflict with each other, and ambitions to push the limits of what is possible technologically would have to be balanced against the need to create the longest possible, climate-quality data records for addressing key issues facing us today, such as climate change.

In 1921 Sir C.V. Raman, in an article in *Nature*, argued that it was optical processes in the ocean, rather than the reflection of sun light, that gave sea water its characteristic blue colour (Raman, 1921), refuting the earlier suggestion by Lord Rayleigh that sea was blue due to reflection of the blue of the sky. It then appeared a solved problem, and of not much further scientific interest, until the influence of phytoplankton on the colour of the sea was articulated (e.g., Morel and Prieur, 1977). That led to the launch of the first ocean-colour sensor, the Coastal Zone Colour Scanner. Since then, the biological applications have been the driver for ocean-colour research. Yet, it has always been physics that drives the theoretical work to understand the optical processes responsible for variations in ocean colour. Interestingly, Raman came to the forefront again when the importance of Raman scattering in determining the marine light field was discovered well after the launch of the CZCS (e.g., Stavn and Weidemann, 1988). Ocean-colour science is interdisciplinary in nature, and it is the constant interplay between physics and biology that has made ocean colour a fascinating subject of study. It will surely remain so for several more decades to come.

Acknowledgements This work was supported by the Canadian Space Agency through its GRIP and EOPI programmes, and by NERC (UK) through its 2025 and NCEO programmes. We thank Vittorio Barale for all his patience as the editor, and Nicolas Hoepffner for his helpful comments on an earlier version of the manuscript.

References

Devred E, Sathyendranath S, Platt T (2007) Delineation of ecological provinces using ocean colour radiometry. Mar Ecol Prog Ser 346:1–13

Doerffer R, Schiller H (1998) Determination of Case 2 water constituents using radiative transfer simulation and its inversion by neural networks. In: Ackleson SG, Campbell J (eds.) Proceedings of the Ocean Optics XIV, Office of Naval Research, Washington, DC

GCOS (2004) Implementation plan for the global observing system for climate in support of UNFCCC. GCOS-92, WMO/TD No 1219

IOCCG (1999) Status and plans for satellite ocean-colour missions: considerations for complementary missions. In: Yoder JA (ed.) Reports of the International Ocean-Colour Coordinating Group, No. 2, IOCCG, Dartmouth, Canada

IOCCG (2000) Remote sensing of ocean colour in coastal, and other optically-complex, waters. In: Sathyendranath S (ed.) Reports of the International Ocean-Colour Coordinating Group, No. 3, IOCCG, Dartmouth, Canada

IOCCG (2006) Remote sensing of inherent optical properties: fundamentals,tests of algorithms, and applications. In: Lee ZP (ed.) Reports of the International Ocean-Colour Coordinating Group, No. 5, IOCCG, Dartmouth, Canada

IOCCG (2007) Ocean-colour data merging. In: Gregg W (ed.) Reports of the International Ocean-Colour Coordinating Group, No. 6, IOCCG, Dartmouth, Canada

IOCCG (2008) Why ocean colour? The societal benefits of ocean-colour technology. In: Platt T, Hoepffner N, Stuart V, Brown C (eds.) Reports of the International Ocean-Colour Coordinating Group, No. 7, IOCCG, Dartmouth, Canada

IOCCG (2009) Remote sensing in fisheries and aquaculture. In: Forget MH, Stuart V, Platt T (eds.) Reports of the International Ocean-Colour Coordinating Group, No. 8, IOCCG, Dartmouth, Canada

Longhurst A (2007) Ecological geography of the sea, 2nd ed, Elsevier, San Diego, California, USA

Morel A, Prieur L (1977) Analysis of variations in ocean colour. Limnol Oceanogr 22:709–722

Nair A, Sathyendranath S, Platt T, Morales J, Stuart V, Forget M-H, Devred E, Bouman H (2008) Remote sensing of phytoplankton functional types. Rem Sens Environ 112:3366–3375

Platt T, Sathyendranath S (1999) Spatial structure of pelagic ecosystem processes in the global ocean. Ecosystems 2:384–394

Platt T, Sathyendranath S (2008) Ecological indicators for the pelagic zone of the ocean. Rem Sens Environ 112:3426–3436

Platt T, Sathyendranath S, Forget MH, White GN, Caverhill C, Bouman H, Devred E, Son S (2008) Operational mode estimation of primary production at large geographical scales. Rem Sens Environ 112:3437–3448

Raman CV (1921) The colour of the sea. Nature 108:367

Sathyendranath S, Platt T, Horne EPW, Harrison WG, Ulloa O, Outerbridge R, Hoepffner N (1991) Estimation of new production in the ocean by compound remote sensing. Nature 353:129–133

Sathyendranath S, Prieur L, Morel A (1989) A three-component model of ocean colour and its application to remote sensing of phytoplankton pigments in coastal waters. Int J Remote Sens 10:1373–1394

Stavn R, Weidemann A (1988) Optical modeling of clear ocean light fields: raman scattering effects. Appl Opt 27:4002–4011

Vountas M, Dinter T, Bracher A, Burrows JP, Sierk B (2007) Spectral studies of ocean colour with space-borne sensor SCIAMACHY using Differential Optical Absorption Spectroscopy (DOAS). Ocean Sci 3: 429–440

Acronyms

AAOT	Acqua Alta Oceanographic Tower
AATSR	Advanced Along-Track Scanning Radiometer
ACC	Antarctic Circumpolar Current
ACSPO	AVHRR Clear-Sky Processor for Oceans
ADEOS	Advanced Earth Observing Satellite
AF-FOV	Alias-Free Field-of-View
AIS	Automated Information System
ALOS	Advanced Land Observing Satellite
AMSR	Advanced Microwave Scanning Radiometer
AMSR-E	Advance Microwave Scanning Radiometer – EOS
AO	Announcement of Opportunity
AOML	Atlantic Oceanographic and Meteorological Laboratory
APC	Antenna Pattern Correction
ARGO	buoy, not an, acronym named for Greek mythological ship
AS Mode	Aperture Switching Mode (of TerraSAR-X)
ASAR	Advanced SAR (Instrument on ENVISAT)
ASCAT	Advanced Scatterometer
ATI	Along-Track (Radar) Interferometry
ATSR	Along-Track Scanning Radiometer (ERS-1)
AVHRR	Advanced Very High Resolution Radiometer
AVISO	A Main Distributor of Multimission Satellite Altimetry Data
AXBTs	Airborne eXtpendable Bathy-Thermographs
BAW	Bundesanstalt für Wasserbau (D)
bb	Backscattering
BEC	Barcelona Expert Centre
BODAS	BLUElink Ocean Data Assimilation System
BoM	Australian Bureau of Meteorology
BSH	Bundesamt für Seeschifffahrt und Hydrographie (D)
CBE	Current Best Estimate
CCMP	Cross Correlated Multi-Platform
CDOM	Colored Dissolved Organic Matter
CDOP	Measured Doppler Shifts at C Band Algorithm
CDR	Climate Data Record
CDTI	Centro para el Desarrollo Tecnológico Industrial
CEOS	Committee on Earth Observation Satellites
CERSAT	Centre d'Archivage et de Traitement
CFAR	Constant False Alarm Rate
CHAMP	ChAllenging Minisatellite Payload Geoscience Application
chl-*a*	Chlorophyll a

V. Barale et al. (eds.), *Oceanography from Space*,
DOI 10.1007/978-90-481-8681-5, © Springer Science+Business Media B.V. 2010

CIRIMS	Calibrated InfraRed In-Situ Measurement System
CLIVAR	Climate Variability and Predictability
CLS	Collect Localisation Satellite
CMC	Canadian Meteorological Centre
CNES	Centre National d'Etudes Spatiales
COAMPS	Coupled Ocean/Atmosphere Mesoscale Prediction System
COMS	Communication Ocean and Meteorological Satellite
CONAE	Comisión Nacional de Actividades Espaciales
CPC	Circular Polarization Coherence
CryoSat-2	Cryosphere Satellite 2
CSA	Canadian Space Agency
CSIRO	Commonwealth Scientific and Industrial Research Organisation
CZCS	Coastal Zone Color Scanner
DAAC	Distributed Active Archive Center
DLR	Deutsches Zentrum für Luft- und Raumfahrt
DMSP	Defense Meteorological Satellite Program
DNSC	Danish National Space Center
DOISST	Daily Optimally Interpolated SST
DOMSAT	Domestic Communications Satellite
DopRIM	Numerical Model for Doppler Simulations
DRA mode	Dual Receive Antenna Mode (of TerraSAR-X)
DV	Diurnal Variability
ECCO	Estimating the Circulation and Climate of the Ocean
ECMWF	European Center for Medium-Range Weather Forecast
EEZ	Exclusive Economic Zone
EGM	Earth Gravity Model
EKE	Eddy Kinetic Energy
EKWC	East Korean Warm Current
EMSA	European Maritime Safety Agency
ENL	Equivalent Number of Looks
ENSO	El Niño – Southern Oscillation
ENVISAT	Environmental Satellite (ESA)
EOS	Earth Observing System
EOSDIS	EOS Data and Information System
EPA	Environmental Protection Agency
ERIM	Environmental Research Institute of Michigan
ERM	Exact Repeat Mission
ERS-1	ESA Remote Sensing Satellite 1
ERS-2	ESA Remote Sensing Satellite 2
ESA	European Space Agency
ESAC	European Space Astronomy Centre
ESMR	Electrically Scanning Microwave Radiometer
ESSF	Earth System Science Pathfinder
ESTAR	Electronically Steered Thinned Array Radiometer
ESTEC	European Space Technology Centre
ETM+	Enhanced Thematic Mapper Plus
EU	European Union
EUMETSAT	European Organisation for the Exploitation of Meteorological Satellites
FANNs	Artificial Neural Networks
FCDR	Fundamental Climate Data Record
FOAM	Forecasting Ocean Assimilation Model
FOV	Field-of-View
FRAC	Full Resolution Area Coverage

FTIR	Fourier Transform InfraRed
G1SST	Global 1 km SST
GAC	Global Area Coverage
GAMSSA	Global Multi-Sensor Sea Surface Temperature Analysis
GBR	Great Barrier Reef
GBRMPA	Great Barrier Reef Marine Park Authority
GCOM-C	Global Change Observation Mission – Climate
GCOM-W	Global Change Observation Mission – Water
GCOS	Global Climate Observing System
GDS	GHRSST Data Specification
GEOS-3	Geodynamics Experimental Ocean Satellite – 3
Geosat	US Navy Altimeter Satellite
GFDL	Geophysical Fluid Dynamics Laboratory
GHRSST	Group for High- Resolution SST
GHRSST-PP	GODAE High-Resolution SST Pilot Project
GIS	Geographic Information Systems
GLI	Global Imager
GLORYS	GLobal Ocean ReanalYses and Simulations
GM	Geodetic Mission
GMF	Geophysical Model Function
GMF-S	Geophysical Model Function – Stress
GMF-W	Geophysical Model Function – Wind
GMPE	GHRSST Multi Product Ensemble
GMS	Geosynchronous Meteorological Satellite
GOCE	Gravity Field Steady-State Ocean Circulation Experiment
GOCI	Geostationary Ocean Color Imager
GOCINA	Geoid and Ocean CIrculation in the North Atlantic
GOCINO	GOCe IN Ocean Modelling
GODAE	Global Ocean Data Assimilation Experiment
GOES	Geostationary Operational Environmental Satellite
GOOS	Global Ocean Observing System
GPM	Global Precipitation Measuring Mission
GPS	Global Positioning System
GRACE	Gravity Recovery and Climate Experiment
GRDC	Global Runoff Data Centre
GSFC	Goddard Space Flight Center
GTS	Global Telecommunications System
GUT	GOCE User Toolbox
HadISST	Hadley Centre Sea Ice Sea Surface Temperature Data Set
HDF	Hierarchical Data Format
HH	Horizontal (Radar) Co-polarisation
HLT	Helsinki Lighthouse Tower
HR-DDS	High Resolution Diagnostic Data Set
HRPT	High Resolution Picture Transmission
HRV	High Resolution Visible
ICOADS	International Comprehensive Ocean-Atmosphere Data Set
IFREMER	Institut francais de recherché et d'Exploitation de la Mer
IMO	International Maritime Organization
IMOS	(Australian) Integrated Marine Observing System
InSAR	Interferometric Synthetic Aperture Radar
IOCCG	International Ocean Colour Coordinating Group
IOS	Institute of Ocean Sciences (Canada)
IPCC	Intergovernmental Panel on Climate Change

IR	InfraRed
IRSO	Indian Space Research Organization
IRS-P3	Indian Remote Sensing Satellite
ISAR	Infrared Scanning Autonomous Radiometer
ISDGM	Instituto per lo Studio della Dinamica delle Grandi Masse
ISMAR	Instituto di Scienze Marine
ITF	Indonesian Throughflow
IUCRM	Inter-Union Commission on Radio Meteorology
IWs	Internal Waves
JASON-1	US-French (Dual Frequency) Altimeter Satellite, 2001
JASON-2	US-French (Dual Frequency) Altimeter Satellite, 2008
JAXA	Japan Aerospace Exploration Agency (Formally NASDA)
JMA	Japan Meteorological Agency
JPL	Jet Propulsion Laboratory
JRC	Joint Research Center
KMA	Korean Meteorological Administration
KUSTWAD	Dutch Wadden Sea Numerical Circulation Model
L0	Level 0
L1	Level 1 – Geolocated Radiances Over a Single Swath
L1b	Level 1b
L2	Level 2 – Geolocated SSTs Over a Single Swath
L2OP	Level 2 Ocean Salinity Processor
L2P	Level 2 – Pre-processed (GHRSST Format L2 File)
L3	Level 3 – Composite, Gridded, SST Product from L2 Data
L4	Gridded, Gap-Free, SST Analyses (by Interpolation)
LAC	Local Area Coverage
Landsat-7	Latest Satellite of the USGS/NASA's Landsat Program
LECT	Local Equator Crossing Times
LICEF	Lightweight Cost-Effective Front-End
L-RIT	Long-Range Identification and Tracking
LUT	Look Up Table
M-AERI	Marine-Atmospheric Emitted Radiance Interferometer
MCI	Maximum Chlorophyll Index
MDB	Matchup Database
MDT	Mean Dynamic Topography
MERCATOR	Numerical Global Ocean Circulation Model
MERIS	Medium Resolution Imaging Spectrometer
MERSEA	Marine Environment and Security for the European Area
MetOp	Meteorological Operational Satellite
METOP	European METeorological OPerational Satellite
MGDSST	Merged Satellite and In-Situ Data Global Daily SST
MIRAS	Microwave Imaging Radiometer using Aperture Synthesis
MIS	Microwave Imager Sounder
MODIS	Moderate Resolution Imaging Spectroradiometer
MOS	Modular Optoelectrical Scanner
MOSE	Modulo Sperimentale Elettromeccanico
MOVE	Multivariate Ocean Variational Estimation
MRI	Meteorological Research Institute(Japan)
MSG	Meteosat Second Generation
MSG	METEOSAT Second Generation
MSS	Mean Sea Surface
MTG	METEOSAT Third Generation
MTSAT	Meteorological Geostationary Satellite of Japan

MVISR	Multispectral Visible and Infra-Red Scan Radiometer
MW	Microwave
MWIR	Mid Wave Infra Red
MWR	Microwave Radiometer
N	North
NAO	North Atlantic Oscillation, a Climatic Index
NASA	National Aeronautics and Space Administration
NAVOCEANO	NAVal OCEANographic Office (USA)
NCDC	National Climatic Data Center (USA)
NCEP	National Centers for Environmental Protection (USA)
NCODA	Navy Coupled Ocean Data Assimilation Analysis (USA)
NCOM	Naval Coastal Ocean Model
NE	North–East
NEDT	Noise Equivalent Temperature Difference
NESDIS	National Environmental Satellite Data Information Service
NESS	National Environmental Satellite Service (Now NESDIS)
NESZ	Noise Equivalent Sigma Zero
NET	Nimbus Experiment Team
NetCDF	Network Common Data Format
NIR	Near InfraRed
NIRST	New InfraRed Scanner Technology
NIST	National Institute of Standards and Technology
NLSST	Non-Linear Sea Surface Temperature
NMEFC	National Marine Environment Forecast Centre (China)
NNE	North–North–East
NNW	North–North–West
NOAA	National Oceanic and Atmospheric Administration
NODC	National Oceanographic Data Center
NOSS	National Ocean Satellite System (USA)
NPOESS	National Polar-Orbiting Operational Environmental Satellite System
NPP	NPOESS Preparatory Program
NRCS	Normalised Radar Cross Section
NRIO	National Research Institute for Oceanology
NRL	Naval Research Laboratory
NRT	Near Real Time
NSCAT	NASA Scatterometer
NSIDC	National Snow and Ice Data Center (USA)
NWP	Numerical Weather Prediction
OBPG	Ocean Biology Processing Group
OceanMAPS	Ocean Model, Analysis and Prediction System
OCS	Ocean Color Scanner
OCTS	Ocean Color and Temperature Sensor
ODYSSEA	Ocean Data analYsis System for merSEA
OISST	Reynolds Optimally Interpolated SST
ONR	Office of Naval Research
OPeNDAP	Open-Source Project for a Network Data Access Protocol
OSCAR	Ocean Surface Current Analyses – Realtime
OSI-SAF	Ocean and Sea Ice Satellite Application Facility
OSMM	Ocean Salinity Measurement Mission
OST	Ocean Surface Topography
OSTIA	Operational Sea Surface Temperature and Sea Ice Analysis
OSWG	Oceans Science Working Group
PAR	Photosynthetically Available Radiation

PATMOS	Pathfinder Atmospheres
PDO	Pacific Decadal Oscillation
PHAM	Pelagic Habitat Analysis Module
PIRATA	Pilot Research Moored Array in the Tropical Atlantic
PMEL	Pacific Marine Environmental Laboratory
PMW	Passive MicroWave
PO.DAAC	Physical Oceanography Distributed Active Archive Center
POAMA	Protective Ocean Atmosphere Model for Australia
POES	Polar Operational Environmental Satellite
POLDER	Polarization and Directionality of the Earth's Reflectances
PSS	Practical Salinity Scale
QAA	Quasi-Analytical Algorithm
QuikSCAT	Quick Scatterometer
RADARSAT	Canadian Satellite Carrying a SAR
RAMSES	Radiométrie Appliquée Mesure Salinité et Eaudans le Sol
RFI	Radio Frequency Interference
RMS/rms	Root-Mean-Square
RSMAS	Rosenstiel School of Marine and Atmospheric Science
RSS	Remote Sensing Systems
RTG	Real Time Global
RTM	Radiative Transfer Model
RWS	Rijkswaterstaat (NL)
SAC	Satélite de Aplicaciones Científicas
SAC-D	Satélite de Aplicaciones Científicas-D
SAR	Synthetic Aperture Radar
SASWG	SST Archive Science Working Group
SEASAT	SEA (Ocean Remote Sensing) SATellite, US
SeaWiFS	Sea-Viewing Wide Field-of-View Sensor
SEAWINDS	Scatterometer on the QuikSCAT
SENTINEL-1	(Future) European Remote Sensing Satellite
SEVIRI	Spinning Enhanced Visible and Infrared Imager
SGLI	Second-Generation Global Imager
SHA	Sea Height Anomaly
SIO	Scripps Institution of Oceanography
SISTeR	Scanning Infrared Sea Surface Temperature Radiometer
SLA	Sea Level Anomalies
SLSTR	Sea and Land Surface Temperature Radiometer
SMAP	Soil Moisture Mapping
SMMR	Scanning Multi-channel Microwave Radiometer
SMOS	Soil Moisture and Ocean Salinity
SMOSops	SMOS Operational System
SRTM	Shuttle Radar Topography Mission
SSALTO	(-DUACS) Multimission Altimeter Data Processing System
SSES	Single Sensor Error Statistics
SSH	Sea Surface Height
SSHA	Sea Surface Height Anomaly
SSIWG	Salinity and Sea Ice Working Group
SSM/I	Scanning Multichannel Microwave/Imager
SSMIS	Special Sensor Microwave Imager/Sounder
SSMR	Scanning Multichannel Microwave Radiometer
SSS	Sea Surface Salinity
SST	Sea Surface Temperature
STD	Standard Deviation

SVP	Surface Velocity Program
SWC	Soya Warm Current
SWG	Science Working Group
SWIR	Short Wave Infra Red
SWOT	Surface Water Ocean Topography
T/P	Topography Experiment/Poseidon
TA	Antenna Temperature
TanDEM-X	TerraSAR-X Add-On Digital Elevation Measurements
TAO	Tropical Atmosphere Ocean
TAO/TRITON	Tropical Atmosphere Ocean/TRITON
TB	Brightness Temperature
T_B	Brightness Temperatures
TC	Tropical Cyclone
TCHP	Tropical Cyclone Heat Potential
TerraSAR-X	German Satellite Carrying an X Band SAR
TIR	Thermal Infra-Red
TIROS	Television and Infrared Observation Satellite
TMI	TRMM, Microwave Imager
TMI TRMM	Microwave Imager
TOGA	Tropical Ocean Global Atmosphere Program
TOPAZ	Towards an Operational Prediction System for the North Atlantic European Coastal Zones
TOPEX	(-Poseidon) TOPography Experiment, US-F, 1992
TOVS	TIROS Operational Vertical Sounder
TRITON	Triangle Trans-Ocean Buoy Network
TRMM	Tropical Rainfall Measuring Mission
TSM	Total Suspended Matter (Mass)
TV	Television
TXR	Transfer Radiometer
UNCLOS	United Nations Convention on the Law of the Sea
UnTRIM	Numerical Circulation Model (Name)
US	United States
USA	United States of America
USGS	US Geological Survey
UTC	Universal Time Coordinate
VIIRS	Visible Infrared Imaging Radiometer Suite
VIRR	Visible and Infrared Scan Radiometer
VIRS	Visible Infrared Radiometer Sounder
VIS	Visible
VMS	Vessel Monitoring System
VV	Vertical (Radar) Co-polarisation
WISE	Wind and Salinity Experiment
WMO	World Meteorological Organisation
WNW	West–North–West
WOCE	World Ocean Circulation Experiment
WSOA	Wide Swath Ocean Altimeter
XBT	eXpendable Bathy Thermograph
XTI	Cross-Track (Radar) Interferometry

Subject Index

This Volume has been prepared and published with the support of the Joint Research Centre, European Commission.

JRC
EUROPEAN COMMISSION

The mission of the JRC is to provide customer-driven scientific and technical support for the conception, development, implementation and monitoring of EU policies. As a service of the European Commission, the JRC functions as a reference centre of science and technology for the Union. Close to the policy-making process, it serves the common interest of the Member States, while being independent of special interests, whether private or national.